DATE DUE

AP 7 '95			
NO 17 '95			
MY 24 '96			
SE 26 '96			
DE 8 '97			
MR 8 '98			
NO 30 '00			
JE 9 '03			
NO 6 '07			

DEMCO 38-296

ANNALS OF
THE NEW YORK ACADEMY
OF SCIENCES

Volume 646

EDITORIAL STAFF

Executive Editor
BILL BOLAND

Managing Editor
JUSTINE CULLINAN

Associate Editor
ANGELA C. FINK

The New York Academy of Sciences
2 East 63rd Street
New York, New York 10021

RECOMBINANT DNA TECHNOLOGY I

ANNALS OF THE NEW YORK ACADEMY OF SCIENCES
Volume 646

RECOMBINANT DNA TECHNOLOGY I

Edited by Aleš Prokop and Rakesh K. Bajpai

The New York Academy of Sciences
New York, New York
1991

Library of Congress Cataloging-in-Publication Data

Recombinant DNA technology I / edited by Aleš Prokop and Rakesh K. Bajpai.
    p. cm. — (Annals of the New York Academy of Sciences, ISSN 0077-8923; v. 646)
    ISBN 0-89766-673-9 (cloth: alk. paper). — ISBN 0-89766-674-7 (pbk.: alk. paper)
    1. Genetic engineering. 2. Recombinant DNA. I. Prokop, Aleš. II. Bajpai, Rakesh K. III. Title: Recombinant DNA technology I. IV. Title: Recombinant DNA technology one. V. Series.
    [DNLM: 1. Biotechnology—congresses. 2. DNA, Recombinant—congresses. 3. Genetic Engineering—congresses. W1 AN626YL v. 646 / QU 58 R31246]
Q11.N5 vol. 646
[TP248.6]
500 s—dc20
[600'.65]
DNLM/DLC
for Library of Congress
                                      92-5973
                                      CIP

SP
Printed in the United States of America
ISBN 0-89766-673-9 (cloth)
ISBN 0-89766-674-7 (paper)
ISSN 0077-8923

ANNALS OF THE NEW YORK ACADEMY OF SCIENCES

Volume 646
December 27, 1991

RECOMBINANT DNA TECHNOLOGY I[a]

Editors
ALEŠ PROKOP AND RAKESH K. BAJPAI

Organizing Committee
ROGER N. BEACHY, ARNOLD L. DEMAIN, ARTHUR E. HUMPHREY,
ERNEST G. JAWORSKI, GEORGE PAVLAKIS, AND K. VENKAT

CONTENTS

[a]This volume is the result of a conference entitled Progress in Recombinant DNA Technology and Applications, which was sponsored by the Engineering Foundation and held on June 3–8, 1990 in Potosi, Missouri.

Biological Considerations: Eukaryotic Systems

Engineering Considerations

Financial assistance was received from:

Major Funders
- ENGINEERING FOUNDATION
- NATIONAL SCIENCE FOUNDATION
- US ARMY RESEARCH OFFICE

Supporters
- AMGEN, INC.
- ANHEUSER-BUSCH COMPANIES, INC.
- ANONYMOUS
- BEHRINGWERKE AG
- BRISTOL MYERS SQUIBBS CO.
- COORS BIOTECH, INC.
- CONNAUGHT LABS, INC. (USA)
- GENENTECH, INC.
- GLAXO GROUP RESEARCH, LTD. (UK)
- H. J. HEINZ
- JAPANESE SOCIETY OF ENZYME ENGINEERING
- KYOWA HAKKO KOGYO CO., LTD.
- MONSANTO COMPANY
- ORTHO PHARMACEUTICAL CORP.
- PHARMACIA LKB (USA)
- SMITHKLINE BEECHAM
- UNITED ENGINEERS AND CONTRACTORS
- US BIOCHEMICAL CORP.

Preface

ALEŠ PROKOP[a] AND RAKESH K. BAJPAI[b]

[a] *Chemical Engineering Department*
Washington University
1 Brookings Drive
St. Louis, Missouri 63130

[b] *Chemical Engineering Department*
University of Missouri-Columbia
1030 Engineering Building
Columbia, Missouri 65211

Recombinant DNA technology has taken major strides in the last decade. A number of products of this technology are already on the market and many more are at the threshold of commercialization. This success has been achieved by involving personnel from several different disciplines, geneticists, molecular biologists, biochemists, physicians, veterinarians, cell biologists, and engineers. To continue and to enhance the rate of progress, it is necessary to promote effective communication between these diverse groups. A symposium, titled "Progress in Recombinant DNA Technology and Applications," was held from June 3–8, 1990, to provide an avenue for improving communications between the different groups of professionals, under the umbrella of the Engineering Foundation Conferences. Over 100 scientists and engineers from around the world gathered at Trout Lodge (Potosi, Missouri) in the scenic Ozarks for five days to exchange views in the formal settings of the sessions and in the informal discussion groups concerning various aspects of this technology, ranging from cloning to facility design. These proceedings are a compilation of several papers presented at the conference. For organization, the papers have been grouped under the classical categories of biological and engineering considerations.

We would like to thank the members of the organizing committee of the symposium consisting of Roger N. Beachy (Washington University), Harold A. Comerer (Engineering Foundation), Arnold L. Demain (MIT), Arthur E. Humphrey (Lehigh University), Ernest G. Jaworski (Monsanto), George Pavlakis (NIH), and K. Venkat (H. J. Heinz Company, now at Genmap, Inc.). In addition, Greg Casey (Anheuser-Busch Companies), James F. Kane (Monsanto Company), Gwen G. Krivi (Monsanto Company), and Allen I. Laskin (Laskin & Associates) also provided substantial input in the organization of this symposium. The true success of the symposium was due to the efforts of the chairpersons and the co-chairpersons of the different sessions: Arnold L. Demain (MIT); Claude Vézina (Institut Rosell-Montreal); Martin Rosenberg (SmithKline Beecham); Janet Westpheling (University of Pennsylvania); Lois K. Miller (University of Georgia); Gerald H. Rank (University of Saskatchewan); Robert E. Hammer (Howard Hughes Medical Institute-Dallas); Herbert Heynecker (Genencor); Raymond L. Rodriguez (University of California-Davis); Robert B. Horsch (Monsanto Company); Jan Desomer (Rijksuniversiteit-Gent); David Fischoff (Monsanto Company); Czaba Horváth (Yale University); John Frenz (Genentech); Charles L. Cooney (MIT); Matthias Reuss (Technische Universität-Stuttgart); Barry C. Buckland (Merck Sharp & Dohme); Daniel Omstead (Ortho Pharmaceuticals Corp.); E. T. Papoutsakis (Northwestern University); and George Georgiou (University of Texas-Austin).

A number of reviewers provided valuable input regarding the manuscripts published in these proceedings. The reviewers' dedication is a guarantee of the quality of any publication, and we are much obliged for their efforts. These reviewers, in addition to the two editors, are: Donald G. Bergmann, SmithKline Beecham; Barry C. Buckland, Merck Sharp & Dohme; Gary S. Ditta, University of California-Davis; Alan M. Easton, Monsanto Company; D. B. Finkelstein, Panlabs Inc.; Julian B. Fleishman, Washington University School of Medicine; Mark E. Gustafson, Monsanto Company; C. Richard Hutchinson, University of Wisconsin-Madison; James F. Kane, Monsanto Company; Gwen G. Krivi, Monsanto Company; Sally A. Leong, University of Wisconsin-Madison; Verne A. Luckow, Monsanto Company; Margaret (Peggy) Marino, Monsanto Company; Joseph K. McLaughlin, Monsanto Company; Victor Morales, Plant Biotechnology Institute-Saskatoon; Gerald H. Rank, University of Saskatchewan; John A. Ryals, Ciba-Geigy Corp.; Robert Schleif, The John Hopkins University; and Roger Weigand, Monsanto Company.

The next symposium, "Recombinant DNA Technology II," will be held in January 1993. To participate, please contact the editors.

Prospectives and Challenges in Genetic Engineering

WILLIAM S. REZNIKOFF

Department of Biochemistry
College of Agricultural and Life Sciences
University of Wisconsin
Madison, Wisconsin 53706

Genetic engineering is the technology by which man chooses and selectively breeds mutants in order to use the resulting organisms for his own benefit. This enterprise started with the first domestication of animals and plants or, perhaps, with the first maintenance and use of microorganisms for food and beverage processing. Although the associated technology and the uses of genetic engineering have changed dramatically since the early 1970s, it is important to realize that genetic engineering is not a new endeavor.

During the last decade we have seen an explosion in the use of (and interest in) genetic engineering. This is largely the result of two developments: the basic scientific discoveries of the structure and functioning of DNA (what genes are, how they are expressed, and how the expression is regulated), and the development of new techniques for analyzing and manipulating DNA molecules (and other biological macromolecules). Although particular research accomplishments can be pointed out as highlights in these developments, the whole process was, as I shall attempt to point out, the cumulative result of the efforts of many scientific projects, the great majority of which were not designed to aid in the development of this enterprise.

Several basic science discoveries were critical in the development of the new technology. Perhaps the first critical findings were that DNA was the genetic material and that DNA acted, for the most part, by encoding the primary sequence of proteins.[1-3] This led to work that elucidated (1) the mechanisms by which this information transfer occurs (usually DNA to RNA to protein),[4] (2) the nature of the code that relates nucleotide sequence to amino acid sequence,[5] and (3) the controlling mechanisms that define individual units of expression and regulate their activity.[6] Of course, the motivation for much of this work may not have been to develop the technology of genetic engineering; rather, it was an attempt to understand life and, for some scientists, it was for personal enjoyment and aesthetic satisfaction. In addition, many discoveries that were made should have a basic impact on our philosophical systems. (What is the real consequence of discovering that life is governed by biochemical reactions?[7] The ubiquitous existence of transposable elements[8] implies that perhaps a basic level of evolution is "competition" between DNA sequences and that each organism carries within its own genome the instructions for changing that genome. The discovery of introns[9,10] has challenged, for awhile, the notion that nature tends toward greater economy.) Nonetheless, the discoveries have given us guidelines that we as "engineers" can use to plan the alteration of selected organisms in specific ways. For instance, if we wish to have an organism make a particular protein, we now know what the structure of the gene should be and what signals are needed for controlling its synthesis.

Knowledge is not enough. Engineers also need tools to apply the knowledge. Although my list may be exhausting, it will not be exhaustive. Of critical importance

1

was the development of the field of nucleic acid enzymology. Many point to the discovery of restriction enzymes[11] as a watershed event, because it enabled us to dissect DNA molecules into defined fragments. However, being able to fragment DNA molecules without knowing how to link various fragments together would not have been very useful; therefore, the discovery and analysis of DNA ligases[12] were important. Likewise, due to the efforts of many unheralded scientists, we have enzymes that will degrade DNA, synthesize DNA, and phosphorylate and dephosphorylate the ends of polynucleotides, and so forth.[13] Other current techniques required the development of DNA sequencing methodologies,[14,15] chemical techniques for synthesizing DNA,[16] and methods for purifying and handling various macromolecules. Most of the methodologies also make use of basic microbiological protocols as well as the discoveries associated with bacteriophages, plasmids, and so on.

Which technologies will have a particularly big impact in the next 10–20 years? My three favorites are: (1) nucleic acid hybridization technology as a diagnostic tool, (2) antisense RNAs or ribozymes as pharmaceuticals, and (3) designer monoclonal antibodies made in *Escherichia coli.*

The use of nucleic acid hybridization technology[17] to identify the presence of specific DNA sequences (and consequently the presence of specific organisms or specific rearranged genes in a given organism) is commonplace in the research laboratory. We already know its use in the forensic laboratory and in paternity testing. The only questions are not if, but rather, why has this technology not been used already, when will it be used, and what detection modalities will be used.

Antisense RNAs[18] and/or ribozymes[19] block gene expression through the same molecular principle, that is, sequence-specific nucleic acid hybridization. Just one example suffices to show its potential usefulness. Attacking a virus during an infection process is problematical because the host cell can be killed as well. In principle, these techniques eliminate this problem by virtue of their specificity. Although these techniques can work right now in the research laboratory, the jump to a practical pharmaceutical application may not be simple. We are faced with major questions of cell targeting and efficient introduction into the cell.

Finally, designer antibodies. Antibodies have the fantastic ability to recognize and bind to specific molecular structures. The potential applications are endless. With the recent development of *E. coli* systems for producing such monoclonal antibodies,[20,21] we now can envision inexpensive and abundant sources for any antibody we wish.

If current controversies and newspaper headlines are any measure, the public is equivocal about accepting modern genetic engineering as a productive technology. In a sense, this is paradoxical, because almost all of our food is a direct product of genetic engineering, and the new technologies are likely to provide, if anything, a more controlled application of genetics. My guess is that our school systems do not educate (and have not educated) our children adequately in genetics (How many people really know that biological properties are inherited?) or in biochemistry (Are life processes really controlled by chemical reactions?). The problem may reflect a general need for upgrading K through 12, but I worry that it may also reflect the fact that the science of genetics and biochemistry challenges some generally accepted philosophical norms. After all, is it not the basis for the creation-evolution controversy? Perhaps there is an almost deliberate wish by our educators to ignore these sciences. When one takes this perspective and adds in a distrust for high technology, is it surprising that the public might be suspicious when genetics and biochemistry are used so obviously? I think not. My only solution is that the scientific and industrial community must start taking education more seriously. A very important mechanism by which the private sector can further public education is through judicious product selection. The choice of products that address perceived needs is likely to encourage a higher comfort level with the technology. An important part of this process should be public safety concerns.

My final concern has to do with the fiscal health of the academic research community. The current situation is guarded and the prognosis is not good. Almost all academic scientists believe that the key to a healthy academic research community is the individual investigator-initiated research grant system. In NIH terminology, these are RO1 grants. The situation is close to desperate. Although the stated numbers vary (and this is symptomatic of one of the problems), the huge majority of competing grant proposals, including the majority of competitive renewal proposals, are not being funded. Why is this important? Without a healthy grant system, most academic research will cease, most graduate student and postdoctoral training will cease, and most undergraduates will get the message that scientific careers are not viable. This is at a time when 30% of our faculties will retire in the next 10–15 years.

How will this crisis affect the biotechnology industry? In the short run, it will probably be an asset. Talented scientists will be encouraged to drop out of academia, and those that stay will seek industrial ties. But in the long run, it will be a disaster. Five years from now, the employee pipeline will start to dry up. In addition, almost all of the basic scientific discoveries and technological developments described before were the products of individual investigator-initiated research projects. We will need new discoveries, but the research will not have been done.

What can we do? I suggest that two initiatives are needed. Number one, increase funding of individual investigator academic research. Little more needs to be said than "give me a Stealth bomber and help me to provide employment to technical help, secretaries, dishwashers, etc." Number two, let's get some leadership at NIH and let's have NIH (and Congress, if necessary) review its priorities. NIH first has to admit that there is a crisis. I haven't heard it yet. Second, NIH has to review the cost and validity of high indirect costs and of completely funding some faculty salaries. Let's have some good old-fashioned decisions about what NIH is supposed to be doing. Third, NIH has to realize that behind big science (e.g., the genome initiative) is small science—good old-fashioned molecular genetics. Finally, NIH has to have an equal playing field, that is, the same quality standards for large targeted grants as for individual grants.

Let me close by being honest. We have a potentially bright future, but if we are going to realize it, we need private industry's help, help in education, help in research funding, and help in lobbying.

REFERENCES

1. AVERY, O. T., C. M. MacLEOD & M. McCARTY. 1944. J. Exp. Med. **79:** 137.
2. WATSON, J. D. & F. H. C. CRICK. 1953. Nature **171:** 737–738.
3. GARROD, A. E. 1923. Inborn Errors of Metabolism. Oxford University Press. Washington, DC.
4. CRICK, F. H. C. 1970. Nature **227:** 561–563.
5. Cold Spring Harbor Symp. Quant. Biol. 1963 (28) and 1966 (31).
6. JACOB, F. & J. MONOD. 1961. J. Mol. Biol. **3:** 318.
7. BUCHNER, E. 1897. Dtsch. Chem. Ges. Ber. **30:** 117–124.
8. BERG, D. E. & M. M. HOWE. 1989. Mobile DNA. Am. Soc. Microbiol.
9. BERGET, S., C. MOORE & P. SHARP. 1977. Proc. Natl. Acad. Sci. USA **74:** 3171–3175.
10. CHOW, L., R. GELINAS, T. BROKER & R. ROBERTS. 1977. Cell **12:** 1–8.
11. SMITH, H. O. & K. W. WILCOX. 1970. J. Mol. Biol. **51:** 379–391.

12. OLIVERA, B. M. & I. R. LEHMAN. 1967. Proc. Natl. Acad. Sci. USA **57:** 1426–1433.
13. KORNBERG, A. 1974. DNA Synthesis. W. H. Freeman & Co. San Francisco, CA.
14. MAXAM, A. & W. GILBERT. 1977. Proc. Natl. Acad. Sci. USA **74:** 560–564.
15. SANGER, F., S. NICKLEN & A. COULSON. 1977. Proc. Natl. Acad. Sci. USA **74:** 5463–5467.
16. KHORANA, H. G. 1979. Science **203:** 614–625.
17. MARMUR, J., R. BOWND & C. L. SCHILDKRAUT. 1963. *In* Progress in Nucleic Acid Research. N. Davidson & W. E. Cohn, eds.: 231–300. Academic Press. New York.
18. WEINTRAUB, H., J. G. IZANT & R. M. HARLAND. 1985. Trends Genet. **1:** 22–25.
19. HASELOFF, J. & W. L. GERLACH. 1988. Nature **334:** 585–591.
20. WARD, E. S., D. GUSSOW, A. D. GRIFFITHS, P. T. JONES & G. WINTER. 1989. Nature **341:** 544.
21. HUSE, W. D., L. SASTRY, S. A. IVERSON, A. S. KOMG, M. ALTING-MEES, D. R. BURTON, S. J. BENKOVIC & R. A. LERNER. 1989. Science **246:** 1275–1281.

Role Played by International Meetings of Genetics of Industrial Microorganisms

MARIJA ALACEVIĆ

University of Zagreb
Faculty of Food Technology and Biotechnology
Pierottijeva 6
Zagreb, Yugoslavia

The late 1960s was a time when antibiotic-producing strains, especially tetracycline, were important and much work was done, mostly in industrial laboratories, to improve the technology, strain productivity, and biochemistry of compounds. Mutagenesis was used to improve strains, and auxotrophic and nonproductive blocked mutants were isolated, both techniques helping a great deal to determine the biosynthetic pathways of desired metabolites. Furthermore, marked strains were used to construct genome maps at a time when little was known about either the mechanisms of mutagenesis of different mutagens or the repair of DNA lesions.

Prof. Giuseppe Sermonti, who helped me to learn the different methods available, and I discussed at length the necessity of establishing more direct contact and exchange of experience between scientists in industry and those in academic laboratories. Because virtually no contact existed between scientists from academic institutions and those from industrial laboratories, we decided to organize an international meeting on *streptomycetes* and antibiotics, the most interesting industrial subjects at that time. The meeting in Dubrovnik, Yugoslavia, took place from May 31 to June 2, 1968. I would like to quote from remarks made by the President of the Meeting, Prof. Sermonti, and its Chairman, Prof. Hopwood.

After ten years we can make a first balance of the progress achieved both in the genetics and the breeding of *Streptomyces*. Pure genetics of *Streptomyces* is no longer in its infancy and has made considerable advancements mainly thanks to the work of the Glasgow group. The biochemistry of antibiotic biosynthesis by *Streptomyces* has also recorded great achievements, especially in the field of the tetracyclines. Considerable success has been made in the breeding of various antibiotic-producing *streptomycetes,* although I must admit that I am only vaguely informed about them.

There has been, in general, a lack of osmosis between theoretical genetics and practical microbiology. It must be said that Dr. M. Demerec was an exception to this rule, having been the first to introduce the mutagenic treatment in the selection of high rated strains of Penicillium.

Interesting attempts to fill the gap will be reported in this meeting, and I hope that some relevant information will emerge from the discussion. It is the chief aim for this Symposium to develop a fruitful dialogue between the two sides of the field, the theoretical and the practical. It would thus be desirable if the speakers dealing with pure genetics and biochemistry could emphasize the possible practical implications of their findings, and, vice versa, if the speakers dealing with breeding could put emphasis on the problems that they think would require a more refined theoretical approach.

(Giuseppe Sermonti)

This morning's session is devoted almost exclusively to a particular strain, A3(2), of *Streptomyces coelicolor*. This is, of course, not an antibiotic-producer, and has no known industrially interesting attributes. It simply happens to be the organism with which most of the work on streptomycete genetics has been done. Clearly it would be inefficient to

5

switch the major effort at this stage to some strain of typical industrial interest, which may not even be sustained. It is surely better to concentrate the main effort in "academic" research on A3(2)—just as most of the work on the genetics of enteric bacteria has been done with *Escherichia coli* K12 and *Salmonella typhimurium* LT2—but at the same time to compare genetic phenomena in other streptomyces with the A3(2) model to see to what extent techniques have to be modified to succeed with other strains. Some of the contributions later in this meeting will illustrate this approach.

(D. A. Hopwood)

We now know that many essential differences exist among *Streptomyces,* especially their productive strains, such as differences in map arrangement (empty zones); the role of different parts of the genome; gene amplification and instability; plasmid structure and stability; interspecific recombination; and the position of genes included in antibiotic biosynthesis. On the other hand, much data show a great similarity between parts of the genome included in primary metabolism that have led to different speculations on the development of the *Streptomyces* genome.

At the Dubrovnik meeting, as cited by the President and the Chairman, a significant advance in the research of radiation and chemical mutagenesis was shown as well as an already well-defined map of *S. coelicolor,* much data on the mutation and selection of productive strains and screening programs from different laboratories, and recombination processes, genetic analysis, and mapping of genomes of *Streptomyces,* primarily *S. coelicolor.* Also discussed were the mutation in *Streptomyces* from research into the effects of radiation and chemical agents on DNA, the possibility of selecting better producers in different screening programs, and the role of block mutants in the biosynthesis of tetracyclines (chlortetracycline). The study of actinophages was also presented in one lecture.

At this meeting a group of scientists, including Sermonti, Hopwood, Alikhanian, McCormick, Vanek, and myself, discussed the necessity of organizing regular meetings on the genetics of industrial microorganisms with the aim of improving collaboration between academic and industrial scientists.

The first general international meeting was organized in Prague in 1970. Vanek, in his review of the meeting, remarked that throughout the decades microbiology and genetics have been the foundation for the great development of the fermentation industry which not only has produced extremely valuable drugs and chemicals but has also become important in the production of food feed and in solving ecological problems.

With progress in the antibiotic industry and with more funds for basic research into molecular biology and formal genetics, great advances occurred in our knowledge through joint research.

The most characteristic feature of industrial microbiology was the proliferation of studies into molecular genetics, the results of which brought us to the fundamentals of biochemical and genetic control of microbial metabolism. This gave rise to what was to be called molecular biology and the newest methods for tailoring and constructing new strains. The rapid development of industrial microbiology and genetics proceeded almost independently, although their basis as well as their results led to similar ends. The start was determined by the work of industrial microbiologists, whereas genetics remained almost entirely basic and theoretically oriented. Those working with industrial microorganisms were to profit from a knowledge of genetics and new methods, acquiring a rationale for their work and hence intensifying their search for less empirical but quicker results for improving their strains. Geneticists were able to refocus their research from formal genetics and the control of primary metabolic pathways to the level of formation of special products and to extend the range of models for research. The mutually coupled investigation in this direction thus opened the way for other contributions of microbiology in solving problems of fundamental biology, for example.

In other words, only a complete understanding of the physiological metabolic and genetic laws could help uncover how and why different products are formed in an organism during cell differentiation. The subjects of the Prague meeting were similar to those of the Dubrovnik meeting, but the number of presentations and participants and the breadth of knowledge were greater.

To achieve excess production of the desired compound, industrial scientists invented many ingenious tricks for overcoming the regulatory mechanisms. This demonstrated that industry employed outstanding scientists who could convert the basic research data to practical ends in a remarkably short time.

There are many examples to document that through the use of fundamental knowledge in strain selection, backward pressure was exerted on pure science, stimulating the solution of further questions of fundamental importance. Beadle and Tatum used the "industrially unsuitable" model of *Neurospora crassa* to study the "industrially uninteresting" auxotrophy that later became the basis of molecular biology, and Kellner first observed photoreactivation in *S. griseus* while working on an industrially interesting strain that explained one of the repair processes in cells.

The whole area, however, was mutually criticized. Doubt arose concerning the suitability of model organisms that were studied, and feedback from industrial microbiologists was almost completely lacking.

The first indisputable achievement of the Prague symposium was that the intended bringing together of "applied" microbiologists with those scientists who contributed to our knowledge of genetic regulation was fully appreciated. This made it possible to confront our present fundamental knowledge with the needs of practical industrial application.

The goal of bringing together the theoretical and the practical scientists at the Prague meeting was achieved, but how was the collaboration to proceed? This problem was very neatly addressed in the final lecture given by A. L. Demain entitled, "Marriage of Genetics and Industrial Microbiology—After a Long Engagement, a Bright Future." Demain was not sure when the bright future would start: "The partners have not yet fully committed themselves to working toward the mutual benefit of the pair, each going down a separate path to a considerable extent. If the union is ever consummated, we can look forward to a bright future for both."

Vanek concluded that it would be immodest to expect miracles from a meeting of this type; still, it cannot be overlooked that the symposium succeeded in focusing attention on theoretical problems, defining fields of future study and indicating some methodical approaches to be used. It was, above all, obvious that a successful solution would require a complex combined approach by geneticists, physiologists, enzymologists, chemists, bioengineers, and physicists. The conclusion that the symposium would represent just the beginning of a regular series of symposia was unanimously accepted.

Simplicity of the mutation techniques and the ease with which microorganisms could be changed by mutation attracted industrial microbiologists and managers, and extensive programs of mutation and selection in strain improvement still exist today in industrial laboratories throughout the world. Fifty years later, it is clear that mutation has been the major factor in up to a thousandfold increase in the production of microbial metabolites, without there being much knowledge about the mechanisms of their action. Today, at a time when geneticists are faced with possibilities unimaginable 50 years ago, we still need to rely occasionally on the old methods, especially selective ones.

Unfortunately, the randomly selected mutants used to increase the production in industry were less attractive for geneticists, and although changes in colonial and cellular morphology were very useful in the selection of active strains, there was no real understanding of what was happening in the cell and colony.

A combined approach based on a knowledge of fundamental research and mutagenesis, life cycle, and physiology presents us with the best approach to strain improvement.

To illustrate this approach, in the selection of alkaloid-producing strains we used strains that were totally nonproductive in a submerged culture but productive in the parasitic stage. First, we carefully studied the cellular and mycelial morphology of the parasitic strain in its productive phase and then we used several mutagens with known mechanisms and conditions. We cultivated the mutated cells together to allow contact of survivals to form multinucleate or diploid cells as they are in sclerotial phase (productive phase) in the parasitic strain. The result was excellent and quick.

To choose the best mutagen, a knowledge of the mechanisms of mutagenesis was needed. Particularly interesting was the discovery that mutation methodology could be used to produce entirely new molecules. A further use of mutants was to elucidate secondary metabolic pathways. The isolation of a good producer, however, has raised many questions, and numerous scientists have tried to explain what really happens in the cell despite many years of observing the phenomenon and its successful use in production.

The use of genetic recombination to improve yields of fermentation products was disappointing, despite the fact that recombination is supposed to be the major source of genetic variation and that many types of recombination occur in microorganisms. The failure of classical recombination with the exchange of large parts of the genome in strain improvement may be explained genetically by the instability of diploids, the recessive nature of genes that lead to superior production, and the lack of methods to select the desired high-producing recombinants. The results show that mutation is the source of the variability and that natural recombination is the way of stabilizing the natural relationship between features in the cell and its existing environment.

Presumably because of the need to maintain chromosomal integrity, recombination *in vivo* requires nearly perfect homology and is abolished by as little as 10% sequence divergence. This may influence the recombination frequency between divergent strains. The mismatch repair system might prevent recombination between partially homologous sequences and provide a functional barrier to interspecies recombination. The authors have now tested the prediction that mutants defective in mismatch repair can become proficient in interspecies recombination by using members of two related bacterial genera, *E. coli* and *S. typhimurium,* which are about 80% homologous in the DNA sequence. Mutations in the *mutL, mutS,* or *mutH* mutator genes inactivate methyl-directed mismatch repair of errors in DNA replication. Rayssiquier *et al.* showed that these mutations can increase intergeneric recombination in conjugational and transductional crosses up to a thousandfold. Hence, mismatch recognition and the mismatch repair system can act as a barrier to recombination between DNAs of different species. The use of technologies that allow interspecific recombination may be wide ranging in future biotechnology.

At the Sheffield meeting of Genetics of Industrial Microorganisms (GIM) in 1974, Pontecorvo stated:

> One thing is clear to the outsider. The advances in the applications of genetics to the improvement of strains of industrial microorganisms are trifles compared to the advances in the fundamental genetics of microorganisms.
>
> With a general basis of knowledge and techniques as formidable as those in microbial genetics it is very disappointing to see how far behind are the applications to the improvement of industrial microorganisms.

The main technique used is still a prehistoric one: mutation and selection. In 1940 independently Demerec, in the USA, and myself in Scotland proposed this technique only as a war-time emergency measure for improving penicillin yields. Penicillin was desperately needed then and even an approach intellectually crude and, a priori, not very likely to be successful was worth a trial. The success was so unexpectedly good that, unfortunately, since then most industrial laboratories have been contented with its exclusive use.

Sexual reproduction, combined with diploidy, is the most highly elaborate of these mechanisms. It is at least 25 years since Müller showed in a simple graphic way how mutation and selection by themselves are so much less effective than in combination with processes of transfer of genetic information.

If industrial laboratories were doing successfully something along these lines, we would not hear about it.

Why does this enormous gap exist between basic knowledge and its application? Pontecorvo mentioned two reasons: fragmentation of effort and predominance of chemical outlook in the microbiological industries. Let me deal with this second point first. Most of the fermentation industries are offshoots of the chemical and, more particularly, pharmaceutical industries. Naturally in these mother industries the predominant outlook was that of the organic chemist. This idea was carried over into the fermentation industries, later only partially infiltrated by biochemistry.

The second reason for the disappointing state of the application of genetics to the improvement of industrial microorganisms is the fragmentation of effort. Every industrial concern has its own miniteam working in secrecy and trying to produce more desirable strains. The absurdity of this situation can be illustrated by an analogy: "agriculture in Great Britain is a highly fragmented, privately run industry vastly more important than all the microbiological industries taken together. Yet, the improvement of varieties of crop plants, which was mainly an individual concern up to a century ago, is now concentrated in three main plant-breeding stations. They are quite successful and the farming community seems to be happy about them. No farmer in his senses, no matter how large his farm, would dream of breeding his own varieties of crop plant and, in addition, keeping them jealously for himself. Recruitment of first class geneticists to the plant-breeding stations was no problem at all. The charge that academic geneticists are reluctant to turn to full-time applied work, quite correct in the case of the microbiological industries, is wrong in the case of agriculture. What should we do about all this? The problem is how to have industrial plants and know-how directly connected with teams of first-class geneticists working on strain improvement. In analogy to plant breeding, this improvement can be carried out efficiently only in a few highly specialized centres."

"I realize that I have dropped a few huge bricks, but I hope they will lead to a serious reappraisal of the situation," Pontecorvo concluded.

Pontecorvo's words coincided with the explosion of new knowledge, after which closer relations between the two unexpected partners began to develop. However, the problem was much more profound than the foregoing technical difficulties indicate. Given that more was known about the mechanisms of mutagenesis and recombination, the approach to strain improvement through classical recombination could have been more successful. Progress in microbial specialty molecular genetics over the last 30 years was truly fantastic and it has shown how knowledge may be directly applied when conditions on both sides are ripe. I believe that this signaled the long-awaited marriage that Demain had referred to earlier.

It has now been demonstrated that the division in industrial or academic strains is no longer pertinent. The most academic strain, *E. coli*, has become industrial and every industrial species, such as yeast and some *Streptomyces*, has become academic.

Although the two GIMs of 1974 and 1978 dealt with similar problems, it was evident that the accumulation of basic knowledge was about to explode with something new. The Kioto meeting, with its subjects or rather its focus on methods used, was totally different from the previous ones.

In vitro genetic engineering of industrial microorganisms, which was the main subject in both Kioto and Split meetings, stems from a vast knowledge of *in vivo* genetics of *E. coli* K-12 and its phages and plasmids. The applications of gene cloning in *E. coli* include natural genetic phenomena as well as unnatural ones. Other "academic" strains, such as *Bacillus subtilis* and *Saccharomyces cerevisiae*, also were immediately included in new research and became important hosts for recombinant DNA because they were well established in other genetic work. In *Streptomyces* strains the situation was slightly different because *S. lividans*, instead of *S. coelicolor*, became the first experimental organism in genetic engineering. Protoplast fusion, alone, although unsuitable for fundamental research, was useful in the selection of some types of production, and naturally the knowledge of methods for protoplast and mass regeneration helped in recent DNA experiments.

The importance of traditional mutation and recombination methods for improving industrial strains was emphasized in the panel discussion convened by the chairman of GIM. Progress in these methods as well as in the techniques of site-directed mutagenesis and chemical synthesis of DNA was referred to frequently in many presentations.

The GIM International Symposia have become very important in the field of biotechnology, and the topics reported at GIM-82 and GIM-86 drew great attention worldwide.

The last 20 years have shown the dramatic development of new methods for effecting genetic changes in industrial microorganisms. We are already applying all these methods. Their more useful results, however, will come from combinations of different procedures. We can now obtain desirable new combinations of nonhomologous DNA segments from the two species, and with a sufficiently powerful selection, we can hope to produce a complex new genotype that cannot be constructed rationally because its components cannot be identified in advance.

We are indeed living in an exciting time for the genetics of industrial microorganisms with its endless possibilities. The close collaboration between academic and industrial microbiologists has taken a long time to be realized. Pontecorvo, however, would be gratified to witness the proliferation of highly specialized centers like those he admiringly made reference to in his plant-breeding analogy.

REFERENCES

1. SERMONTI, G. & M. ALACEVIĆ. (Eds). 1968. Genetics and Breeding of *Streptomyces*. Proceedings of International Symposium. Dubrovnik, 1968.
2. VANEK, Z., Z. HOSTALEK & J. CUDLIN. (Eds). 1973. Genetics of Industrial Microorganisms, Parts A and B. Proceedings of International Symposium. Prague, 1973.
3. MCDONALD. (Ed). 1976. Genetics of Industrial Microorganisms, Proceedings of the Second International Symposium on Genetics of Industrial Microorganisms. Sheffield, 1974.
4. SEBEK, O. K. & A. I. LASKIN. (Eds). 1978. Genetics of Industrial Microorganisms. Proceedings of the Third International Symposium on Genetics of Industrial Microorganisms. Madison, 1978.
5. IKEDA, Y. & T. BEPPU. (Eds). 1982. Genetics of Industrial Microorganisms. Proceedings of the Fourth International Symposium. Tokyo, 1982.

6. ALACEVIĆ, M., D. HRANUELI & Z. TOMAN. (Eds). 1986. Genetics of Industrial Microorganisms. Proceedings of the Fifth International Symposium. Split, 1986.
7. RAYSSIQUIER, C., D. S. THALER & M. RADMAN. 1989. The barrier to recombination between *Escherichia coli* and *Salmonella typhimurium* is disrupted in mismatch-repair mutants. Nature **342** (6248): 396–401.
8. DIDEK-BRUMEC, M., A. PUC, H. SOCIC & M. ALACEVIĆ. 1987. Isolation and characterization of a high-yielding *Claviceps purpurea* strain producing ergotoxines. Prehrambenotehnolo. biotehnol. revija **25:** 103–109.

Ethics, Trust, and the Future of Recombinant DNA

THOMAS H. MURRAY

Center for Biomedical Ethics
School of Medicine
Case Western Reserve University
Cleveland, Ohio 44106

There is a tendency, as unfortunate as it is understandable, to focus on the most dramatic and direct ethical issues raised by recombinant DNA research and applications. So much has been written about such issues as human gene therapy[1] and so much intellectual energy diverted to them that other issues that may be more important in the long run get insufficient attention. In a conference on applications of genetic engineering, such as the one reported in this volume, a paper on ethics can take either of two directions. It can rehearse a set of well-known and exhaustively discussed issues or it can attempt to address issues that may be every bit as important to the long-term fate of genetic engineering, but less frequently addressed in forums such as this one. This paper takes the latter course. I discuss two relatively neglected but potentially significant issues: the centrality of trust and its implications, and the need to be attentive to indirect, unintended impacts of genetic engineering applications.

WHY TRUST IS IMPORTANT

Psychologists and psychiatrists emphasize the fundamental significance of trust in human development and human relationships.[2] Trust may be no less important to developing a science and technology or to sustaining a mutually beneficial and supportive relationship between social institutions than it is to individual development or personal relationships. My claim here is that scientists' willingness to deal openly (more or less) with the first significant moral issue in recombinant DNA research—laboratory safety—was important in allowing the early development of the science, that continuing openness on particular applications has helped ensure their public acceptability, and that the future health of genetic engineering depends upon continuing, indeed improved, openness as a prerequisite to trust between scientists and the public.

An important distinction must be made here between basic research and the development of applications, and how these two endeavors relate differently to values. In general, basic research serves a value that is shared by scientists and the public—knowledge. Like all values, knowledge is not absolute. We accept restrictions on the things we can subject other people, or animals, to in the quest for knowledge. We insist that experiments not pose avoidable and direct dangers to surrounding communities. There may be some subjects—knowledge that would improve the effectiveness of torture or the construction of human-ape chimera, for example—that we think ought not to be pursued at all. But for the most part, scientists and nonscientists alike value the pursuit of new knowledge very highly.

Where scientists and nonscientists differ dramatically is in their ability to judge what research opportunities are most likely to result in significant new knowledge.

In contrast, new applications may affect a number of different values. The decisions of whether and in what manner to permit a new application to be made available require consideration and balancing of many values, some of which will be enhanced and some threatened. Scientists and engineers, expert as they may be on technical aspects of the application, hold no special expertise on the value choices that ought to be made in deciding on the desirability and shape of a particular application. Technical knowledge is essential in making applications such as biotechnology, but other values supersede when the question is what do we do with this new tool?

TRUST AND MISTRUST

Mistrust thrives on misunderstanding and disrespect. Scientists are prone to regard public apprehension about research familiar to the scientists as just so much ignorance or superstition. Public fears may be based on misinformation. Nevertheless, scientists make a mistake when they dismiss such fears for at least two reasons.

First, failing to take the public's concerns seriously is imprudent, not in one's own interest. If people believe that their deep-felt fears are being dismissed, they may react with anger. Should there be widespread sentiment among the public that scientists are unsympathetic and arrogant, the harm that might be done to science could be enormous. In its early years, recombinant DNA research flirted with this possibility. Precisely how it was averted will be for historians to tell, but two factors were probably important: scientists themselves were the first to identify the possible danger from genetic engineering research, and the passage of time without any of the horrific scenarios predicted by critics of the research reassured the public that the research was indeed safe.

In addition to the concern about the public's general response, it must be acknowledged that any human enterprise as vast and diverse as recombinant DNA research in this country is bound to experience both errors and abuses. Errors are just that—unintentional mishaps that cause harm or merely the possibility of harm. Documented abuses may be more difficult to find, but when they occur, they shake everyone's confidence. The US military's effort to slip a P-4 laboratory through an appropriations procedure intended for minor and uncontroversial changes was either an instance of remarkable ignorance or an outright abuse.[3]

If errors and abuses are virtually inevitable, must they necessarily destroy public trust in the larger enterprise? The answer is no, if the public has a basic attitude of trust towards the scientists conducting the research. Against a background of trust, occasional errors and even abuses will be seen as warnings perhaps, but not proof that the worst will happen or that scientists are fundamentally untrustworthy. If basic trust is lacking, then even innocent errors may be given the most dire interpretations.

The second reason that scientists should not dismiss public worries is that the public may be right—in its general concerns, if not in the details. Scientists and engineers are likely to be enthusiasts for their work; we would not wish it otherwise. The nature of enthusiasm is such that it highlights the advantages and possibilities and takes less notice of disadvantages and dangers.

An example of such an enthusiastic idea in genetics began over 50 years ago with a suggestion by J. B. S. Haldane that prospective workers in the potters' trade might be screened to identify those with genetic susceptibilities to bronchitis, a common ailment among pottery workers.[4] Those especially susceptible could be steered to another industry, Haldane suggested, and his idea was picked up several decades later,[5] leading eventually to two studies by the US Congress Office of Technology Assessment.[6] Neither Haldane nor later proponents of the idea saw beyond the hoped-for public health benefits (of fewer workers with occupational illnesses) to the potential misuses of such genetic screening programs such as using screening to exclude racial or ethnic minorities and using screening to save money for employers, irrespective of its effect on those seeking work.[7] (It is essential to stress the word *potential* here, because the best available evidence indicates that little such screening is actually done.) Public discussion revealed potential harms of workplace genetic screening that apparently were never contemplated by its scientific proponents.

THE ASILOMAR CONFERENCE: TWO INTERPRETATIONS

The conference held in Asilomar, California in 1975 to develop guidelines for containing the risks of recombinant DNA research is well known and well documented[8] as is the 1974 letter by Paul Berg and other distinguished scientists that urged a moratorium on certain lines of recombinant DNA research[9] and was one of the factors leading to the Asilomar conference. This article is not the place to discuss the Asilomar conference in general, but rather as a signal event in the trust, or mistrust, between science and the public.

The context within which Asilomar and related events occurred is important in understanding how and why they happened. The late 1960s and early 1970s were a time in which authority of any type was more readily questioned. Science did not escape this increased scrutiny. Vietnam War related research on campuses sparked demonstrations, confrontations, and debates. The contemporary bioethics movement was born. Articles on the public accountability of scientists appeared.[8] Qualms about scientists' social responsibility for atomic weapons contributed to a growing public willingness to question the trust with which science and scientists had tended to be regarded. Asilomar took place in the midst of this ferment, a situation ripe for a major breach of trust between science and the public.

Two interpretations of Asilomar have been offered. On the one hand, it can be regarded as an applaudable example of scientists' responsiveness to the public, and as an example of responsible behavior, in which scientists created self-imposed limits on scientific inquiry. One implication of this interpretation is that scientists themselves made an implicit but unambiguous rejection of the proposition that scientific inquiry must be unfettered by any values other than the search for new knowledge.

Another interpretation is possible. In a fair-minded review of Asilomar, Grobstein[10] describes the point of view taken by critics: that Asilomar was "public decision making without appropriate public access, in extreme terms, a covert exercise of power by a special interest group." One critical account of the conference argues that:

> To secure the goal of disciplinary autonomy, the organizers of Asilomar had accomplished two objectives: (1) they defined the issues in such a way that the expertise remained the monopoly of those who gain the most from the technique, and (2) they chose to place authority for regulating the use of the technique in the agency that is the major supporter of biomedical research in the United States. (p. 153)[8]

A relative handful of nonscientists were present at Asilomar, mostly lawyers and journalists, but their numbers paled in comparison to the 150 scientists who participated. The discussion focused narrowly on the immediate question of laboratory safety, and did not expand to include several other questions about risk and value that soon emerged as important.

Both interpretations of Asilomar are plausible, partial portraits. I believe Grobstein[10] is correct in his characterization of its virtues and flaws:

> The public image of science was humanized and thereby enhanced. Asilomar is not interpreted generally as a brilliant conspiracy but as a conscientious effort by well-intentioned people who did not know what was best to do but did their best despite conflicting personal interests and inclinations. Asilomar also protected the extension of knowledge for a smaller price than might otherwise have been paid. (p. 7)

Asilomar, the Berg letter, and other events that demonstrated that scientists were willing to consider their responsibility to the public critically and openly (at least to a degree) closed the distance between science and the public, and strengthened trust.

Some scientists regret Asilomar and the extensive public involvement over recombinant DNA research. Their regret is understandable: the process is time-consuming, exhausting, and frustrating; one feels misunderstood and misrepresented.

The current state of recombinant DNA research is eloquent testimony to the long-term value of openness. The research thrives, the safety record is exemplary, and although public debate continues—appropriately, I believe—on a wide range of ethical and value issues concerning such research and technology development, the relationship between science and the public is generally a good one.

The best assurance that the future of recombinant DNA research will be both scientifically vigorous and publicly responsible is a continuing and improved openness to public deliberation about those issues that are not the sole province of scientific experts. Asilomar may have been a success as a demonstration of good intentions and a willingness to be more open, but as an example of sound public policy-making it left much to be desired. Grobstein again:

> In an open society public decision making requires access by all interests who will later be called upon to support the decision. Private caucuses have their place, but they must not be confused with public decision arenas. Nor should private caucuses be allowed to set the public agenda. This subverts democratic processes of policy making. The scientific community should be seeking additional ways to cooperate with the larger community in making policy . . . (p. 10)[10]

HUMAN GENE THERAPY

The first approved human trials using recombinant DNA have just begun. They stand, I suspect, as the most rigorously and exhaustively reviewed experiments in the history of scientific inquiry. The first approved experiment, tagging tumor-infiltrating lymphocytes (TILs) with a bacterial gene for neomycin resistance, went through 16 levels of review before being approved by the Director of NIH.[11] The very first genuine gene therapy experiment, an attempt to develop a treatment for a lethal immune defect caused by deficiency in the enzyme adenosine deaminase, has just begun after years of effort and public review.

Human gene therapy, more than any other manifestation of recombinant DNA research, has inspired public awe and fear. It inspired a 1980 letter to then-President Carter by leaders of the three largest religions in the United States,[12] a report by a Presidential Commission,[13] and countless articles by scholars and journalists. This public scrutiny has made the road to the first trials long and, undoubtedly for the scientists, tedious. Yet, as with Asilomar, the result is clear: the public seems more intrigued and hopeful than afraid. However painful the process may have seemed to the scientists involved, the ethics of human gene therapy have been exhaustively and publicly reviewed, and the path to future research looks smooth. We have progressed from fantasies of "flying nuns" to hope for improved treatments for cancers, congenital immune deficiencies, and other dreaded diseases.[14]

OPENNESS AND TRUST

The examples of Asilomar and human gene therapy suggest that a relationship exists between the openness with which scientists treat an issue and the public's (eventual) willingness to trust those scientists. Trust is an exceptionally important value, sturdy enough to withstand temporary suspicions, but difficult to rebuild once it is destroyed. The trust between scientists and the public serves both parties well, as long as both not only remain trustworthy, but also increase their mutual openness and understanding.

SUBTLE CHOICES, MAJOR CONSEQUENCES

Public attention has tended to be focused on the direct effects of genetic engineering, especially potential or imagined threats to human health. Beginning with "andromeda strains" and moving to the possible deleterious impact of field releases of recombinant organisms, the debates over the wisdom of various applications of genetic engineering have been less attentive to more subtle and indirect, but potentially significant, consequences. For example, there are increased or new capacities to manipulate human physiology as rare bioactive molecules are produced in quantity by inserting human genes into microbes or cell cultures.

New Capacities to Manipulate Human Form and Function. People have always, so far as we can tell, attempted to manipulate their appearance in a myriad of ways. From ancient cosmetics to modern "body sculpting," people have sought to mold their bodies to desired forms. With equal ingenuity, they have sought to control its function. In the ancient Olympiads, athletes used herbs or mushrooms in the belief that they could enhance their performance in the games, gaining an edge over their competitors.

Recombinant DNA applications have already had a major impact on our ability to manipulate both form and function. Biosynthetic human growth hormone (hGH), available since the mid-1980s, has made it possible to treat many more children who do not produce sufficient bioactive hGH to sustain normal physical growth. But, the same hGH, it is suspected, can boost the ultimate height that might be attained by a child with normal levels of hGH. It may also increase the bulk and perhaps the muscle mass and strength of adults.

In like manner, biosynthetic erythropoietin (EPO) is now available in quantity. EPO stimulates the production of erythrocytes, increasing the oxygen-carrying capacity of the blood. Individuals afflicted with chronic anemia can derive enormous benefit from EPO, as their red cell count approaches normality. Such has been the case with patients on chronic renal dialysis, who typically experience chronic anemia with a resulting lack of energy. But EPO may also increase the number of red cells in the blood of people with a baseline normal count of such cells.

A technique known as blood-doping has been used by athletes to increase their endurance.[15] Blood-doping consists of infusing red cells or whole blood sufficiently in advance of competition so that the excess fluid volume is excreted, but the red cell count remains high. The theory is that the higher proportion of circulating red cells increases the blood's capacity to carry oxygen to exercise-starved tissues, staving off exhaustion. Athletes, particularly those engaged in sports in which endurance is a major factor, may use EPO in the hope that it will give them a competitive advantage, just as they have used other substances or techniques thought to enhance performance, such as anabolic steroids.[16]

Recombinant DNA applications may afford the possibility to perform more powerful alterations of physique or physiology than were imaginable 20 years ago. This power, like all power, is susceptible to misuse. For one, using hGH to add a few inches of height to someone who is not physiologically deficient and would have been within the normal height range anyway is a morally questionable employment of a potent technology. In a society that tends to ascribe desirable attributes to taller people, irrespective of merit, height can be an advantage. But here, the use of a potent drug is directed at solely competitive social ends, not at the treatment or cure of any disease or disability. In addition, because biosynthetic hGH is quite expensive, it will be more readily available to the relatively wealthy. If the wealthy are permitted to use it for their children, then we shall have added height to the advantages money can buy, adding inequality to inequality.[17] If parents are permitted to purchase hGH for their children, potentially large numbers of children could be exposed to the discomfort and risks of such injections.

EPO, which may bring enormous benefit to hundreds of thousands of people with chronic anemia, may also be used by athletes to improve their stamina and gain a competitive advantage. Such a pattern of use is consistent with the experience of this author and other members of the US Olympic Committee's Committee on Substance Abuse Research and Education. Athletes who use EPO in this way risk severe harm and even death from dangerously elevated plasma hemoglobin from a combination of EPO-stimulated erythrocyte production and loss of plasma volume during extensive exercise.[18] No effective means of controlling EPO use by athletes has been devised. Merely finding it in body fluids is insufficient because the biosynthetic form is indistinguishable from the endogenous form.

Society is forced to make choices in order to cope with the indirect, subtle consequences of recombinant DNA applications such as hGH and EPO. Shall we permit parents to buy hGH and arrange for its administration to their normal children? Should we attempt to discourage such use, and by what means? Thus far, manufacturers of hGH have controlled its availability by distributing it only through certain hospital pharmacies. If preliminary research showing that it may have benefits for restoring muscle and strength to elderly persons holds up, the demand for hGH may explode, and the current control strategy prove unworkable. So far, the use of EPO for performance enhancement appears limited to sport. But we can easily envisage other fruits of recombinant DNA research for which there will be socially problematic uses touching other spheres of human activity. Recognizing and dealing with them will be a very difficult challenge.

CONCLUSION

The future vigor of recombinant DNA research and applications, such as those described elsewhere in this volume, depends heavily on the ability of those conducting the research to remain trustworthy in the eyes of the public. Despite the powerful temptations to avoid the effort open communication with the public entails, and the frustration that sometimes occurs, scientists should be open about the potential risks of their research, and they should seek a dialogue with the public when there are value-laden decisions to be made that are not solely the province of scientific expertise. Scientists—and the public—must struggle to create more effective ways to communicate with each other. This approach, with its visible costs to scientists, is both ethically more defensible and socially more robust than is a secretive attitude that may appear to make life easier in the short run.

An attitude of openness will also be important in dealing with the subtle, indirect impacts of recombinant DNA applications. We cannot know the precise shape the challenges to our institutions and ethics of such developments will take. But we can anticipate that there will be serious challenges. The ultimate impact of recombinant DNA, whether it is a boon to humankind or a bane, will depend heavily on our capacity to confront these challenges openly, placing it in the service of our most deeply held and durable values.

REFERENCES

1. NICHOLS, E. K. 1988. Human Gene Therapy. Harvard University Press. Cambridge, MA.
2. ERIKSON, E. H. 1963. Childhood and Society. 2nd ed. W. W. Norton. New York, NY.
3. SMITH, J. R. 1984. New army biowarfare lab raises concerns. Science **226**: 1116–1118.
4. HALDANE, J. B. S. 1938. Heredity and Politics. Allen and Unwin. London.
5. STOKINGER, H. E. & SCHEEL, L. D. 1973. Hypersusceptibility and genetic problems in occupational medicine—a consensus report. J. Occup. Med. **15**(7): 564–573.
6. U.S. Congress Office of Technology Assessment. 1983. The Role of Genetic Testing in the Prevention of Occupational Disease. U.S. Government Printing Office. Washington, DC.; Genetic Monitoring and Screening in the Workplace. 1990.
7. MURRAY, T. H. 1983. Warning: Screening workers for genetic risk. Hastings Center Rep. **13**(1): 5–8.
8. KRIMSKY, S. 1982. Genetic Alchemy: The Social History of the Recombinant DNA Controversy. MIT Press. Cambridge, MA.
9. BERG, P. *et al.* 1974. Potential biohazards of recombinant DNA molecules [letter]. Science **185**: 303.
10. GROBSTEIN, C. 1986. Asilomar and the formation of public policy. *In* The Gene Splicing Wars: Reflections of the Recombinant DNA Controversy. R. A. Zilinskas & B. K. Zimmerman, eds. Macmillan. New York, NY.
11. MARWICK, C. 1989. Preliminary results may open door to gene therapy just a bit wider. JAMA **262**(14): 1909.
12. Letter to President Jimmy Carter, June 20, 1980, signed by general secretaries of the National Council of Churches, Synagogue Council of America, and U.S. Catholic Conference.
13. President's Commission for the Study of Ethical Problems in Medicine and Biomedical and Behavioral Research. 1982. Splicing Life: The Social and Ethical Issues of Genetic Engineering with Human Beings. U.S. Government Printing Office. Washington, DC.
14. FRIEDMANN, T. 1989. Progress toward human gene therapy. Science **244**: 1275–1281.
15. KLEIN, H. G. Blood transfusion and athletics: Games people play. New Engl. J. Med. **312**: 854–856.
16. MURRAY, T. H. 1983. The coercive power of drugs in sports. Hastings Center Rep. **13**: 24–30.
17. MURRAY, T. H. 1987. The growing danger from gene-spliced hormones. Discover Feb.: 88–92.
18. COWART, V. S. 1989. Erythropoietin: A dangerous new form of blood doping? Physician Sports Med. **17**: 115–118.

Long-Distance deoR Regulation of Gene Expression in *Escherichia coli*[a]

GERT DANDANELL,[b] KJELD NORRIS,[c]
AND KARIN HAMMER[d]

[b]*Institute of Biological Chemistry B*
University of Copenhagen
Sølvgade 83
DK-1307 Copenhagen K, Denmark
[c]*Novo Research Institute*
Novo Allé
DK-2880 Bagsværd, Copenhagen, Denmark
[d]*Department of Microbiology*
Technical University of Copenhagen
DK-2800 Lyngby, Copenhagen, Denmark

During the last decade "regulation at a distance" has reportedly been involved in the regulation of several very well-studied operons. In the *gal* operon, for example, a second operator involved in repression by the gal repressor was found within the *galE* structural gene (FIG. 1).[1,2] Also, in the arabinose system, a second operator, *araO₂*, 210 base pairs (bp) from *araI* was required for efficient regulation (repression) of the *araBAD* operon.[3] A periodicity in repression was first discovered in the arabinose system by changing the distance between the operator sites. With these distances, repression of *araBAD* only occurred when the ratio of the distance to the number of base pairs per helical turn was an integer.[3] It was concluded that both operator sites have to be on the same side of the DNA molecule to allow cooperative binding of the araC protein, suggesting a DNA-loop formation. Also in the *deo* operon, regulation at a distance was shown to be responsible for repression of the two promoters *deoP1* and *deoP2*. In this system, the distance between the two operators was unusually long for prokaryotic systems (279, 599, and 878 bp). FIGURE 1 shows some of the operons from *Escherichia coli*, in which regulation at a distance has been studied extensively. This list includes the *lac* operon, which has become the model of prokaryotic gene regulation. Although the three *lac* operators were identified and studied as early as 1974,[4] the importance *in vivo* of the second and third operators was not comprehended because their affinity for lac repressor was 10- and 30-fold lower than that of the primary operator site. Recently, the importance of lac regulation was shown both *in vivo*[5–8] and *in vitro*.[9–11]

The λ repressor is also included in FIGURE 1. This repressor binds to adjacent operators, but by separating the λcI operators on the DNA, λcI repressor was also shown to be active at a distance.[12,13]

This paper reviews the long-distance deoR regulation, the DNA chelate model with some of the data on which it is based, and we present new data supporting this model. This will illustrate how regulation at a distance might be engineered to fulfill special needs in genetic engineering. For previous reviews about DNA looping we

[a]This research was supported by grants from the Danish Natural Science Research Council and the Centre of Microbiology.

19

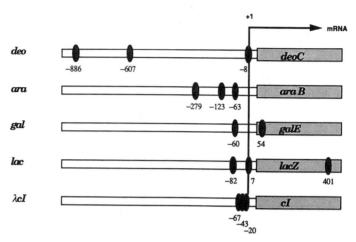

FIGURE 1. Comparison of some of the most extensively studied operons for which regulation at a distance has been demonstrated. *Circles* indicate the position of the natural operators that affect transcription from the promoter initiating at +1 (*arrow*). *Numbers* indicate the centers of the operators relative to the start site at +1 (CRP-activated promoter in *deo, gal,* and *lac* operons, and P_{RM} in case of λcI). In the *deo* operon, operators are $O_E(-886)$, $O_1(-607)$, and $O_2(-8)$.[24,25] In the *araBAD* operon the operators are: $araO_2(-279)$, $araO_1(-123)$, and $araI(-63)$.[3,29] In the *lac* operon the operators are: $lacO_3(-82)$, $lacO_1(7)$, and $lacO_2(401)$.[4,30] In the λcI gene the operators are $O_{R1}(-67)$, $O_{R2}(-43)$, and $O_{R3}(-20)$.[31]

recommend that the reader consult references 14–17. The chelate model is particularly discussed in references 14, 15, and 17.

THE *deo* OPERON

The structure of the *deo* operon, which encodes enzymes that are involved in nucleoside catabolism, is shown in FIGURE 2. Two promoters, *deoP1* and *deoP2,* are regulated by the deoR repressor. When derepressed, both promoters initiate tetracistronic transcripts. *In vivo* derepression (induction) occurs when *E. coli* grows in the presence of deoxyribonucleosides which are catabolized and converted to deoxyribose-5-phosphate, the true internal inducer for the deoR regulation of the *deo* operon. Mutations in the *deoR* gene also cause derepression.

The deoR operators overlap the −10 region in *deoP1* and *deoP2,* respectively, indicating that the mechanism of repression is a steric hindrance of the open-complex formation between the promoter and RNA polymerase. In FIGURE 2, *deoP2* is shown as a complex promoter, regulated by the cytR repressor and the CRP/cAMP protein. This regulation, however, is independent of the deoR system and will not be discussed further here.[14,18,19]

EVIDENCE FOR deoR LONG-DISTANCE REGULATION

The O_E, O_1, and O_2 operators, which are the targets of the deoR repressor, have been defined by (1) footprinting analysis with purified deoR repressor[20,21] (Lisbeth Bech, unpublished data), (2) mutational analysis of O_1,[19,22,23] and (3) construction of synthetic operators[23] (see below). The deoR regulation has been studied *in vivo* using different single-copy systems including the native *deo* operon and various *galK* or *lacZ* fusions carried on λ-phages or single-copy plasmids. (It is necessary to use single-copy systems because multicopy plasmids carrying operators titrate the deoR repressor.) A surprising result was that two operator sites were needed to obtain an efficient deoR regulation of both *deoP1* and *deoP2* (FIG. 3). In λKT33 (FIG. 3, line 1) a fragment containing both P_1O_1 and P_2O_2 is fused to *galK*, and an efficient deoR repression of 34-fold is observed. However, when either of the promoters is present alone, deoR only weakly repressed transcription two- to threefold (lines 2 and 3). The efficient deoR repression can be restored if P_1O_1 is inserted in the opposite orientation upstream of P_2O_2 (line 4), but not when a mutated operator site is used (line 5). Also, transcription from *deoP1* is efficiently repressed when there are two operators (compare lines 2 and 6). Consequently, when two operators are present, a 23- to 115-fold repression is seen (lines 1, 4, and 6). The presence of the third operator further increases the repression two- to threefold.[24]

Increased intracellular concentration of deoR repressor due to the presence of a multicopy plasmid containing the *deoR* gene is sufficient to repress transcription from *deoP1*. This also occurs in constructions containing just one operator site (FIG. 3, line 7). A reduction in *deoP2* expression is also observed, although not as significantly as that in *deoP1* (FIG. 3, line 8). These results imply that the concentration of the repressor is important.

To understand the mechanism of long-distance regulation we changed the interoperator distances. Increasing the distance from 224 and 998 bp had no obvious effect on regulation. Further increases in the interoperator distance gave three clear results (FIG. 4): (1) The two operators work cooperatively even when the distance is more than 4,000 bp; (2) the efficiency of regulation decreases as the interoperator distance increases (FIG. 4, lines 2, 4, and 5); and (3) the second operator can be placed downstream of the promoter and target gene and still function.

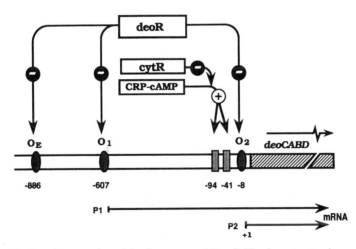

FIGURE 2. Regulatory region of the *deo* operon of *E. coli*. The four structural genes, *deoC, deoA, deoB,* and *deoD,* encode deoxyriboaldolase, thymidine phosphorylase, phosphopentomutase, and purine nucleoside phosphorylase, respectively.[32] *Circles* indicate the deoR operators, which in the case of O_1 and O_2 overlap the −10 region of *deoP1* and *deoP2*, respectively. *Boxes* indicate the two targets, where cAMP receptor protein (CRP) binds with high affinity.[33] The cytR repressor binds to the same region as does CRP.[18,33] *Numbers* indicate the centers of the targets and are relative to the start site of *deoP2*. The expression is indicated to be activated (+) or repressed (−) by the regulatory protein.

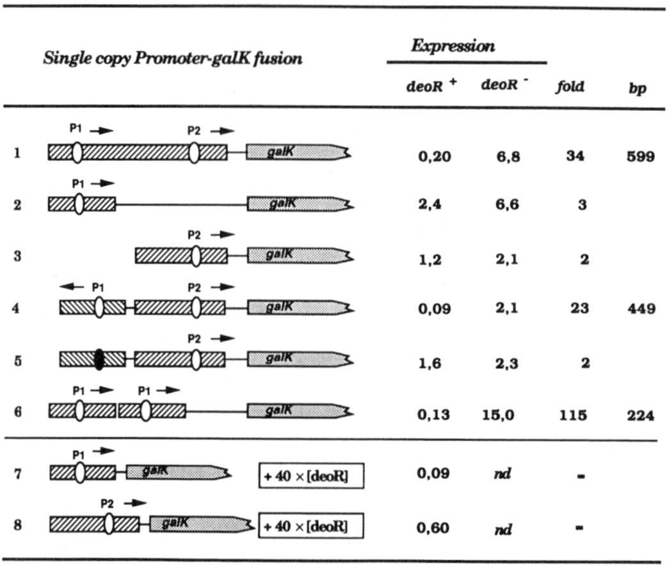
Single copy Promoter-galK fusion
Expression
deoR + deoR - fold bp
1 P1 P2 galK 0,20 6,8 34 599
2 P1 galK 2,4 6,6 3
3 P2 galK 1,2 2,1 2
4 P1 P2 galK 0,09 2,1 23 449
5 P2 galK 1,6 2,3 2
6 P1 P1 galK 0,13 15,0 115 224
7 P1 galK + 40 × [deoR] 0,09 nd -
8 P2 galK + 40 × [deoR] 0,60 nd -

FIGURE 3. DeoR regulation of promoter-*galK* fusions containing one or two operator sites. *White circles* indicate the operator sites, and *arrows* indicate the orientation of the promoters, whereas the *black circle* indicates that the operator is mutated. 40 × [deoR] indicates that the cells contained a multicopy plasmid expressing the *deoR*$^+$ gene. *In vivo* expression was monitored by measurements of galactokinase in the corresponding *deoR*$^+$ and *deoR*$^-$ strain and is given in nanomoles of substrate converted per minute using a cell density of $OD_{436} = 1$. *Fold* indicates the *deoR*$^-$/*deoR*$^+$ ratio. Data are taken from ref. 25.

CHELATE MODEL

Our results are consistent with the chelate model[14] (FIG. 5) which was suggested for the *deo* operon in 1985.[25] This model requires a regulatory protein or protein complex containing binding sites for at least two operator sites. (For simplicity, the model in FIGURE 5 only involves two operator sites.) In the first step of the binding reaction, free deoR repressor binds to one of the two operators. In the second step, deoR, already bound to one site, binds to the second operator, thus creating a loop in the intervening DNA. Although the intrinsic affinity of deoR for the two operators may be the same, loop formation is favored because of the increased "effective concentration" of deoR for the second step relative to the first step, inasmuch as deoR is already bound close to the second operator. This loop structure is thought to be "breathing" as a result of deoR dissociating and reassociating from one of the operators. In this way, deoR binds with high affinity to DNA containing two operator sites, but preserves a high dissociation rate.

The model is based on the findings that large rate accelerations are observed when two reactants are tethered together rather than being free in solution (i.e., a reduction in three-dimensional diffusion). In FIGURE 6, the anhydrid formation of succinic acid (intramolecular reaction) is compared with the formation of acetic acid anhydrid between two molecules of acedic acid (intermolecular reaction). Although the two reactions are very similar, the ratio of equilibrium constants is 3×10^5 M.[26] In other words, the intramolecular reaction proceeds as if the concentration of the carboxylic acid groups in succinic acid is 3×10^5 M. This physically impossible concentration has been termed the "effective concentration" of one reactant relative to the other.[26] In the chelate model, the initial binding of deoR to the first operator then corresponds to the intermolecular reaction, whereas the loop formation corresponds to the intramolecular reaction. According to this model, the cooperative effect of having two operators should be mimicked with one operator simply by increasing the deoR concentration, which was indeed shown experimentally (FIG. 3, lines 7 and 8).

CONSTRUCTION OF A SYNTHETIC OPERATOR

To define the size of the operator and to define what is needed for loop formation, we constructed a synthetic operator, O_S, that consisted of 23 bp of P_1O_1 DNA containing the 16-bp palindrome (FIG. 7B). This operator has 19 bp identical to P_2O_2 and 20 bp identical to O_E. When we inserted O_S upstream of the *galK* gene in a promoter-cloning vector and transformed this plasmid into a *galK⁻* strain, the transformants formed red colonies on MacConkey galactose plates, indicating that a promoter had been created (P_{SYN}). This may be expected because the -10 region of *deoP1* is inherited in the 16-bp palindrome. As the "-35 region" must reside in the vector, we could change this region by filling in the *Hind*III cloning site using DNA polymerase (Klenow), thereby increasing the distance between the expected -10 and -35 regions by 4 bp (FIG. 7). *GalK⁻* strains containing this plasmid produced white colonies, indicating that P_{SYN} had been inactivated.

To test if the 16-bp core sequence is sufficient to act as the second operator, we first inserted O_S upstream of a P_2O_2-*galK* fusion at a distance of 139 bp from O_2. The construction was transferred to a λ phage and inserted onto the chromosome of a *deoR⁺*, *galK⁻* and a *deoR⁻*, *galK⁻* strain. Table 1 indicates that when O_S is inserted

		Expression			
		deoR⁺	deoR⁻	fold	bp
1		191	389	2,0	
2		17	357	21	1245
3		210	361	1,7	
4		49	427	8,8	2540
5		85	444	5,2	4076

FIGURE 4. Cooperativity as a function of interoperator distance. *In vivo* expression from *deoP2* was monitored by measurements of the deoC gene product (from a single-copy plasmid) in *deoR⁺* and *deoR⁻* strains both carrying a deletion of the chromosomal *deo* operon. See also legend to FIGURE 3. Data are taken from ref. 34.

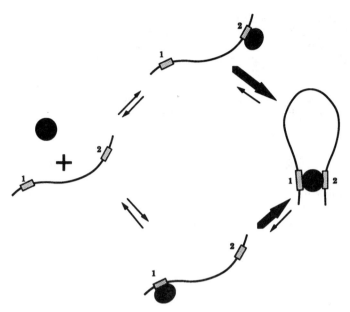

FIGURE 5. The DNA chelate model. *Circle* indicates the deoR protein that has the ability to bind two (or more) operators at the same time. *Boxes* indicate *deo* operator sites. (See text for explanation.)

upstream of P_2O_2, the level in *deoR*$^+$ strains decreases from 8.9 units/OD_{436} to 0.21 units/OD_{436}, whereas the derepressed level is unchanged. In λGD34, which only contains the P_2O_2-*galK* fusion, deoR can repress transcription from *deoP2* 1.5-fold. The deoR regulation in λGD28 is 53-fold, and we concluded that O_S can fully act as a second operator for repression of *deoP2*.

deoR REGULATION OF A SYNTHETIC PROMOTER

Because the weak P_{SYN} promoter contains the 16-bp palindrome, it is expected to be regulated by deoR. It should therefore be possible to construct an entirely synthetic system in which O_S is used as a second operator for P_{SYN}. In such a construction, only the palindrome placed in tandem originates from the *deo* operon (FIG. 7). O_S was inserted 162 bp upstream of $P_{SYN}O_S$ which was fused to *galK*, and the constructions carrying O_S, $P_{SYN}O_S$ and O_S-162-$P_{SYN}O_S$ were transferred to λ phages and inserted into the chromosome of *deoR*$^+$ and *deoR*$^-$ cells as already described. Under derepressed conditions (*deoR*$^-$) the *in vivo* promoter strength can be compared. It was found that P_{SYN} is a weak promoter with only about 3–4% of the activity of the very strong *deoP2* promoter. Under repressed conditions (*deoR*$^+$) P_{SYN} is weakly repressed by deoR (4.4-fold), but in λGD64, with O_S inserted upstream of $P_{SYN}O_S$, P_{SYN} is almost completely repressed by deoR with a regulation of 18-fold. To demonstrate that this regulation depends on deoR, we increased the intracellular concentration of deoR by transforming the cells with a multicopy plasmid expressing the *deoR*$^+$ gene. As was found for *deoP1*, P_{SYN} is completely repressed when the deoR concentration is increased, despite the fact that it contains only one operator (compare FIG. 3, line 7 with TABLE 1, line 6).

TITRATION OF deoR REPRESSOR

When multicopy plasmids carrying deoR operator sites are introduced into a *deoR*[+] strain, an increase in the expression of the chromosomal *deo* operon is observed[22] (Dandanell and Hammer, in preparation). This derepression is a result of a titration of deoR repressor, which is present in a very low concentration in the cell. We tested the ability of the synthetic operator to titrate the deoR repressor and compared it to that of the natural *deo* operators. The results in TABLE 2 show the expression of the chromosomal *deoA* gene in different multicopy plasmid-containing strains. When the expression is normalized to the strain carrying the vector plasmid which is unable to titrate deoR, it is seen that O_S and $P_{SYN}O_S$ titrate 1.8- and 4.6-fold, respectively. This derepression is very similar to that obtained with P_1O_1 and P_2O_2 (compare lines 2 and 3 with lines 6 and 7). When O_S and $P_{SYN}O_S$ are both present on the plasmid, a much stronger titration occurs, which is similar to that obtained with the P_1O_1-599-P_2O_2 construction. The synthetic operators titrate as well as do the natural sites (P_2O_2 titrates more efficiently than does P_1O_1 because it also titrates the cytR repressor, and in wild-type cells the *deo* operon is repressed by both deoR and cytR).

DISCUSSION

Regulation at a distance (according to the chelate model) only requires a protein that can bind to two or more targets at the same time and a DNA molecule

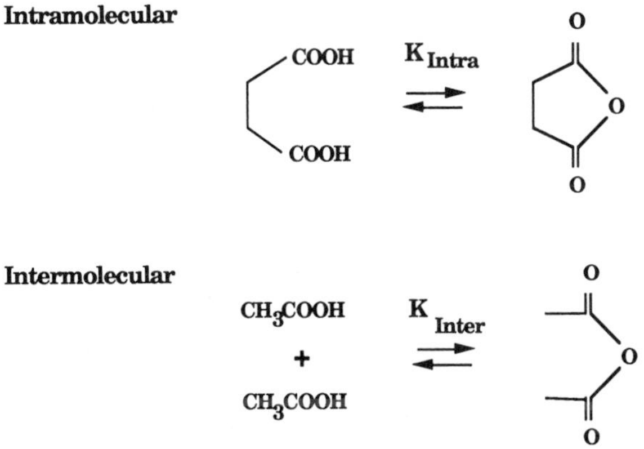

Intramolecular

$$\frac{K_{Intra}}{K_{Inter}} = 3 \times 10^5 \text{ M}$$

FIGURE 6. Chelate effect. See text for explanation.

containing two or more of these targets in the vicinity of each other. Accordingly, it is also possible to construct entirely synthetic promoters that are regulated at a distance like the P_{SYN} promoter described in this paper. With the correct -35 region this promoter can be made very strong, and it can even be combined with other

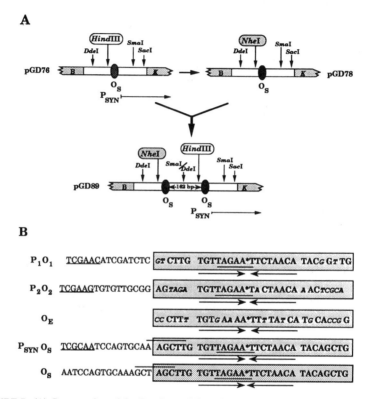

FIGURE 7. (A) Construction of O_S, $P_{SYN}O_S$, and O_S-162-$P_{SYN}O_S$. The synthetic DNA fragment O_S, shown in the gray box in FIGURE 7B, was ligated into the HindIII and BamHI sites of pGD52 (a galK promoter cloning vector derived from pKO500[22]), resulting in pGD76 ($P_{SYN}O_S$). pGD78 (O_S) was constructed by filling in the HindIII cloning site in pGD76 using Klenow polymerase, thereby creating an NheI site. pGD89 was constructed by isolating the 160-bp DdeI-SacI fragment of pGD76, filling in the DdeI site with Klenow polymerase and substituting this fragment with the SmaI-SacI fragment of pGD78. Constructions in pGD76, pGD78, and pGD89 were transferred to a λ phage by in vivo recombination, resulting in λGD63, λGD27, and λGD64, respectively.[35] (B) Comparison of promoter sequences. The sequence in the gray box of O_S and $P_{SYN}O_S$ is the synthetic operator inserted between HindIII and BamHI sites of pGD52. In the gray boxes of O_1, O_2, and O_E, the bases that are identical to the O_S sequence are indicated by bold type. Arrows indicate the palindrome. Underlined sequences indicate -35 and -10 regions (the exact transcription start of P_{SYN} has not been mapped). Overlined sequences indicate HindIII ($P_{SYN}O_S$) and NheI (O_S) restriction sites.

regulatory systems like CRP/cAMP activation. (See refs. 14 and 19 for a more extensive discussion.)

In many other operons that are regulated at a distance, a strong dependency on the interoperator distance has been reported when the distances are below 200

TABLE 1. Single-Copy Expression of Operator/Promoter-galK Fusions[a]

| | | | Active | $galK$ (units/OD$_{436}$) | | |
	λ phage	Operator	Promoter	$deoR^+$	$deoR^-$	Fold
1	λGD34	O$_2$	$deoP2$	8.9	13.1	1.5
2	λGD28	O$_S$-139-O$_2$	$deoP2$	0.21	11.1	53
3	λGD27	O$_S$	−	0.06[b]	0.15[b]	2.5
4	λGD63	P$_{SYN}$O$_S$	P_{SYN}	0.11[b]	0.48[b]	4.4
5	λGD64	O$_S$-162-P$_{SYN}$O$_S$	P_{SYN}	0.03[b]	0.54[b]	18
6	λGD63	P$_{SYN}$O$_S$ + 40 × [deoR]	P_{SYN}	<0.01	<0.01	−

[a]*E. coli* strains SØ3431 ($deoR^+$, $galK^-$) and SØ3432 ($deoR^-$, $galK^-$) were lysogenized with the different λ phages, and the single copy expression of galactokinase was measured in cell-free extracts after exponential growth in glycerol minimal medium, as described previously.[25] The activities are the average of two independent experiments, each determined in triplicate with a standard deviation less than 12%.

[b]The activity was measured in SØ3426 and SØ3428, respectively (grown in fructose minimal medium). The reason for this is that the background of 0.10 units/OD$_{436}$ (subtracted from the numbers in lines 1 and 2) is eliminated in *cya* strains, and therefore the low expression is more accurately measured in *cya* strains. (Similar results were obtained in *cya*$^+$ strain.) 40 × [deoR] indicates that the strain carrying λGD63 also contained a multicopy plasmid expressing the *deoR* gene (see also FIG. 3). λGD28 and λGD34 have been described in ref. 22.

bp.[3,7,9,12,27] Such dependency of the position on the DNA helix has become diagnostic of loop formation. In the *deo* operon we also studied regulation at short distances, but we have not seen a strong periodic variation (Dandanell and Hammer, in preparation). The most likely explanation for this deviation is that deoR can bind three or four operators at the same time (deoR is a hexamer or an octamer[21,28]), and because only two operators on the DNA are used for these studies, there will be alternative ways by which loops can be formed. That deoR can bind more than two operators was clearly shown when purified deoR protein was incubated with DNA containing all three *deo* operators and studied by electron microscopy. A double-

TABLE 2. Titration of deoR Repressor in Wild-Type Cells[a]

	Plasmid	Operator	deoA (units/OD$_{436}$)	Fold of Titration
1	pGD52	− (vector)	1.9	1.0
2	pGD78	O$_S$	3.4	1.8
3	pGD76	P$_{SYN}$O$_S$	8.8	4.6
4	pGD89	O$_S$-162-P$_{SYN}$O$_S$	21	11
5	pGD69	P$_1$O$_1$-599-P$_2$O$_2$	23	12
6	pGD17	P$_1$O$_1$	3.2	1.7
7	pGD70	P$_2$O$_2$	7.8	4.1
8	pGD79	O$_S$-139-P$_2$O$_2$	34	18

[a]SØ3430 ($deoR^+$, $cytR^+$, $deoCABD^+$) was transformed with the plasmids indicated, and cells were grown exponentially in glycerol minimal medium supplemented with 100 μg/ml ampicillin for five generations at 37°C. Expression of the chromosomal *deoA* gene was measured in cell-free extracts.[25,36] Also, expression of the plasmid-borne *bla* gene was measured as an indication of changes in the plasmid copy number. No significant variations were seen. Similar results for O$_S$ and P$_{SYN}$O$_S$ are obtained if glucose is used as a carbon source or if the experiment is carried out in an isogenic *cya* strain with fructose as the carbon source. Fold of titration is the *deoA* expression normalized to that of the vector plasmid pGD52. pGD17 has been described in ref. 25, whereas pGD52, pGD69, pGD70, and pGD79 are described in ref. 22.

looped structure in which one deoR binds all three operators was seen, and the loop sizes exactly corresponded to the position of the operators.[28] As shown in FIGURE 1, the *ara, lac,* and λcI systems each contain three operators, but araC, lacR, and λcI proteins can only bind two operators at a time. In the *araCABD* and λcI systems the alternative single loops are formed under different regulatory conditions.

We have not discussed any other models for regulation at a distance such as protein condensation, sliding, or induction of conformational changes in the DNA. Protein condensation can be excluded as a mechanism for deoR regulation, because 25 deoR molecules per cell cannot condense the DNA at a distance of 4,000 bp. These mechanisms, however, may contribute to modulate the overall efficiency of the chelate effect in other regulatory systems.

ACKNOWLEDGMENT

We thank Klara Zelei Kovacs for excellent technical assistance.

REFERENCES

1. IRANI, M. H., L. OROSZ & S. ADHYA. 1983. A control element within a structural gene: The *gal* operon of *Escherichia coli.* Cell **32:** 783–788.
2. FRITZ, H.-J., H. BICKNÄSE, B. GLEUMES, C. HEIBACH, S. ROSAHL & R. EHRING. 1983. Characterization of two mutations in the *Escherichia coli galE* gene inactivating the second galactose operator and comparative studies of repressor binding. EMBO J. **2:** 2129–2135.
3. DUNN, T. M., S. HAHN, S. OGDEN & R. F. SCHLEIF. 1984. An operator at −280 base pairs that is required for repression of *araBAD* operon promoter: Addition of DNA helical turns between the operator and promoter cyclically hinders repression. Proc. Natl. Acad. Sci. USA **81:** 5017–5020.
4. REZNIKOFF, W. S., R. B. WINTER & C. K. HURLEY. 1974. The location of the repressor binding sites in the *lac* operon. Proc. Natl. Acad. Sci. USA **71:** 2314–2318.
5. MOSSING, M. C. & J. M. T. RECORD. 1986. Upstream operators enhance repression of the *lac* promoter. Science **233:** 889–892.
6. EISMANN, E., B. V. WILCKEN-BERGMAN & B. MÜLLER-HILL. 1987. Specific destruction of the second *lac* operator decreases repression of the *lac* operon in *Escherichia coli* fivefold. J. Mol. Biol. **195:** 949–952.
7. BELLOMY, G. R., M. C. MOSSING & M. T. RECORD. 1988. Physical properties of DNA *in vivo* as probed by the length dependence of the *lac* operator looping process. Biochemistry **27:** 3900–3906.
8. OEHLER, S., E. R. EISMANN, H. KRÄMER & B. MÜLLER-HILL. 1990. The three operators of the *lac* operon cooperates in repression. EMBO J. **9:** 973–977.
9. KRÄMER, H., M. NIEMÖLLER, M. AMOUYAL, B. REVET, B. V. WILCKEN-BERGMAN & B. MÜLLER-HILL. 1987. lac repressor forms loops with linear DNA carrying two suitably spaced *lac* operators. EMBO J. **6:** 1481–1491.
10. KRÄMER, H., M. AMOUYAL, A. NORDHEIM & B. MÜLLER-HILL. 1988. DNA supercoiling changes the spacing requirements of two *lac* operators for DNA loop formation with lac repressor. EMBO J. **7:** 547–556.
11. FLASHNER, Y. & J. D. GRALLA. 1988. DNA flexibility and protein recognition: Differential stimulation by bacterial histone-like protein HU. Cell **54:** 713–721.
12. GRIFFITH, J., A. HOCHSCHILD & M. PTASHNE. 1986. DNA loops induced by cooperative binding of λ repressor. Nature **322:** 750–752.

13. HOCHSCHILD, A. & M. PTASHNE. 1988. Interaction at a distance between λ repressors disrupt gene activation. Nature **336:** 353–357.
14. HAMMER, K. & G. DANDANELL. 1989. The deoR repressor from *E. coli* and its action in regulation-at-distance. *In* Nucleic Acids and Molecular Biology. F. Eckstein & D. M. J. Lilley, eds.: 79–91. Springer-Verlag. Berlin-Heidelberg.
15. SCHLEIF, R. 1988. DNA looping. Science **240:** 127–128.
16. ADHYA, S. 1989. Multipartite genetic control elements: Communication by DNA loop. Ann. Rev. Genet. **23:** 227–250.
17. BELLOMY, G. R. & M. T. RECORD, JR. 1990. Stable DNA Loops *in vivo* and *in vitro:* Roles in gene regulation at a distance and in biophysical characterization of DNA. *In* Progress in Nucleic Acids Research and Molecular Biology, vol 39: 81–128. Academic Press.
18. VALENTIN-HANSEN, P. 1985. DNA sequences involved in expression and regulation of deoR, cytR and cAMP/CRP controlled genes in *Escherichia coli. In* Gene Manipulation and Expression. R. E. Glass & J. Spizek, eds.: 273–288. Croom Helm. London.
19. DANDANELL, G. & K. HAMMER. 1990. *In vivo* CRP-cAMP regulation of the deoP1 and deoP2 promoter of *E. coli* K12. Life Science Advances Molecular Genetics **9:** 41–45.
20. MORTENSEN, L. 1988. Model system for the mechanism regulating the transcription initiation: Purification and characterization of the deoR repressor of *Escherichia coli* by *in vitro* measurements. Dissertation, University of Copenhagen.
21. MORTENSEN, L., G. DANDANELL & K. HAMMER. 1989. Purification and characterization of the deoR repressor of *Escherichia coli.* EMBO J. **8:** 325–331.
22. DANDANELL, G. 1989. Control of transcription initiation: Long-range deoR regulation in *Escherichia coli.* Ph.D. thesis, University of Copenhagen.
23. HAMMER, K., L. BECH, G. DANDANELL & P. HOBOLTH. 1990. DNA sequence specificity of *deoP1* operator-deoR repressor recognition in *Escherichia coli.* Proceedings of the 5th European Congress on Biotechnology. Munksgaard, Copenhagen. **1:** 468–471.
24. VALENTIN-HANSEN, P., A. BJARNE & J. E. L. LARSEN. 1986. DNA-protein recognition: Demonstration of three genetically separated operator elements that are required for the repression of the *Escherichia coli deoCABD* promoters by the deoR repressor. EMBO J. **5:** 2015–2021.
25. DANDANELL, G. & K. HAMMER. 1985. Two operator sites separated by 599 base pairs are required for deoR repression of the *deo* operon of *Escherichia coli.* EMBO J. **4:** 3333–3338.
26. PAGE, M. I. & W. P. JENCKS. 1971. Entropic contributions to rate accelerations in enzymatic and intramolecular reactions and the chelate effect. Proc. Natl. Acad. Sci. USA **68:** 1678–1683.
27. LEE, D.-H. & R. F. SCHLEIFF. 1989. *In vivo* DNA loops in *araCBAD:* Size limits and helical repeat. Proc. Natl. Acad. Sci. USA **86:** 476–480.
28. AMOUYAL, M., L. MORTENSEN, H. BUC & K. HAMMER. 1989. Single and double loop formation when deoR repressor binds to its natural operator sites. Cell **58:** 545–551.
29. OGDEN, S., D. HAGGERTY, C. STONNER, D. KOLODRUBETZ & R. SCHLEIF. 1980. The *Escherichia coli* L-arabinose operon: Binding sites of the regulatory proteins and a mechanism of positive and negative regulation. Proc. Natl. Acad. Sci. USA **77:** 3346–3350.
30. GILBERT, S., J. GRALLA, J. MAJORS & A. MAXAM. 1975. Lactose operator sequences and the action of lac repressor. *In* Protein Ligand Interactions. H. Sund and G. Blauer, eds.: 193–210. de Gruyter. Berlin.
31. GUSSIN, G. N., A. D. JOHNSON, C. O. PABO & R. T. SAURE. 1983. Repressor and cro protein: Structure, function, and role in lysogenization. *In* Lambda II. R. W. Hendrix, J. W. Roberts, F. W. Stahl & R. A. Weisberg, eds.: 93–121. Cold Spring Harbor Laboratory. Cold Spring Harbor. NY.
32. HAMMER-JESPERSEN, K. 1983. Nucleoside catabolism. *In* Metabolism of Nucleotides, Nucleosides and Nucleobases in Microorganisms. A. Munch-Petersen, ed.: 203–258. Academic Press. London.
33. VALENTIN-HANSEN, P., H. AIBA & D. SCHÜMPERLI. 1982. The structure of tandem regulatory regions in the *deo* operon of *Escherichia coli* K12. EMBO J. **1:** 317–322.

34. DANDANELL, G., P. VALENTIN-HANSEN, P. E. LØVE-LARSEN & K. HAMMER. 1987. Long-range cooperativity between gene regulatory sequences in a prokaryote. Nature **325**: 823–826.
35. MCKENNEY, K. 1982. A vector system for the molecular analysis of transcription initiation and termination in *Escherichia coli*. Ph.D. thesis, The John Hopkins University.
36. MUNCH-PETERSEN, A. 1968. On the catabolism of deoxyribonucleosides in cells and cell extracts of *Escherichia coli*. Eur. J. Biochem. **6**: 432–442.

Ribosomal RNA and the Site Specificity of Chloramphenicol-Dependent Ribosome Stalling in *cat* Gene Leaders

PAUL S. LOVETT

Department of Biological Sciences
University of Maryland
Catonsville, Maryland 21228

Regulation of mRNA translation as a primary control mechanism for gene expression is uncommon; however, specific examples have been described. Translational attenuation is one such system that regulates the expression of at least two types of antibiotic-resistant genes in gram-positive bacteria.[1-3] These genes, *cat* and *erm*, specify resistance to chloramphenicol and erythromycin, respectively, and each is induced by the corresponding antibiotic. The regulatory principle of translational attenuation is readily seen by using the *cat* gene system as a model. *cat* transcripts are continuously synthesized.[4] Each has a characteristic structure at the 5' end that imparts the regulated phenotype. The ribosome binding site (RBS) for the *cat* structural gene, designated RBS-3 in FIGURE 1, is within a 14-nucleotide sequence that is repeated in the inverted form 12 nucleotides upstream. The consequence is that RBS-3 is sequestered in a stable stem-loop structure and is presumably unavailable for translation initiation. Upstream from the RNA stem-loop is a short

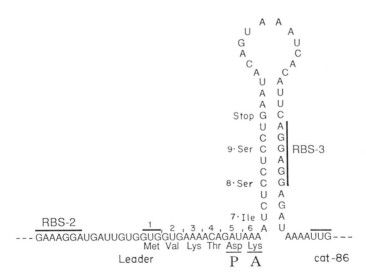

FIGURE 1. Regulatory region for *cat-86* RBS-3 is the ribosome binding site for the *cat-86* coding sequence, and RBS-2 is the ribosome binding site for the leader. A and P refer to the location of the aminoacyl and peptidyl sites of a chloramphenicol-stalled ribosome. Leader codons 2 through 5 comprise the *crb* box.

31

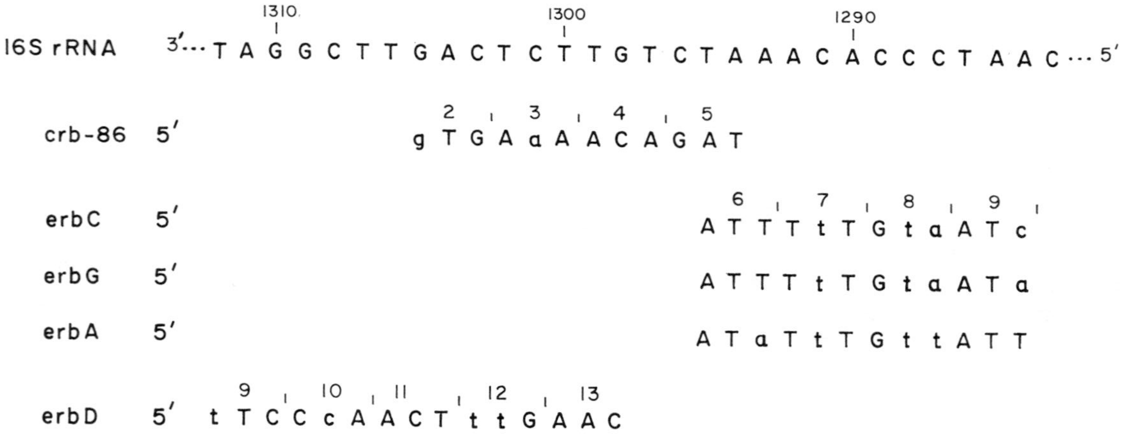

FIGURE 2. Complementarity of crb-86 and erbC, G, A, and D with a 30-nucleotide region of *B. subtilis* 16S rRNA. Small letters designate mismatches. Codons are numbered according to their location in the respective leaders.

open reading frame (the regulatory leader) that has its own RBS, RBS-2. The regulatory leader is continuously translated, whereas the *cat* coding sequence is not. The addition of chloramphenicol to cells containing the regulated *cat* gene stalls a ribosome in the leader, which destabilizes the RNA stem-loop, allowing translation of *cat*.

A remarkable feature of the *cat* induction system is the specificity of the site of ribosome stalling that is brought about by chloramphenicol. The antibiotic stalls a ribosome with its aminoacyl site at leader codon 6.[5] This site of stalling is dictated by both the antibiotic and the nature of leader codons 2 through 5 which we designate the *crb* box.[6] In the *erm* system an analogous sequence exists, the *erb* box, which is composed of *ermC* leader codons 5 through 9. Ribosome stalling at *erbC* is due to an action of erythromycin. How *crb* and *erb* act in stalling ribosomes remains a mystery; however, the resulting specificity of drug-induced stalling must be a function of the *crb* and *erb* nucleotide sequence and/or the corresponding amino acid sequence of the leader peptide.

Interaction of ribosomes with a specific mRNA sequence is driven by rRNA pairing with a complementary mRNA sequence.[8,9] We have identified that *crb* and *erb* sequences from all *cat* and *erm* sequences show complementarity with an internal region of *Bacillus subtilis* 16S rRNA[10] (FIG. 2). Evidence provided elsewhere[10] supports the idea that *crb* interacts with *B. subtilis* rRNA during antibiotic induction of *cat in vivo*.

Both *cat* and *erm* regulatory signals have been shown to effectively modulate heterologous gene expression in *B. subtilis*, but at the level of mRNA translation rather than transcription. Both, therefore, have the potential to provide an alternative gene control system in the biotechnologic applications of *B. subtilis*. We suggest that these regions of complementarity, *crb* and *erb*, guide a ribosome to the correct stall site needed to induce the regulated gene.

REFERENCES

1. LOVETT, P. S. 1990. Translational attenuation as the regulator of inducible *cat* genes. J. Bacteriol. **172:** 1–6.
2. DUBNAU, D. 1984. Translational attenuation. The regulation of bacterial resistance to the macrolide-lincosamide-streptogramin B antibiotics. Crit. Rev. Biochem. **16:** 103–132.
3. WEISBLUM, B. 1983. Inducible resistance to macrolides, lincosamides, and streptogramin B type antibiotics: The resistance phenotype, its biological diversity, and structural elements that regulate expression. *In* Gene Function in Procaryotes. J. Beckwith, J. Davis & J. A. Gallent, eds.: 91–121. Cold Spring Harbor Laboratory. Cold Spring Harbor, NY.
4. DUVALL, E. J. & P. S. LOVETT. 1986. Chloramphenicol induces translation of the mRNA for a chloramphenicol-resistance gene in *Bacillus subtilis*. Proc. Natl. Acad. Sci. USA **83:** 3939–3943.
5. ALEXIEVA, Z., E. J. DUVALL, N. P. AMBULOS, JR., U. J. KIM & P. S. LOVETT. 1988. Chloramphenicol induction of *cat-86* requires ribosome stalling at a specific site in the leader. Proc. Natl. Acad. Sci. USA **85:** 3057–3061.
6. ROGERS, E. J., U. J. KIM, N. P. AMBULOS, JR. & P. S. LOVETT. 1990. Four codons in the *cat-86* leader define a chloramphenicol-sensitive ribosome stall sequence. J. Bacteriol. **172:** 110–115.
7. MAYFORD, M. & B. WEISBLUM. 1989. *ermC* leader peptide. Amino acid sequence critical for induction by translational attenuation. J. Mol. Biol. **206:** 69–79.
8. SHINE, J. & L. DALGARNO. 1974. The 3′-terminal sequence of *Escherichia coli* 16S ribosomal RNA: Complementarity to nonsense triplets and ribosome binding sites. Proc. Natl. Acad. Sci. USA **71:** 1342–1346.

9. HUI, A. & H. A. DEBOER. 1987. Specialized ribosome system: Preferential translation of a single mRNA species by a subpopulation of mutated ribosomes in *Escherichia coli.* Proc. Natl. Acad. Sci. USA **84:** 4762–4766.
10. ROGERS, E. J., N. P. AMBULOS, JR. & P. S. LOVETT. 1990. Complementarity of *Bacillus subtilis* 16S rRNA with sites of antibiotic-dependent ribosome stalling in *cat* and *erm* leaders. J. Bacteriol. **172:** 6282–6290.

Horseradish Peroxidase Gene Expression in *Escherichia coli*

A. M. EGOROV,[a] I. G. GAZARYAN,[a] S. V. SAVELYEV,[a]
V. A. FECHINA,[b] A. N. VERYOVKIN,[b] AND B. B. KIM

[a]*Chemical Department*
Moscow State University
Moscow 119899 GSP, USSR

[b]*A. N. Bach Institute of Biochemistry*
USSR Academy of Sciences
Moscow, USSR

Plant peroxidase (EC 1.11.1.7) plays an important role in several physiological functions such as removal of hydrogen peroxide, oxidation of toxic reductants, biosynthesis and degradation of lignin, and defense response against wounding and virus infections. Peroxidase is widely used for practical purposes. It is one of the most available enzymes for enzyme immunoassay and diagnostics because of a high sensitivity and possibility of visual detection. For example, the reaction of enhanced chemiluminescence is of great importance for diagnostics as an ultrasensitive analytical technique.[1-3]

The development of expression systems for plant peroxidases, as it has already been done for yeast[4] and microbial[5] peroxidases, might be helpful in clarifying the physiological roles and catalytic functions of plant peroxidases.

Horseradish peroxidase isozyme C is a monomeric glycohemeprotein (MW 44,000) containing ferriprotoporphyrin IX as the noncovalently bound prosthetic group. According to Welinder,[6] its amino acid chain consists of 308 residues with 4 disulfide bridges (MW 33,890). Isozyme C also contains 8 neutral carbohydrate chains bound to asparagine residue and 2 calcium ions per molecule. On the basis of the amino acid sequence of isozyme C, two genes encoding horseradish peroxidase have been synthesized by Amersham Int plc[7] and British Biotechnology Ltd.[8]

Ortlepp *et al.*[9] reported previously that expression level for 33-kD peroxidase protein in the case of *Escherichia coli* JM101 transformants varied between 0.5 and 2% of the total protein. Expression levels in the deletion mutant (deleting 72 C-terminal amino acids, MW 26 kD) were as high as 40%. The expression of the synthetic gene from British Biotechnology Ltd using *tac*-promoter was investigated in *E. coli* HW110 strain.[10]

We have studied the possibility of expression in *E. coli* of the horseradish peroxidase synthetic gene (Amersham) and properties of the refolded peroxidase in an enhanced chemiluminescence reaction.

MATERIALS AND METHODS

Plasmids pSA233, pSA261, and pSA262 with the synthetic gene of horseradish peroxidase isozyme C (FIG. 1A and B) were synthesized (kindly provided by S. A. Ortlepp[7]). Plasmid pSA233 has no promoter in front of the peroxidase gene, whereas

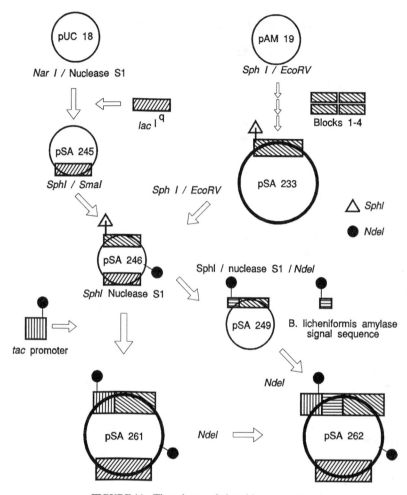

FIGURE 1A. The scheme of plasmid construction.

the gene in pSA261 is under the *tac*-promoter. pSA262 was derived from pSA261 by the addition of the alpha-amylase signal sequence.

E. coli JM 103, JM 109, HB101, and TG1 strains grown in LB medium[11] were used as the major host strains throughout these experiments. Cells containing recombinant plasmids were grown in LB with ampicillin (100 μg/ml). Strains were made competent and transformed by the method of Mandel and Higa.[12]

We used 100 μM IPTG concentration to induce peroxidase expression in the middle of the log-phase of cell growth. Transformants were grown in 300 ml LB medium at 37°C. The cell biomass was harvested by centrifugation after a 4–6 hour induction. The expression was tested by means of SDS-PAGE electrophoresis[13] in sonicated E. coli/pSA preparations. Nontransformed E. coli cells were treated the same way and used as the control.

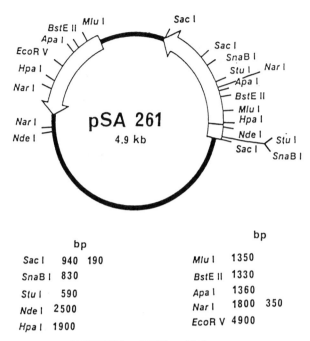

	bp	
Sac I	940	190
SnaB I	830	
Stu I	590	
Nde I	2500	
Hpa I	1900	

	bp	
Mlu I	1350	
BstE II	1330	
Apa I	1360	
Nar I	1800	350
EcoR V	4900	

FIGURE 1B. pSA261 restriction map.

Solubilization experiments were carried out with pellets after cell biomass sonication. The pellets were resuspended in 50 mM Tris.HCl buffer, pH 7.0, with the addition of solubilizers (5% Triton X-100, 5% Tween 20, 2 M NaCl, 2% SDS, 2% 2-mercaptoethanol, 8 M urea, 5 M guanidine, or combinations of 2 M NaCl, 2% SDS, and 8 M urea with mercaptoethanol). The solubilization effects were studied by SDS-PAGE/electrophoresis.

Recombinant protein refolding was carried out in 50 mM Tris.HCl buffer, pH 7.0, containing 5 μM hemin, 0.02% Tween 20, 1 mM oxidized glutathione, and 1 mM CaCl₂. The solubilized protein in a 5 M urea solution was diluted by its addition drop by drop to the buffer just described (dilution to a final urea concentration of about 1–2 M). The mixture obtained was incubated for 48 hours at 4°C. The refolded protein was then purified by fast protein liquid chromatography (FPLC) gel filtration on Superose 6 HR 10/30 column, Pharmacia LKB Biotechnology.

Isoelectrofocusing was carried out using a MiniF device (BioRad, Austria). Peroxidase activity staining was performed using 0.2 mg/ml diaminobenzidine in 50 mM Tris.HCl buffer solution, pH 7.2, in the presence of 0.02% CoCl₂.

Peroxidase activity was measured using a phenol-4-aminoantipyrine probe[14] and enhanced chemiluminescence with luminol-*p*-iodophenol substrate pair.[2]

RESULTS AND DISCUSSION

Plasmid pSA233 (without the *tac*-promoter) was replicated without expression. For plasmids pSA261 and pSA262, expression but not excretion of the recombinant protein was observed as it could be expected for pSA262 transformants. Therefore, for subsequent investigations we chose only pSA261 transformants.

Most of the recombinant protein was found forming inclusion bodies and was tested in pellets after cell sonication. The highest level of expression of recombinant peroxidase (33 kD) was found in *E. coli* JM109/pSA261 (30% of the total protein in cell pellets, FIG. 2, lanes a, b, and c). In *E. coli* HB 101/pSA261 and TG1/pSA261 expression was lower (10%), and in *E. coli* JM103/pSA261, it was only 2%. Data obtained led to the conclusion that the expression level strictly depends on the host strain.

Pellets after sonication *E. coli* JM109/pSA261 cells were used to solubilize the recombinant protein. Treatment with 8 M urea was the best solubilization method (FIG. 3, lanes a and c). The presence of mercaptoethanol did not significantly increase the solubilization effect of urea (compare lanes a, b, and c in FIG. 3). The decrease in urea concentration to the 4–5 M level did not decrease the yield of solubilized peroxidase. Similar results were obtained for 3–5 M guanidine (data not shown). Surfactants did not solubilize the protein (lanes e and f in FIG. 2 and lane e in FIG. 3) as well as NaCl (lane f in FIG. 3). However, 2 M NaCl treatment resulted in solubilization of all the *E. coli* proteins except the recombinant one. This observation led to the development of a simple procedure for recombinant peroxidase solubilization. After cell sonication the pellet was suspended in 2 M NaCl and 2% mercapto-ethanol, incubated at room temperature for 30 minutes, and centrifuged. The supernatant solution was removed, the pellet was washed with 50 mM Tris.HCl buffer solution, pH 7.0, or water and incubated for 30 minutes in 5 M urea solution for total solubilization of the recombinant protein. The purity of the solubilized protein was 90–95%.

The insoluble recombinant protein could be detected by an enhanced chemilumi-nescence reaction. To visualize the recombinant peroxidase activity, we performed refolding experiments after protein solubilization. After 48 hours of incubation the peroxidase activity with phenol-antipyrine was about 10 μmol/min per milligram of protein. In the absence of oxidized glutathione the reactivation resulted in peroxi-

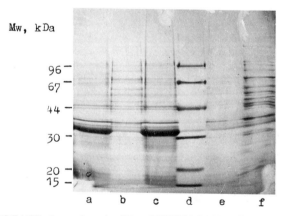

FIGURE 2. SDS-PAGE electrophoresis of *E. coli* JM109/pSA261 cell proteins. *Lanes a–c* are cells (a), supernatant solution after cell sonication (b), and pellet after cell sonication (c), boiled with 1% SDS and 2% mercaptoethanol. *Lane* d contains molecular weight markers. *Lanes e and f* are cell pellets after sonication, solubilized with 5% Triton X-100 (e) and Tween 20 (f).

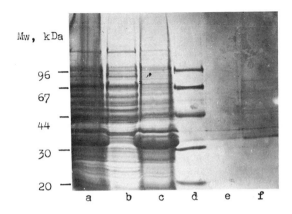

FIGURE 3. SDS-PAGE electrophoresis of proteins solubilized from the pellets after *E. coli* JM109/pSA261 sonication. Incubation with 8 M urea (a), 1% SDS (b), 8 M urea + 1% SDS (c), 5% Triton + 2% mercaptoethanol (e), and 2 M NaCl (f). Molecular weight markers (d).

dase specific activity of about 40 nmol/min per milligram of protein. In the absence of calcium, reactivation did not proceed at all, indicating the importance of disulfide bond formation and calcium binding in peroxidase reactivation.

FPLC gel filtration revealed two bands corresponding to monomeric (major) and dimeric peroxidase (minor) forms. Isoelectrofocusing of reactivated peroxidase monomer (33 kD) demonstrated the presence of three isozymes (9.05—major, 8.2, and 7.5) as was observed for the peroxidase type IX preparation (Sigma, USA). In the case of a one-gene product the existence of three isozymes could be explained only by posttranslational modification including hemin and calcium binding.

The reactivated peroxidase was studied in the reaction of enhanced chemiluminescence. The reactivated peroxidase exhibited catalytic properties close to those of the native enzyme. It could be subjected to lyophilization without the loss of its specific activity.

To obtain more promising results in refolding experiments, promoter optimization or low temperature expression for *E. coli* expression systems must be performed. Another way to solve the problem of active peroxidase expression may be cloning the gene in yeast, plant tissue culture, or the plant itself. Anionic tobacco peroxidase has been cloned and expressed in transgenic tobacco plants in active form.[15] These experiments open a new stage in plant physiology and peroxidase biochemistry.

REFERENCES

1. WHITEHED, T. P., G. H. G. THORPE, T. J. N. CARTER, C. CROUCUTT & L. J. KRICKA. 1983. Enhanced luminescence procedure for sensitive determination of peroxidase-labelled conjugates in immunoassay. Nature **305:** 158–159.

2. VLASENKO, S. B., A. A. AREFIEV, A. D. KLIMOV, B. B. KIM, E. L. GOROVITS, A. P. OSIPOV, E. M. GAVRILOVA & A. M. EGOROV. 1989. An investigation of the catalytic mechanism of enhanced chemiluminescence: Immunochemical application of this reaction. J. Biolum. Chemilum. **4:** 164–176.

3. GOROVITS, E. L., B. B. KIM, S. B. VLASENKO, E. M. GAVRILOVA & A. M. EGOROV. 1989. ELISA with enhanced chemiluminescence detection. Biotechnology (USSR) **5:** 233–239.

4. FISHEL, L. A., J. E. VILLFRANCA, M. MAURO & J. KRAUT. 1987. Yeast cytochrome C peroxidase: Mutagenesis and expression in *Escherichia coli* show tryptophan-51 is not the radical site in compound I. Biochemistry **26:** 351–360.

5. LOPRASERT, S., I. URABE & H. OKADA. 1990. Overproduction and single-step purification of *Bacillus stearothermophilus* peroxidase in *Escherichia coli*. Appl. Microbiol. Biotechnol. **32:** 690–692.
6. WELINDER, K. G. 1979. Amino acid sequence studies of horseradish peroxidase. Eur. J. Biochem. **96:** 483–502.
7. CHISWELL, D. J. & S. A. ORTLEPP. 1989. DNA sequence coding for HRP enzyme. European patent no. EP 0299682.
8. EDWARDS, R. M. & J. F. BURKE. 1989. Synthetic gene. International patent no. WO 89/03424.
9. ORTLEPP, S. A., D. POLARD-KNIGHT & D. J. CHISWELL. 1990. Expression and characterization of a protein specified by a synthetic horseradish peroxidase gene in *Escherichia coli*. 1989. J. Biotechnol. **11:** 353–364.
10. BURKE, J. F., A. SMITH, N. SANTAMA, R. C. BRAY, R. N. F. THORNELEY & S. DACEY. 1989. Expression of recombinant horseradish peroxidase C in *Escherichia coli*—artificial gene construction. Biochem. Soc. Trans. **17:** 1077–1078.
11. MANIATIS, T., E. F. FRITSCH & J. SAMBROOK. 1982. Molecular cloning. Laboratory manual. Cold Spring Harbor Laboratory. Cold Spring Harbor. New York.
12. GLOVER, D. M. (Ed.) 1985. DNA Cloning. A Practical Approach. Vol. 1. IRL Press. Oxford, Washington DC.
13. LAEMMLI, U. K. 1970. Cleavage of structural proteins during the assembly of the head of bacteriophage T4. Nature **227:** 680–685.
14. KEESEY, J. (Ed.) 1987. Biochemica Information, 1st Ed. Boehringer Mannheim Biochemicals. Indianapolis, IN.
15. LAGRIMINI, L. M., S. BRADFORD & S. ROTHSTEIN. 1990. Peroxidase-induced wilting in transgenic tobacco plants increased synthesis of peroxidase in transgenic *Nicotiana tabacum* and *N. sylvestris*. Plant Cell **2:** 7–18.

Cloning of a β-Glucosidase Gene from *Ruminococcus albus* and Its Expression in *Escherichia coli*

KUNIO OHMIYA,[a] MASAYUKI TAKANO,[b] AND
SHOICHI SHIMIZU[c]

[a]*Faculty of Bioresources*
Mie University
Tsu 514, Japan

[b]*Biochemical Research Laboratories*
Kanegafuchi Chemical Industry Co. Ltd.
Takasago 676, Japan

[c]*Professor Emeritus*
Nagoya University
Chikusa, Nagoya 464-01, Japan

Ruminococcus albus, a rumen anaerobe, is a potent cellulolytic bacterium dominant in the rumen cattle. Because cattle supply a large amount of high quality meat, milk, and leather by consuming mainly cellulosic materials, they are economically important animals. One of the growth-limiting factors of cattle is the cellulose-degrading rate of rumen microorganisms. Therefore, it is desirable to enhance utilization of cellulose by the rumen microorganisms. In addition, several cellulolytic enzymes are present in a microorganism, making it difficult to obtain pure preparations of a specific cellulase. As a result, trace contaminants in the purified enzyme preparations would obscure the mechanism of cellulose degradation. Therefore, cloning of the genes encoding the cellulases and their expression individually in suitable hosts should contribute to our understanding of the properties of each cellulase. The molecular biology of cellulase genes and their products has developed rapidly, opening new fields of investigation such as structure and properties of cellulase genes, their regulation at the molecular level, and the structural features required for enzymatic activity. There are many reports about cloning and expression of the gene encoding endo- and exoglucanases and β-glucosidases from cellulolytic microorganisms.[1,2] Ohmiya *et al.*[3] cloned an endoglucanase gene of *R. albus,* determined the nucleotide sequence of the gene, and purified the enzyme from the transformant. This enzyme was modified by gene truncation.[4] However, the gene encoding β-glucosidase from *R. albus* has not been isolated and characterized. If the structure of the gene and the properties of the enzyme were understood, a more effective way for enhancing cellulose utilization would be found. For this purpose, more details of the enzyme are required. The present study describes the cloning of β-glucosidase from *R. albus* into *Escherichia coli.* The structure of the gene and some of the properties of the enzyme are also described.

MATERIALS AND METHODS

Pure cellulose (3% suspension of KC flock W-300, Sanyo Kokusaku Pulp Co., Tokyo) was used after ball-milling for 3 days as a main carbon source of the medium for *R. albus.* Yeast extract and tryptone as medium components were products of Difco Laboratories (Detroit). RNase, lysozyme, and ampicillin were purchased from Sigma Chemicals (St. Louis). Restriction endonucleases, T4 ligase, and bacterial alkaline phosphatase, obtained from Boehringer Mannheim GmbH (Tokyo), Takara Shuzo (Kyoto), and Toyobo (Osaka), respectively, were used under the conditions recommended by the suppliers. Cellooligomers for high performance liquid chromatography (HPLC) were obtained from Seikagaku Kogyo (Tokyo). All other reagents used were commercial products with the highest quality.

E. coli HB101 and E. coli JM109 were cultivated in Luria broth or M9 broth[5] in shaking test tubes or flasks at 37°C. The competent cells of these organisms were prepared according to the method of Mandel and Higa.[6] Ampicillin was added to each medium after filter sterilization to a final concentration of 50 μg/ml. Plasmids pBR322, pUC118, and pUC119 were used as cloning vectors. *R. albus* F-40, an anaerobic cellulolytic rumen bacterium, was cultivated in ball-milled cellulose or cellobiose medium at pH 6.5 and 37°C for a given period.[7]

Chromosomal DNA of *R. albus* was partially digested with *Hin*dIII. The digest was separated on an 0.7% agarose gel. Digested fragments with the molecular size of 4–10 kbp were extracted from the gel and ligated into pBR322 at *Hin*dIII site. The chimera plasmids were used for transformation of competent *E. coli* HB101. The transformants were cultivated overnight on LB agar plates containing ampicillin at 37°C. All colonies on the plates were covered with 1% agar containing 1 mM of 4-methylumbelliferyl-β-D-cellobiopyranoside (MUC) at 50°C and incubated for several hours at 37°C. The 4-methylumbelliferone released from MUC by β-glucosidase was detected by an ultraviolet transilluminator.

The nucleotide sequence of DNA was determined by the dideoxy chain termination method[8] using [α-^{35}S]dCTP. DNA sequences were determined by analyzing the autoradiograms.

β-Glucosidase activity was determined by measuring the release of *p*-nitrophenol (PNP) from *p*-nitrophenyl-β-D-glucoside (PNPG). PNPG solution (2 mM) and enzyme aliquot were added to the sodium phosphate buffer to the final concentration of 50 mM, at pH 6.5, and incubated at 30°C for 60 minutes. After the reaction was stopped by the addition of Na_2CO_3 to a final concentration of 0.4 M, the absorbance of the released PNP was measured at 405 nm. One unit of enzymatic activity was defined as the amount of enzyme catalyzing the release of 1 μmol of PNP per minute.

The purification of the β-glucosidase from transformed *E. coli* was carried out at 20°C with fast protein liquid chromatography (FPLC, Pharmacia Fine Chemicals, Tokyo). Culture broth was centrifuged at 13,000 × *g* at 4°C for 20 minutes to remove cells. Ammonium sulfate was added to the supernatant at 70% saturation. The precipitated proteins were collected by centrifugation at 13,000 × *g* for 15 minutes and dissolved in 10 mM sodium phosphate buffer (pH 6.5, containing 10 mM 2-mercaptoethanol in the sodium phosphate buffer). This solution was desalted by gel filtration (Sephacryl S-200 HR, Pharmacia, 1.6 × 100 cm) with 0.5 M NaCl in the phosphate buffer at a flow rate of 0.4 ml/min. The fractions with high activity of β-glucosidase were applied to a DEAE Bio-Gel A (BioRad) column (2.6 × 30 cm) equilibrated with the phosphate buffer. The column was eluted with a 1,200 ml linear gradient of NaCl (0–1.0 M) in equilibration buffer at a flow rate of 1.5 ml/min. Mono Q column (Pharmacia, 0.5 × 5 cm) chromatography was performed as the next step. The fractions with activity were eluted out (flow rate: 1 ml/min) with a linear gradient of NaCl (0–0.6 M) in the phosphate buffer and pooled. Purification of β-glucosidase from *R. albus* was performed by the method of Ohmiya *et al.*[9]

The properties of the purified β-glucosidase were studied using cellooligomers. Cellooligomer solutions (20 mM; 10 μl) of cellobiose (G2), cellotriose (G3), cellotetraose (G4), cellopentaose (G5), or cellohexaose (G6) were incubated separately with 1 μl of 2-mercaptoethanol and 8 mU of β-glucosidase. Aliquots of the enzymatic hydrolysates were analyzed with an HPLC system (JASCO model VL-614, JASCO, Tokyo) equipped with an Ultron NH_2 column (Shinwakako Co., Kyoto) at room temperature. The solvent was composed of acetonitrile and water at a ratio of 55:45, and the flow rate was 0.7 ml/min.

NH$_2$-terminal amino acid sequence of β-glucosidase was determined with an ABI 477A/120A sequence analyzer. The purified enzyme was spotted onto a polyvinylidene difluoride (PVDF) membrane,[10] stained with Coomassie blue R-250, and sequenced directly. Quadral buffer (0.1 M, pH 9.0) was used as a coupling buffer. Polyprene (4 mg) was used as a carrier. Phenylthiohydantoin (PTH)-amino acid derivatives were identified by reverse-phase HPLC using Develosil ODS-5 column (4.6 × 250 mm; Nomura Chemicals Co., Seto, Aichi) with an isocratic elution system (10 mM ammonium acetate:methanol:acetonitrile = 15:9:1, v/v/v).

A 200-μg sample of β-glucosidase purified from the transformant was mixed well with 150 μl of complete Freund's adjuvant and injected into a mature BALB/c mouse. A booster shot was injected in the same manner 4 weeks after the first injection. The serum was collected 1 week after the booster injection from an eye blood vessel and stored at 4°C in NaN$_3$ (0.01%) until used.

RESULTS AND DISCUSSION

Cloning of a β-Glucosidase Gene

Positive clones were detected by fluorescence of the degraded product (FIG. 1) from approximately 20,000 colonies examined. All of the positive clones carried the same recombinant plasmid DNA, and it was designated as pRA2. This plasmid contained a 3.3 kbp insert in vector pBR322. The fragment in pRA2 was partially digested by *Eco*RI and subcloned into the *Eco*RI site of pUC119. Plasmid pURA2 in the β-glucosidase-positive subclones harbored a 3.1 kpb *Eco*RI fragment that was used for DNA sequence and for generation of the translation product. A restriction map of pURA2 is shown in FIGURE 2. Three *Hin*dIII, two *Pst*I, two *Eco*RI, and three *Pvu*II sites were found in the inserted fragment. The fragment inserted in pURA2 was shown to be chromosomal DNA by a Southern hybridization method using the ^{32}P-labeled 1.3 kbp and 0.9 kbp *Hin*dIII fragments contained in pRA201 as probes (data not shown).

DNA Sequence

The DNA sequence strategy for the *R. albus* β-glucosidase gene is shown in FIGURE 3. The DNA sequence and the deduced amino acid sequence of the cloned DNA fragment are shown in FIGURE 4. The deduced NH$_2$-terminal amino acid sequence (the underlined amino acid residues) was identical to that of purified enzyme. The GGAGG sequence (nucleotide No. 197-201) in the upstream of the translational initiation codon ATG (nucleotide No. 208-210) would provide a strong Shine-Dalgarno sequence, typical of the ribosome binding site of gram-positive bacteria.[11] The open reading frame of 2,844 bp (including a stop codon) extending downstream of the initiation codon encodes a polypeptide of 947 amino acids with a molecular weight of 104,276 daltons. Palindromic sequences were found both

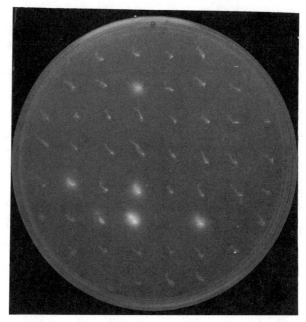

FIGURE 1. β-Glucosidase-producing (fluorescence-emitting) colonies of *E. coli* transformed with a plasmid containing a DNA fragment encoding β-glucosidase from *R. albus*.

upstream and downstream of the β-glucosidase structural gene (nucleotide No. 10-33 and No. 3066-3075). These structures, corresponding to the mRNA hairpin loop with ΔG of −13.20 kcal/mol and −14.70 kcal/mol,[12] respectively, are supposed to act as a transcriptional terminator.

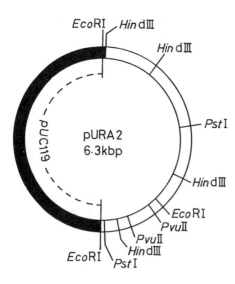

FIGURE 2. Restriction map of the *R. albus* β-glucosidase gene in pRA2. Dark and light areas indicate vector pBR322 and the integrated DNA fragment, respectively.

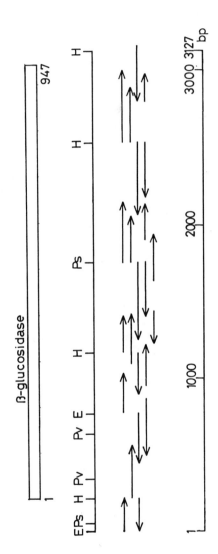

FIGURE 3. Sequence strategy for the *R. albus* β-glucosidase gene. *Arrows* indicate the starting point and direction of individual sequence reading. E, Ps, H, and Pv are restriction sites for *Eco*R1, *Pst*I, *Hind*III and *Pvu*II, respectively.

```
      10        20        30        40        50        60
GAATTCATCAAGGTGTGATGTTGATTATACCTTCGTGAAGTTTGTGAAGAAGCCTGCAGG
         ————————→   ←————————    ————————


      70        80        90       100       110       120
CTCGAAGCCAGGTATGGTGATATTCACACATAATTATACGATAACCGTAAAGCCTGACGA
             ————————              ——————————
              -35                    -10


     130       140       150       160       170       180
AGAAACAGTACTGGGACTGATGCACAAAAGCTGAATCCCTGTGACGTTTCTTTGTCAATT


     190       200       210       220       230       240
GACCTGACTGTGATACGGAGGTAAGATATGATAAAGCTTGATTGGAACGAATATCTCGAA
               ————————            MetIleLysLeuAspTrpAsnGluTyrLeuGlu
                 SD


     250       260       270       280       290       300
AAGGCAGCAGAGGTAAACGCTGAGGGCGCAGTGCTGCTTGTGAACAACGGCGTCCTGCCG
LysAlaAlaGluValAsnAlaGluGlyAlaValLeuLeuValAsnAsnGlyValLeuPro


     310       320       330       340       350       360
CTGGATAAAAATGCCGTTACGCAGGTTTTCGGACGTATACAGCTGGATTATTATAAAAGC
LeuAspLysAsnAlaValThrGlnValPheGlyArgIleGlnLeuAspTyrTyrLysSer


     370       380       390       400       410       420
GGTACGGGCTCTGGCGGAATGGTGAATGTTGCAAAGGTCACGGGAATAACCGATGGCCTT
GlyThrGlySerGlyGlyMetValAsnValAlaLysValThrGlyIleThrAspGlyLeu


     430       440       450       460       470       480
ATAGAAGCAGGTGCAAAACTCAATGAGGATGTGCTGAAGGCTTACAAGGACTATGTTGCT
IleGluAlaGlyAlaLysLeuAsnGluAspValLeuLysAlaTyrLysAspTyrValAla


     490       500       510       520       530       540
GAACATCCCTACGATTACGGCGAGGGCTGGGGCGGCGAGCCCTGGTGTCAGGAGGAGATG
GluHisProTyrAspTyrGlyGluGlyTrpGlyGlyGluProTrpCysGlnGluGluMet


     550       560       570       580       590       600
CCTCTTGATGACAGCCTTGTAAAAAGGGCGGCTGAGAGTTCCGATACAGCGATATGTATT
ProLeuAspAspSerLeuValLysArgAlaAlaGluSerSerAspThrAlaIleCysIle


     610       620       630       640       650       660
ATAGGACGCACCGCAGGCGAGGAACAGGACAACAGCTGCAAGGCAGGTTCTTATCTGCTG
IleGlyArgThrAlaGlyGluGluGlnAspAsnSerCysLysAlaGlySerTyrLeuLeu


     670       680       690       700       710       720
ACAGACGGTGAAAAGGCTATTCTGCGCAAGGTAAGGGATAATTTCAGCAAAATGGTGATA
ThrAspGlyGluLysAlaIleLeuLeuArgLysValArgAspAsnPheSerLysMetValIle


     730       740       750       760       770       780
CTGCTCAATGTGGGCAATATAATCGACATGGGCTTTATCGACGAATTCTCACCCGATGCT
LeuLeuAsnValGlyAsnIleIleAspMetGlyPheIleAspGluPheSerProAspAla


     790       800       810       820       830       840
GTAATGTATGTATGGCAGGGTGGTATGACAGGCGGTACAGGCACTGCAAGGGTGCTGCTG
ValMetTyrValTrpGlnGlyGlyMetThrGlyGlyThrGlyThrAlaArgValLeuLeu
```

FIGURE 4. Legend on page 48.

```
       850        860        870        880        890        900
GGTGAGGTATCTCCCTGCGGCAAGCTGCCCGATACTATCGCTTATGATATCACAGACTAT
GlyGluValSerProCysGlyLysLeuProAspThrIleAlaTyrAspIleThrAspTyr

       910        920        930        940        950        960
CCCTCTGACAAAAATTTCCACAACAGGGATGTGGATATCTATGCTGAAGATATCTTCGTG
ProSerAspLysAsnPheHisAsnArgAspValAspIleTyrAlaGluAspIlePheVal

       970        980        990       1000       1010       1020
GGATACAGATACTTTGATACCTTTGCAAAGGACAGGGTAAGATTCCCCTTCGGATACGGA
GlyTyrArgTyrPheAspThrPheAlaLysAspArgValArgPheProPheGlyTyrGly

      1030       1040       1050       1060       1070       1080
CTTAGCTATACGCAGTTTGAGATAAGTGCCGAGGGCAGAAAGACTGATGACGGTGTTGTC
LeuSerTyrThrGlnPheGluIleSerAlaGluGlyArgLysThrAspAspGlyValVal

      1090       1100       1110       1120       1130       1140
ATAACTGCTAAAGTGAAGAATATCGGCAGTGCGGCAGGCAAGGAAGTCGTGCAGGTATAC
IleThrAlaLysValLysAsnIleGlySerAlaAlaGlyLysGluValValGlnValTyr

      1150       1160       1170       1180       1190       1200
CTTGAAGCGCCCAACTGTAAGCTTGGCAAGGCTGCGCGTGTGCTTTGCGGATTTGAAAAG
LeuGluAlaProAsnCysLysLeuGlyLysAlaAlaArgValLeuCysGlyPheGluLys

      1210       1220       1230       1240       1250       1260
ACAAAGGTACTTGCACCGAATGAAGAACAGACGCTGACGATAGAAGTCACCGAGCGTGAT
ThrLysValLeuAlaProAsnGluGluGlnThrLeuThrIleGluValThrGluArgAsp

      1270       1280       1290       1300       1310       1320
ATAGCTTCCTACGATGACAGCGGCATTACAGGAAATGCCTTCGCATGGGTAGAGGAAGCA
IleAlaSerTyrAspAspSerGlyIleThrGlyAsnAlaPheAlaTrpValGluGluAla

      1330       1340       1350       1360       1370       1380
GGAGAGTACACATTCTATGCAGGCAGTGATGTGCGCAGTGCAAAGGAATGCTTTGCTTTC
GlyGluTyrThrPheTyrAlaGlySerAspValArgSerAlaLysGluCysPheAlaPhe

      1390       1400       1410       1420       1430       1440
ACACTGGATTCTACCAAGGTCATCGAACAGCTTGAACAGGCACTGGCACCTGTTACGCCT
ThrLeuAspSerThrLysValIleGluGlnLeuGluGlnAlaLeuAlaProValThrPro

      1450       1460       1470       1480       1490       1500
TTCAAGAGGATGGTTCGCACCGCAGAGGGTCTTTCCTATGAGGATACCCCTCTTTCAAAG
PheLysArgMetValArgThrAlaGluGlyLeuSerTyrGluAspThrProLeuSerLys

      1510       1520       1530       1540       1550       1560
GTTGACGAAGCTGCACGCAGACTTGGATATCTGCCTGCGGAAACAGCATATACAGGTGAT
ValAspGluAlaAlaArgArgLeuGlyTyrLeuProAlaGluThrAlaTyrThrGlyAsp

      1570       1580       1590       1600       1610       1620
AAGGGTATAGCCCTTTCCGATGTGGCCCATGGTAAGAACACCCTTGATGAGTTCATAGCA
LysGlyIleAlaLeuSerAspValAlaHisGlyLysAsnThrLeuAspGluPheIleAla

      1630       1640       1650       1660       1670       1680
CAGCTTGATGACAATGACCTTAACTGCCTTGTGCGCGGCGAGGGTATGTGTTCTCCAAAG
GlnLeuAspAspAsnAspLeuAsnCysLeuValArgGlyGluGlyMetCysSerProLys
```

FIGURE 4. Legend on page 48.

```
        2530      2540      2550      2560      2570      2580
CTTGGCGAGGACGAGGCTGTTGAAGTTATCAACAAGCCTGCCGAGACCGTTGATGACGGC
LeuGlyGluAspGluAlaValGluValIleAsnLysProAlaGluThrValAspAspGly

        2590      2600      2610      2620      2630      2640
GAGGGCGACAGAGTGTTCCTGCTGGACGGCGACCTGACCATAGATATGAGTGGTGTTAAG
GluGlyAspArgValPheLeuLeuAspGlyAspLeuThrIleAspMetSerGlyValLys

        2650      2660      2670      2680      2690      2700
ACCGAGAGAAATCTCGATTACAGCTTCACTGTAGATGTGGCACAGTTCGGTCAGTACCGC
ThrGluArgAsnLeuAspTyrSerPheThrValAspValAlaGlnPheGlyGlnTyrArg

        2710      2720      2730      2740      2750      2760
ATGGAAATGACAGCAAGCTCCACACAGAGCGAGCTTGCACAGATGCCCGTGACCGTATTC
MetGluMetThrAlaSerSerThrGlnSerGluLeuAlaGlnMetProValThrValPhe

        2770      2780      2790      2800      2810      2820
AGCATGGGTACTGCATGGGGCACATTCACATGGAACGGCACAGGCGGAAAACCCGTGACC
SerMetGlyThrAlaTrpGlyThrPheThrTrpAsnGlyThrGlyGlyLysProValThr

        2830      2840      2850      2860      2870      2880
TTCGCCGTGGAAGAAATGCCCATGTTCTCCCGGTATACTATATTCAGGCTTCACTTCGGT
PheAlaValGluGluMetProMetPheSerArgTyrThrIlePheArgLeuHisPheGly

        2890      2900      2910      2920      2930      2940
CTGGGCGGACTTGATATGGATAAGATAGTATTCAAAAAGATAAGACCCGCCGAGGCACAG
LeuGlyGlyLeuAspMetAspLysIleValPheLysLysIleArgProAlaGluAlaGln

        2950      2960      2970      2980      2990      3000
GTCTGTCGGTTGAGGATATCTGAGAGATGGCTTCAAACGCAGACGTACTTCTGGCTGAAA
ValCysArgLeuArgIleSerGluArgTrpLeuGlnThrGlnThrTyrPheTrpLeuLys

        3010      3020      3030      3040      3050      3060
GCGAACTTTCAAAGTAAAAAGCTGCTTCGAGGACGCCGTGCTTATCGATAAAAGAACGGG
AlaAsnPheGlnSerLysLysLeuLeuArgGlyArgArgAlaTyrArg***      ─────

        3070      3080      3090      3100      3110      3120
CAGAACTTTTCCCGTATGCGATATGTACGCGACCCGAGGGCGCACTGATAAGCAAAGACG
 ─→        ←─────

        3130      3140      3150
AAAGCTTATCGATGATAAGCTGTCAAACATGAGAATTC
  └──→ pBR322
```

FIGURE 4. Nucleotide and deduced amino acid sequences of an *R. albus* β-glucosidase gene. The underlined sequences marked −35 and −10 refer to the RNA polymerase recognition and binding sites, respectively. A Shine-Dalgarno ribosome binding site upstream from the start codon is doubly underlined. The NH$_2$-terminal amino acid sequence of the matured enzyme from the transformant was determined by an amino acid sequencer and underlined in this figure. Nucleotide sequences that may form stem loop structures are found upstream and downstream from the structure gene as shown with a pair of *arrows*. The stop codon is indicated by three *asterisks*.

Production of β-Glucosidase from the Transformant

The transformant harboring pURA2 was cultivated in Luria broth. The time courses of bacterial growth and β-glucosidase formation are shown in FIGURE 5. Unexpectedly, most of the β-glucosidase activity was found in culture supernatant and not in the transformant cells without a decrease in cell density during cultivation for 12 hours, suggesting that cloned β-glucosidase was excreted into the extracellular fraction without autolysis. A similar phenomenon was observed by Shima *et al.*[13] when the cellulase gene from *Clostridium cellobioparum* was expressed in *E. coli* JM109. In their experiment, β-lactamase and malate dehydrogenase, the periplasmic and cytoplasmic enzymes, respectively, were excreted as the extracellular fraction, when a certain 0.5 kbp DNA fragment was harbored in *E. coli* JM109. On the basis of this result, the DNA fragment we cloned may encode, in addition to β-glucosidase of *R. albus,* some signals that facilitate permeability of the membrane of *E. coli* JM109. Therefore, the enzyme was harvested from culture supernatant at 10 hours of cultivation and used for purification.

Enzyme Purification from the Transformant

The proteins in the culture supernatant of *E. coli* JM109 harboring pURA2 were precipitated by ammonium sulfate at 70% saturation, redissolved in sodium phosphate buffer, and gel filtrated. One peak of activity against PNPG was obtained, and the purity increased up to 29 times. Anion exchange chromatography with DEAE Bio-Gel A was the next step. A single peak of enzyme activity was eluted at a NaCl concentration of 0.45 M. The fractions with activity were loaded on an anion exchange resin, Mono Q prepacked column. Enzyme activity was detected in a fraction eluted by NaCl at a concentration of 0.28 M. The specific activity of the fraction was 0.41 units/mg. Quantitative representation for purification of β-glucosidase is summarized in TABLE 1. The specific activity of β-glucosidase purified from the transformant is one tenth that from *R. albus,*[9] suggesting that the production of anaerobic enzyme in *E. coli* under an aerated condition might charge some oxidative effects on the enzyme protein.

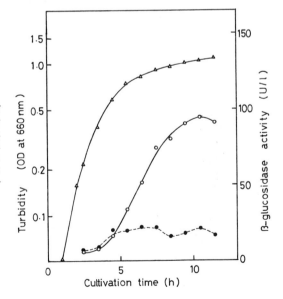

FIGURE 5. Time courses of growth and β-glucosidase activity of *E. coli* JM109 (pRA2) in Luria medium containing 50 μg/ml ampicillin at 37°C on a shaker. △—△: growth of *E. coli;* ○—○: activity in supernatant; ●—●: activity in cells.

TABLE 1. Summary of Purification of β-Glucosidase from *E. coli* JM109 Harboring pURA2

Purification Step	Total Activity (U)	Total Protein (mg)	Specific Activity (U/mg)	Recovery of Activity (%)	Purification (fold)
Culture supernatant	170	27,000	0.0064	100	1
70% (NH₄)₂SO₄ precipitation	140	1,700	0.085	82	13
Sephacryl S-200 HR	62	340	0.18	36	29
DEAE Bio-Gel A	19	96	0.20	11	31
Mono Q	12	31	0.39	6.8	64

Characterization of β-Glucosidase Purified from the Transformant

The purity of β-glucosidase was homogeneous as shown in sodium dodecylsulfate polyacrylamide gel electrophoretogram (SDS-PAGE), and the molecular weight of the purified enzyme was estimated to be 120,000 (FIG. 6). This value is higher than that calculated on the basis of the 947 amino acid residues in the open reading frame of the DNA sequence (FIG. 4), suggesting that the molecular weight estimated from the mobility by SDS-PAGE might be overestimated. Maximum activity of the enzyme was observed at pH 6.5 (data not shown) and at 30°C. The enzyme was stable at pH 6.5 and at a temperature below 30°C, but it was unstable above 37°C (FIG. 7).

The effects of chemical reagents on the activity were studied. The results were shown in TABLE 2. Zn²⁺, Hg²⁺, Cu²⁺ inhibited the activity 80–100%. Glucose and its derivative, glucono-δ-lactone, also inhibited the activity, suggesting that it is a product inhibition. Reducing reagents such as 2-mercaptoethanol, dithiothreitol, glutathione, and cysteine-HCl activated the enzyme, whereas sulfhydryl reagents (*p*-chloromercuribenzoic acid and iodoacetoamide) inhibited the activity, indicating that the purified enzyme may belong to the group of sulfhydryl enzymes. Because the activities of cellulase from *R. albus* are very easily lost in the absence of reducing reagents such as 2-mercaptoethanol and dithiothreitol,⁹,¹⁴ these cellulases including

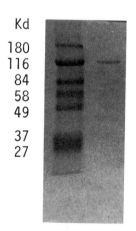

FIGURE 6. SDS-polyacrylamide gel (14%) electrophoretogram of the purified enzyme. *Right lane:* Standard proteins as molecular weight marker with the molecular size of 180,000, 116,000, 84,000, 58,000, 48,500, 36,500, and 26,600 (from top to bottom). *Left lane:* Purified enzyme.

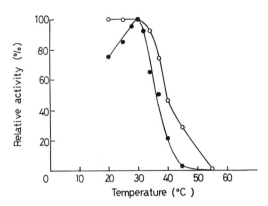

FIGURE 7. Effects of temperature on β-glucosidase activity. ●—●: optimal temperature (enzymatic reaction was carried out at each temperature for 30 minutes). ○—○: temperature stability: residual activity after incubating the enzyme at each temperature for 10 minutes.

β-glucosidase might be oxidatively inactivated. As the transformant was cultivated in an anaerobic condition, it seems that the enzyme produced by the transformant was partially inactivated by oxidation, lowering the specific activity of the enzyme. The activation of the enzyme by reducing reagents (three- to fourfold, TABLE 2) proves it. The K_m value of the enzyme against PNPG was estimated to be 2.4 mM from Lineweaver-Burk plots. When cellobiose was used as a substrate, the value was 96 mM. Each value was comparable to the values of the enzyme purified from *R. albus*.[9] Hydrolysates of cellooligomers by β-glucosidase were analyzed by HPLC. G2, G3, G4, G5, and G6 were hydrolyzed to smaller products and finally to G1, indicating that the enzyme is a β-glucosidase. The NH_2 terminal amino acid sequence of the purified enzyme from the transformant was the same as that of the deduced sequences in FIGURE 4. The purified β-glucosidases from the transformant and that purified from *R. albus* were immunologically identical to each other, as they gave a fused single precipitation line against rat antisurum of β-glucosidase from the transformant.

TABLE 2. Effects of Various Reagents on the β-Glucosidase Activity

Reagents (1 mM)	Relative Activity (%)
2-Mercaptoethanol	273
Dithiothreitol	299
Glutathione	264
Cysteine-HCl	407
p-Chloromercuribenzoic acid	2
Iodoacetoamide	45
Glucose	55
Glucono-δ-lactone	0
KCl	89
$CaCl_2$	90
$CoCl_2$	50
$MgCl_2$	101
$ZnCl_2$	10
$HgCl_2$	0
$FeSO_4$	95
$CuSO_4$	17
None	100

From these results, the cloned gene encodes the *R. albus* β-glucosidase purified in our previous work.[9]

SUMMARY

A *Hin*dIII fragment of *R. albus* DNA encoding β-glucosidase was cloned into *E. coli.* The DNA sequence (3158 bp) was determined, and the longest potential encoding sequence consisted of 2,841 bp (947 amino acids with the calculated molecular weight of 104,276. The deduced NH_2-terminal amino acid sequence from the first (methionine) to the twentieth (glycine) was identical to that of the purified enzyme, suggesting that the gene for β-glucosidase does not encode a signal peptide. The enzyme purified from the culture supernatant of the transformant had a molecular weight of 120,000 and its maximum activity was revealed at pH 6.5 and 30°C. Reducing reagents activated the enzyme, whereas the sulfhydryl group-blocking reagents and reaction products (glucose) inhibited the activity. Hydrolyzates of celloorigomers contained glucose as a major product, indicating that the enzyme acts as β-glucosidase. The enzyme from the transformant revealed similar properties to that from *R. albus,* and both enzyme proteins were immunologically the same to each other, indicating that the cloned gene encodes β-glucosidase from *R. albus.*

REFERENCES

1. KNOWLES, J., P. LEHTOVAARA & T. TEERI. 1987. Cellulase families and their genes. TIBTECH **5:** 255–261.
2. ROBSON, L. M. & G. H. CHAMBLISS. 1989. Cellulases of bacterial origin. Enzyme Microb. Technol. **11:** 626–644.
3. OHMIYA, K., K. NAGASHIMA, T. KAJINO, E. GOTO, A. TSUKADA & S. SHIMIZU. 1988. Cloning of the cellulase gene from *Ruminococcus albus* and its expression in *Escherichia coli.* Appl. Environ. Microbiol. **54:** 1511–1515.
4. OHMIYA, K., H. DEGUCHI & S. SHIMIZU. 1991. Modification of the properties of a *Ruminococcus albus* endo-1,4-β-glucanase by gene truncation. J. Bacteriol. **173:** 636–641.
5. MANIATIS, T., E. F. FRITSCH & J. SAMBROOK. 1982. Molecular Cloning, A Laboratory Manual. Cold Spring Harbor Laboratory. Cold Spring Harbor, NY.
6. MANDEL, M. & A. HIGA. 1970. Calcium-dependent bacteriophage DNA infection. J. Mol. Biol. **53:** 159–162.
7. TAYA, M., T. KOBAYASHI & S. SHIMIZU. 1980. Synthetic medium for cellulolytic anaerobe, *Ruminococcus albus.* Agric. Biol. Chem. **44:** 2225–2227.
8. SANGER, F. 1981. Determination of nucleotide sequence in DNA. Science **214:** 1205–1210.
9. OHMIYA, K., M. SHIRAI, Y. KURACHI & S. SHIMIZU. 1985. Isolation and properties of β-glucosidase from *Ruminococcus albus.* J.Bacteriol.**161:** 432–434.
10. MATSUDAIRA, P. 1987. Sequence from picomole quantities of proteins electroblotted on polyvinylidene difluoride membranes. J. Biol. Chem. **262:** 10035–10038.
11. MCLAUGHLIN, J. R., C. L. MURRAY & J. C. RABINOWITZ. 1981. Unique features in the ribosome binding site sequence of the Gram-positive *Staphylococcus aureus* β-lactamase gene. J. Biol. Chem. **256:** 11283–11291.
12. TINOCO, I., P. N. BRORER, B. DENGLER & M. D. LEVIN. 1973. Improved estimation of secondary in ribonucleic acids. Nature New Biol. **246:** 40–41.
13. SHIMA, S., J. KATO, Y. IGARASHI & T. KODAMA. 1989. Cloning and expression of a *Clostridium cellobioparum* cellulase gene and its expression in *Escherichia coli* JM109. J. Ferment. Bioeng. **68:** 75–78.
14. OHMIYA, K., K. MAEDA & S. SHIMIZU. 1987. Purification and properties of endo-β-1,4-glucanase from *Ruminococcus albus.* Carbohydrate Res. **166:** 145–155.

Construction and Evaluation of a Self-Luminescent Biosensor[a]

L. GEISELHART,[b] M. OSGOOD, AND D. S. HOLMES[c]

Department of Biology
Rowley Labs
Clarkson University
Potsdam, New York 13699

BIOSENSOR DEVELOPMENT

Research and development of biosensors are proceeding extremely rapidly, in part due to advances in microsensor technologies and to remarkable developments in biotechnology. A biosensor uses a biologically derived material immobilized to a suitable transducing system that converts the binding or analyte recognition event into a quantifiable and processable signal. One of the key advantages of using biological components in sensors is their exquisite specificity of molecular recognition. Examples of the range of biological components used in sensors are given in TABLE 1. Another advantage is that biosensor recognition is usually directed towards molecules of biological relevance. Consequently, biosensors are particularly suitable for applications in the pharmaceutical industry, medical diagnostics, food and drug testing, and environmental monitoring.

A central issue in the development of biosensors is how to transduce the molecular recognition event into a detectable and quantifiable signal. Many ways have been investigated, each having its own set of particular advantages and disadvantages (TABLE 2).

Recent advances in luminescence instrumentation, laser miniaturization, fiberoptics research, and biotechnology have permitted the development of a number of types of biosensor based on measurement of light signals.[1] For example, Tromberg and his colleagues[2,3] have developed a fiberoptic biosensor for a benzo(a)pyrene based on the interaction of benzo(a)pyrene with antibody-coated optical fibers. The bound benzo(a)pyrene can then be detected by measuring laser-induced fluorescence.[2,3] A limitation of this approach is that only fluorescence-inducible antigens can be detected.

In an approach developed by Kulp and his colleagues,[4,5] enzymes and fluorescein have been immobilized in polyacrylamide at the tip of optical fibers, creating a device the authors call an enzyme optrode. A signal is produced when the enzyme catalyzes the conversion of the analyte into a product that lowers the microenvironmental pH and consequently lowers the fluorescence intensity of the dye. For example, immobilized glucose oxidase converts glucose to gluconic acid and penicillinase converts penicillin to penicilloic acid. Problems with the enzyme electrode are its lowered sensitivity in buffered solutions and its limited use with enzymes that result in products that alter pH.

[a] This work was supported in part by grant ECS-8915294 from the National Science Foundation and grant H-00014-90-J-1715 from the Office of Naval Research.
[b] PRESENT ADDRESS: Albany Medical College, Albany, New York 12208.
[c] To whom correspondence should be addressed.

53

Another approach involves the use of surface relief gratings and wave guides that respond to the adsorption of molecules, with a change in the intensity of uncoupled light produced by a laser.[6] The sensor responds to changes in the refractive index of the cover medium and to the adsorption of antibodies covalently bound to the grating. More experiments will be directed towards visualization of biological interactions such as those between antigen and antibody, receptor and cell, and inhibitor and enzyme. This approach is limited to measuring biological molecules that undergo such interactions or to reactions that change the refractive index of the medium.

TABLE 1. Biological Components Provide the Molecular Recognition Element for Biosensors[12]

Biological Component	Advantages	Disadvantages
Chemoreceptors		
Intact structures	Use natural structures that are already in the optimal state for detection	Fragility, limited lifetime
Preparations containing isolated chemoreceptors	Can be prepared just before use for freshness, which enhances sensitivity	Difficult to store, few types available, limited selectivity
Antibodies/antigens		
Polyclonal antibodies	Relatively inexpensive	Have several binding constants, limited selectivity
Monoclonal antibodies	Good selectivity, binding constant is uniform	High cost, limited availability
Antibody fragments	Low molecular weight	Limited availability, cost uncertain
Enzyme/antigen conjugates or labeled antibodies	Improved amplification	Tedious to prepare
Biocatalysts		
Isolated enzymes	High selectivity and activity	Limited availability and stability
Enzyme sequences	Expand accessible pathways	Subject to interferences, stringent operating conditions
Microorganisms	Great variety, self-contained, regeneration possible	Poor selectivity, limited lifetime
Plant and animal tissue	High activity, natural configuration	Subject to interferences and contamination

Our laboratory is developing biosensors that are based on the ability of genetically engineered bacteria to emit visible light in response to specific compounds. The ease of measuring light makes this type of biosensor especially attractive. A prototype biosensor has been constructed by fusing the light-emitting genes (*lux* genes) of the marine bacterium *Vibrio fischeri* to the regulatory elements of the mercury detoxification genes (*mer* genes) of the bacterium *Serratia marcescens*. The objective is to develop a genetically engineered microorganism that expresses light in the presence of mercury. Preliminary data on the development and operation of such a biosensor are presented.

TABLE 2. Transducers Provide the Signal-Generating Element for Biosensors[12]

Transducer	Advantages	Disadvantages
Electrochemical		
Potentiometric	Wide response range, commonly available instrumentation, selective	Responses to change in dose are logarithmic, slow response, limited selectivity
Amperometric	Response to change in dose is linear, sensitive	Subject to interferences and protein fouling
Conductimetric	Inexpensive, rugged	Nonselective, poor signal-to-noise ratio
Optical		
Fiberoptics	Small size, rapid response, no reference cell needed, suitable for sensor arrays, calibration stability	Special instrumentation and indicators needed, some problems in turbid samples, losses from photobleaching
Surface plasmon resonance	Suitable for immunoreactions	Geometric factors must be controlled
Interference-effect optodes	Can be used for multiplexing and in array sensors, small size	Require specialized instrumentation
Total internal reflection fluorescence	Can distinguish free and bound fluophor	Multiple reflections required to achieve sensitivity
Field effect transistors	Very small size, potentially low cost, suitable for automated systems	Unstable, leakage problems, slow response, biocompatibility problems
Thermistors	Simple, linear response	Nonselective, subject to drift
Piezoelectric crystals	Operate over a wide temperature range, sensitive	Work best in gas phase, slow recovery

The Lux *System.* The *lux* gene system of *V. fischeri* consists of five structural genes (*lux* A-E) and two regulatory genes (*lux* R and I) organized as an operon, as shown in FIGURE 1. *Lux* A and *lux* B encode two subunits of a heterodimer ($\alpha\beta$) flavin monooxygenase, luciferase, that catalyzes light production by oxidation of a reduced tetradecanal aldehyde, as shown in FIGURE 2. Aldehyde production is catalyzed by enzymes encoded by *lux* C, D, and E.

The Mer *System.* Plasmid-encoded resistance to mercury is widespread in bacteria. Two distinct types of mercury resistance have been described. In the first type, called narrow range, resistance involves reduction of Hg^{+2} to elemental Hg^0, which

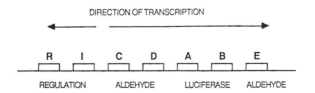

FIGURE 1. *Lux* operon from *V. fischeri.*

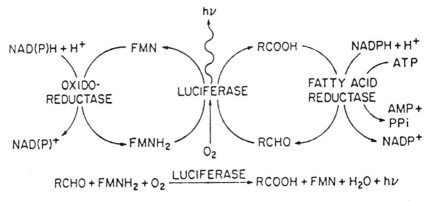

$$RCHO + FMNH_2 + O_2 \xrightarrow{\text{LUCIFERASE}} RCOOH + FMN + H_2O + h\nu$$

FIGURE 2. Substrates, products, and pathways involved in bacterial bioluminescence reaction.

volatilizes from the medium.[7] The reduction is carried out by an intracellular mercuric reductase enzyme (*mer*A, FIG. 3). Another component of the resistance mechanism is an Hg^{+2}-specific transport protein (*mer*T) that carries Hg^{+2} into the cytoplasm where it is detoxified by the reductase enzyme. The mercuric resistance genes are regulated by an operator protein (*mer*R) that, in the absence of exogenous Hg^{+2}, prevents transcription of the *mer* genes by binding to the promoter region (PR). However, in the presence of exogenous mercury, the Hg^{+2} binds to the

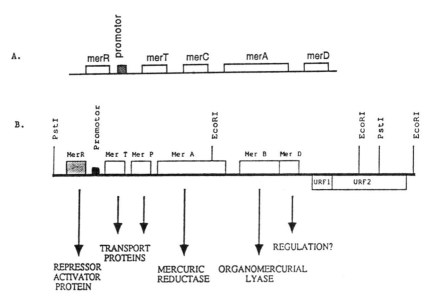

FIGURE 3. (**A**) Narrow range *mer* operon. (**B**) Broad range *mer* operon from pGN21, a plasmid originally isolated from a hospital isolate of *S. marcescens*. The functions of the gene products, as well as a few significant restriction enzyme sites, are noted.

operator protein, and the protein-Hg complex, in turn, binds to the PR, activating the transcription of the *mer* genes.

The second type of resistance, called broad range, also involves a mercury transport protein and a mercuric reductase enzyme. However, it also involves an intracellular enzyme, organomercurial lyase, which cleaves C-Hg bonds to yield Hg^{+2} ions, which are then detoxified by the mercuric reductase. The *mer* resistance genes of this group are activated by a range of organomercurials including phenylmercuric acetate and thimerosal.

MATERIALS AND METHODS

Plasmid pJE202 containing the *lux* operon was a gift from Joanne Engebrecht. Plasmid pGN21 containing the *mer* genes was a gift from Simon Silver. Bacteria were grown in Luria Bertani (LB) broth or on solid media (15 g agarose/liter LB broth) at 30°C or 37°C. Selective media contained 100 μg/ml ampicillin. Plasmid DNA from bacterial cultures was purified by the method of Holmes and Quigley.[8] Restriction endonuclease digestions were carried out according to the manufacturer's instructions (Gibco BRL). Agarose gel electrophoresis was performed as described by Maniatis *et al.*[9] using a tris-acetate EDTA buffer system or a tris-borate EDTA buffer system. Ligation of DNA segments was carried out according to the enzyme manufacturer's directions, and transformation of *Escherichia coli* was performed according to the $CaCl_2$ method described by Maniatis *et al.*[9]

Light emission was measured with a luminometer (Lumac Biocounter 2010, Integrated Biosolutions, Inc., Princeton, NJ) using 0.1 ml sample volume and 30-second internal integration. Light emission was recorded in relative light units which can be converted using a light standard. Filter-sterilized mercuric chloride was added to bacterial cultures at various concentrations and at various times.

RESULTS AND DISCUSSION

The *lux* structural genes were spliced downstream from the *mer* regulating genes and several of the *mer* structural genes as shown in FIGURE 4. This brings the transcription of the *lux* structural genes under the control of the *mer* regulatory genes. Thus, in the presence of Hg^{+2}, the product of the *mer*R gene binds the Hg^{+2}, activating transcription.

Preliminary results show the luminescence activity of a successfully constructed mercury biosensor incorporating the *mer* operon and the *lux* system (FIG. 5). Induction of luminescence occurred within 90 minutes of the addition of mercuric chloride to a culture of *E. coli* that had been transformed with a plasmid (pMLB72) that contained the *mer* genes ligated upstream of the major portion of the *lux* operon and interrupting the *lux* R gene. Without mercury, no luminescence above the background was seen.

The addition of mercuric chloride to *E. coli* cultures transformed with the plasmid pJE202 (which contains all the genes necessary for light production, but no *mer* genes) decreased the normal light output by various amounts according to the bacterial culture growth phase. Mercury did not increase the light output at any stage of growth. We do not yet have data on the dynamic range of mercury concentration necessary for induction or on the specificity of the luminescent response to organic or inorganic mercury compounds.

The transport and reductase genes of the *mer* operon have been included in the construction of the *mer-lux* plasmid, in addition to the promoter region which turns on the *lux system,* for several reasons. The transport proteins are necessary to carry the mercury into the cell. Without such facilitated entry, induction of the *lux* system would be slower and less sensitive. The mercuric reductase and organomercurial lyase proteins enable the bacteria to detoxify the mercury compounds. Excluding these genes in the biosensor construction would result in a hypersensitive organism, that is, one that would allow mercury into the cell, but then be unable to survive.

Experiments with the pJE202 plasmid indicate an absolute requirement for high

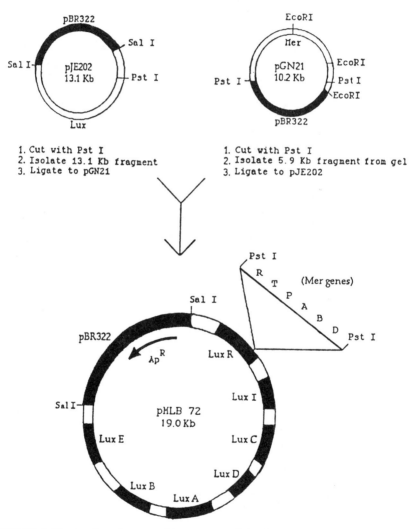

FIGURE 4. Construction of mercury biosensor containing the *mer* genes of plasmid pGN21 and the *lux* genes from plasmid pJE202.

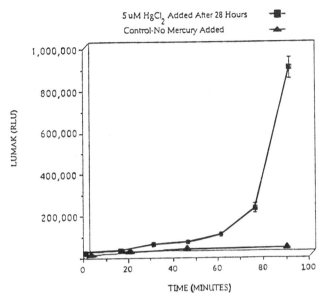

FIGURE 5. Response of pMLB72 to mercury induction.

levels of oxygen for light production (ref. 10 and our own observations). This is not surprising, considering the biochemical reactions involved (FIG. 1). Growth of the microorganisms in a minimal medium, such as M9 medium,[9] results in no light production; however, the necessary constituents present in the rich medium normally used (LB broth) have not yet been elucidated. The production of light is an energy-intensive process that consumes large quantities of ATP and reduced flavin mononucleotide.

A bioluminescent reporter plasmid for naphthalene catabolism, which uses the *lux* genes, has already been described.[11] This construction is induced to luminesce and is responsive to variations in naphthalene exposure. It also was shown to produce sufficient light to act as a biosensor of naphthalene in contaminated soils. This work, as well as our own preliminary results, indicates that this type of microbial biosensor has broad possibilities for environmental use, limited only by the existence of promoters that recognize analytes of interest.

SUMMARY

The genes encoding bioluminescence (*lux* genes), derived from the marine bacterium *V. fischeri,* have been fused next to the genes encoding mercury detoxification (*mer* genes), derived from a clinical isolate of *S. marcescens.* The fusion has been made so that the expression of the light genes comes under the control of the *mer* regulatory gene and promoter. These genetic elements activate the expression of the light genes in the presence of mercury. The light can readily be collected and quantitated, resulting in a biosensor for the detection of mercury.

REFERENCES

1. BORMAN, S. 1987. Optical and piezoelectric biosensors. Anal. Chem. **59:** 1161A–1164A.
2. TROMBERG, B. J., M. J. SEPANIAK, J. P. ALARIE, T. VO-DINH & R. M. SANTELLA. 1988. Development of antibody-based fiber-optic sensors for detection of a benzo(a)pyrene metabolite. Anal. Chem. **60:** 1901–1908.
3. VO-DINH, T., B. J. TROMBERG, G. D. GRIFFIN, K. R. AMBROSE, M. J. SEPANIAK & E. M. GARDENHIRE. 1987. Antibody-based fiberoptics biosensor for the carcinogen benzo(a) pyrene. Appl. Spectroscopy. **41:** 735–738.
4. (a) KULP, T. J., I. CAMINS & S. M. ANGEL. 1987. Polymer immobilized enzyme optrodes for the detection of penicillin. Anal. Chem. **59:** 2849–2853.
   (b) KULP, T. J., I. CAMINS, S. M. ANGEL, C. MUNKHOLM & D. R. WALT. 1987. Polymer immobilized enzyme optrodes for the detection of penicillin. Anal. Chem. **59:** 2849–2853.
5. KULP, T. J., I. CAMINS & J. M. ANGEL. 1988. Enzyme based fiber optic sensors. In Optical Fibers in Medicine III. Proc. Int. Soc. Optical Eng. A. Katzir, ed.: 134–138. SPIE. Washington, DC.
6. SEIFERT, M., K. TIEFENTHALER, K. HEUBERGER, W. LUKOSZ & K. MOSBACH. 1986. An integrated optical biosensor. Analyt. Letter **19:** 205–216.
7. NI'BHRIAN, N. N., S. SILVER & T. J. FOSTER. 1983. Tn5 insertion mutations in the mercury ion resistance genes derived from plasmid R100. J. Bacteriol. **155:** 690–703.
8. HOLMES, D. S. & M. QUIGLEY. 1981. A rapid method for the preparation of plasmids. Anal. Biochem. **114:** 193–197.
9. MANIATIS, T., E. F. FRITSCH & J. SAMBROOK. 1982. Molecular Cloning: A Laboratory Manual. Cold Spring Harbor Laboratory. Cold Spring Harbor. New York.
10. ENGEBRECHT, J., M. SIMON & M. SILVERMAN. 1985. Measuring gene expression with light. Science **227:** 1345–1347.
11. KING, J. M. H., P. M. DIGRAZIA, B. APPLEGATE, R. BURLAGE, J. SANSVERINO, P. DUNBAR, F. LARIMER & G. S. SAYER. 1990. Rapid, sensitive bioluminescent reporter technology for naphthalene exposure and biodegradation. Science **249:** 778–781.
12. RECHNITZ, G. A. 1988. Biosensors. Chem. Eng. News **66:** 24–36.

Rhizobium meliloti Exopolysaccharides

Structures, Genetic Analyses, and Symbiotic Roles[a]

T. L. REUBER, A. URZAINQUI, J. GLAZEBROOK,
J. W. REED, AND G. C. WALKER

Department of Biology
Massachusetts Institute of Technology
Cambridge, Massachusetts 02139

In our attempts to understand the molecular mechanisms of nodulation of legumes by nitrogen-fixing rhizobia, we discovered that exopolysaccharides excreted by *Rhizobium meliloti* play key roles in the invasion and development of nodules on alfalfa and other plant hosts. In a set of genetic analyses of exopolysaccharide biosynthesis by this organism, we have been attempting to understand the biological functions of exopolysaccharides in nodulation. The studies were facilitated by the availability of a wide variety of genetic techniques for *R. meliloti* that permit essentially any genetic strategy used in the study of *Escherichia coli* to also be used in the study of *R. meliloti*.[1] In this paper we discuss our genetic analyses of the regulation and synthesis of exopolysaccharides by *R. meliloti*, possible roles for these exopolysaccharides in nodulation, and examples in which our genetic studies have allowed us to manipulate the regulation of synthesis, structure, or molecular weight of exopolysaccharides.

TWO EXOPOLYSACCHARIDES SYNTHESIZED BY *R. MELILOTI*

Rhizobium meliloti strain Rm1021 excretes two exopolysaccharides. The first of these, succinoglycan (EPS I), is a high molecular weight polymer[2] composed of polymerized octasaccharide subunits. Each octasaccharide consists of a backbone of three glucose and one galactose residue, a side chain of four glucose residues, and 1-carboxyethylidene (pyruvate), acetyl, and succinyl modifications in a ratio of approximately 1:1:1. Succinoglycan is also synthesized by *Alcaligenes faecalis* var. *myxogenes*,[3] *Agrobacterium tumefaciens*,[4] *Agrobacterium radiobacter*,[5] and *Pseudomonas sp.*[6] The structure of succinoglycan was determined through studies in a variety of labs[7-9]; however, the position of the acetyl and succinyl modifications has not yet been unambiguously established. Aman *et al.*[10] established that the succinoglycan produced by strain Rm1021 has the sugar backbone and terminal carboxyethylidene modification shown in FIGURE 1A. This figure also shows possible assignments for the positions of the succinyl and acetyl modifications that are based on studies of succinoglycan produced by other organisms.[7,8,11] Before polymerization, the subunits are assembled on a lipid carrier, and *in vitro* labeling studies show that the sequence of assembly is, first, galactose and β1,3 glucose, then the rest of the glucose residues, and finally the other substituents.[12,13] Succinoglycan has certain resemblances to the commercially important exopolysaccharide xanthan gum of *Xanthomonas campestris* and is of commercial interest in its own right.[14,15] As will be discussed, *R. meliloti*

[a]This work was supported by Public Health Service Grant GM31030. J. W. R., J. G., and T. L. R. were supported by National Science Foundation Predoctoral Fellowships.

strain Rm1021 also has the capacity to make a second exopolysaccharide, EPS II, whose structure is shown in FIGURE 1B.

NODULATION BY *RHIZOBIUM*

The bacterium *Rhizobium* fixes nitrogen in symbiotic association with leguminous plants. In the course of this symbiotic interaction, the bacteria induce the formation of nodules on the roots of the plant, enter these nodules through tubes called infection threads, are encoated in a membrane of plant origin and released into the interior of plant cells, differentiate into morphologically distinct forms termed bacteroids, and then begin to fix nitrogen. (For reviews see refs. 16–20 and 49.) Considerable attention has been devoted to the mechanisms whereby a *Rhizobium* strain and its host plant might recognize each other and whereby the bacteria enter the nodules that they induce. The products of the rhizobial *nod* genes were shown to

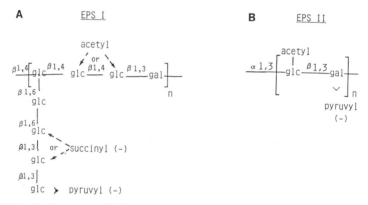

FIGURE 1. (A) Structure of EPS I (succinoglycan) of *R. meliloti* strain Rm1021 as established by Aman *et al.*[10] The possible assignments for the positions of the succinyl and acetyl modifications that are shown are based on studies of succinoglycan produced by other organisms.[7,8,11] (B) Structure of EPS II of *R. meliloti* strain Rm1021.[39,44]

be central to this process and are under active investigation in numerous laboratories.[18,21] A recent study showed that the *R. meliloti nod* genes determine the synthesis of a sulfated, acylated glucosamine oligosaccharide that seems to function as an extracellular signal to the plant during nodulation.[22] In addition, most rhizobia produce a variety of polysaccharides,[16,23] and over the years it has been hypothesized that these polysaccharides play important roles in bacteria-plant interactions, for example, that of being responsible for the host specificity of various *Rhizobium* strains.

THE SUCCINOGLYCAN EXOPOLYSACCHARIDE OF *R. MELILOTI* IS REQUIRED FOR NODULE INVASION

Our laboratory recently obtained strong genetic evidence that succinoglycan, the acidic Calcofluor-binding exopolysaccharide of *R. meliloti* strain Rm1021,[9,10] is required for nodule invasion and possibly for later events in nodule development.[24-26] We isolated a set of mutants (*exo*) of *R. meliloti* Rm1021 on the basis of their failure to fluoresce under ultraviolet light on medium containing Calcofluor and showed that these mutants did not synthesize succinoglycan.[26] Alfalfa seedlings inoculated with these *exo* mutants formed ineffective (non-nitrogen-fixing) nodules that contained few, if any, bacteria and no bacteroids.[26] More detailed characterizations of nodules elicited by an *exoB* mutant revealed that in one genetic background no infection threads are formed after inoculation with this mutant and that the plant cells in the interior of nodules elicited by this strain do not contain bacteria or bacteroids.[24] Subsequent studies (A. Hirsch, personal communication) of a series of *exo* mutants generated by insertion of the transposon Tn5 in *R. meliloti* strain Rm1021 showed that root hair curling is significantly delayed and that infection threads can form, but abort at a very early stage, so that the bacteria are never able to reach the interior of the nodule. Furthermore, these *exo* mutants are only able to elicit the synthesis of 2 of the nodule-specific plant proteins, termed nodulins, as opposed to the 17 that are elicited by infection with a wild-type *R. meliloti* (refs. 25, 27, 28 and A. Hirsch, personal communication). Consequently, these *R. meliloti exo* mutants uncouple the ability of the bacteria to signal the plant to initiate nodule formation from the ability of the bacteria to invade the nodule and establish an effective symbiosis.

Similar findings implicating *Rhizobium* exopolysaccharides in normal nodule development were reported for the nodulation of clover by *R. trifolii*,[29] the nodulation of *Leucaena* by *Rhizobium* sp. strain NGR234,[30] and the nodulation of pea by *R. leguminosarum*.[31] In the case of *Rhizobium* sp. strain NGR234 and *R. trifolii* strain ANU843, Djordjevic *et al.*[32] reported that it is possible to suppress the symbiotic phenotypes of these strains by the addition of exogenous polysaccharide purified from the homologous wild-type strain. We and others[33] have been unable to suppress the symbiotic deficiencies of *R. meliloti exo* mutants by the addition of purified high molecular weight exopolysaccharide isolated from the *exo*$^+$ parent Rm1021, suggesting that the polysaccharide may play a different or an additional role(s) in the case of *R. meliloti*.

GENETIC ANALYSES OF THE SYNTHESIS OF SUCCINOGLYCAN BY *R. MELILOTI*

We subdivided our initial set of *R. meliloti exo* mutants into five different genetic classes.[26] Three of these classes were located on the second symbiotic megaplasmid, pRmeSU47B, of *R. meliloti* and the other two classes (*exoC* and *exoD*) on the chromosome.[34] A detailed genetic analysis of the cluster of *exo* genes located on the second symbiotic megaplasmid has led to the identification of a number of loci that are required for the synthesis of the *R. meliloti* succinoglycan.[35] Mutations in *exoA*, *exoB*, *exoF*, *exoL*, *exoM*, *exoP*, *exoQ*, and *exoT* completely abolish production of the succinoglycan and result in mutants that form Fix⁻ nodules.[26,35,36] Analyses of the properties of fusions of various *exo* genes to alkaline phosphatase that were generated using Tn*phoA*[37] suggest that the *exoF*, *exoP*, *exoQ*, and *exoA* gene products are membrane proteins.[38] To determine the roles of the *exo* gene products in the synthesis of succinoglycan, we initiated *in vitro* studies[12,13] on the synthesis of succinoglycan by these various mutants.

In addition, we found that mutations in *exoG*, *exoJ*, and *exoN*[35] diminish the production of Calcofluor-binding material. *exoG* and *exoJ* mutants form effective nodules with decreased efficiency, whereas plants inoculated with *exoN* mutants fix nitrogen normally. The *exoG* mutants are of particular interest because they produce no detectable high molecular weight exopolysaccharide; yet the low molecular weight Calcofluor-binding material they do produce seems to be sufficient to allow some amount of nodule invasion and development to proceed. This mutant may represent an example of a derivative of *R. meliloti* that excretes a biologically active oligosaccharide or low molecular weight polysaccharide but no high molecular weight polysaccharide. Another class of *exo* mutation, *exoK,* that causes a decrease in exopolysaccharide production, also causes a delay in the appearance of a fluorescent halo on Calcofluor plates but no nodulation defect.[25,35]

R. MELILOTI MUTANTS THAT FAIL TO SUCCINYLATE THEIR SUCCINOGLYCAN ARE DEFECTIVE IN NODULE INVASION

A very interesting class of *exo* mutants that synthesize a structural variant of succinoglycan was originally identified by the failure of colonies of such mutants to form a fluorescent halo under ultraviolet light when grown on medium containing Calcofluor.[25] These mutations defined a locus termed *exoH* that mapped in the middle of a cluster of *exo* genes on the second symbiotic megaplasmid. Alfalfa seedlings inoculated with *exoH* mutants form ineffective nodules that do not contain intracellular bacteria or bacteroids. Root hair curling is significantly delayed, and infection threads abort in the nodule cortex. In other words, even though these mutants make Calcofluor-binding material, their behavior on plants is the same as that of *exo* mutants that produce no succinoglycan. NMR analyses of exopolysaccharide secreted by *exoH* mutants have shown it to be identical to the Calcofluor-binding exopolysaccharide secreted by the parental strain except that it completely lacks the succinyl modification. *In vitro* translation of total RNA isolated from nodules induced by an *exoH* mutant has shown that, as in exopolysaccharide-deficient *exo* mutants, only 2 of the plant-encoded nodulins are induced compared to the 17 nodulins induced by the wild-type strain. These observations suggest that succinylation of the bacterial exopolysaccharide is important for its role(s) in nodule invasion and possibly in nodule development.[25]

Our ability to isolate *exoH* mutants suggests that the biosynthetic pathway for the synthesis of succinoglycan is sufficiently flexible to allow the polymerization of nonsuccinylated octasaccharides to yield high molecular weight polysaccharide. The resulting polysaccharide has lower charge density. We are interested in obtaining mutants that synthesize variants of succinoglycan lacking the acetyl and carboxyethylidene modifications, but have not yet been successful.

REGULATION OF SUCCINOGLYCAN BIOSYNTHESIS IN THE FREE-LIVING STATE

We initially observed that, as for the synthesis of several other bacterial heteropolysaccharides,[39] the synthesis of succinoglycan by *R. meliloti* is greatly increased if the cells are limited for nitrogen, phosphorus, or sulfur in the presence of a good carbon source. We have identified two new unlinked loci, *exoR* and *exoS,* whose products play a role in regulating the synthesis of succinoglycan by *R. meliloti* strain Rm1021.[40] Tn5-generated mutations in these loci are recessive and lead to substantial increases in the amount of exopolysaccharide synthesized, indicating that the *exoR* and *exoS* gene products play negative roles in regulating exopolysaccharide synthesis. Introduction of an *exoR95*::Tn5 or *exoS96*::Tn5 mutation into strains containing various *exo*::Tn*phoA* fusions results in a two- to fivefold increase in the level of expression of alkaline phosphatase activity, suggesting that *exoR* and *exoS* negatively regulate *exo* expression. A fundamental difference between the *exoR95*:: Tn5 and *exoS96*::Tn5 mutants is that the *exoR95*::Tn5 mutant synthesizes its succinoglycan at a high constitutive level regardless of the presence or absence of nitrogen in the medium, whereas the *exoS96*::Tn5 mutant undergoes a further increase in the rate of synthesis on nitrogen starvation. It seems that the *exoR* gene product is either involved directly in sensing the level of nitrogen in the medium or acts later in a putative regulatory cascade than the element(s) that actually does the sensing. The relationship of *exoS* action to *exoR* action is not yet clear.

Despite the high constitutive level of synthesis of succinoglycan by *exoR* mutants, we have not found these strains to be useful for production of succinoglycan. These mutants are extremely slow growing, so that larger cultures of *exoR* mutants become overgrown by derivatives that have acquired secondary mutations that reduce the level of synthesis of succinoglycan.

In examining the regulation of *exo*::Tn*phoA* fusions by *exoR* and *exoS,* we found that five of the eight classes of *exo* mutations that produce a complete block of succinoglycan biosynthesis (*exoL, M, P, Q,* and *T*) are lethal in *exoR* or *exoS* backgrounds.[36] These double mutants are viable, however, when a plasmid complementing the *exo* mutation is present. This implies that blocking succinoglycan biosynthesis at certain stages results in an accumulation of intermediates that are toxic when overproduced. One possibility is that buildup of lipid-linked intermediates may deplete the available pool of lipid carriers and therefore block lipopolyysaccharide and peptidoglycan biosynthesis. Availability of lipid carriers has been proposed to be a factor influencing polymer biosynthesis.[41]

REGULATION OF SUCCINOGLYCAN SYNTHESIS DURING NODULATION

We found that it is possible to stain nodules specifically for the alkaline phosphatase activity present in the inducing bacteria by staining at pH 9. At this pH, nodules induced by a Pho⁻ strain show no staining.[36] Nodules induced by a strain carrying the *exoF369*::Tn*phoA* fusion and a plasmid that complements the *exoF* mutation to allow normal nodulation show staining primarily in the early symbiotic or invasion zone of the nodule where the bacteria are invading the plant cells.[36] In the late symbiotic zone, which contains mature bacteroids, little staining was seen. Nodules induced by strains carrying *exoP*::Tn*phoA* and *exoA*::Tn*phoA* fusions showed a much lower degree of staining than did nodules induced by the *exoF*::Tn*phoA* strain, but faint staining of the invasion zone was seen after long incubation. These results suggest that little or no new EPS I synthesis is needed after nodule invasion.

exoS96::Tn5 mutants formed Fix⁺ nodules on alfalfa. In contrast, we found that on alfalfa, *exoR95*::Tn5 mutants formed both empty Fix⁻ nodules and also Fix⁺ nodules that contained widely varying numbers of bacteria and bacteroids. All the bacteria we isolated from the Fix⁺ nodules induced by the *exoR95*::Tn5 strain had acquired unlinked suppressors that reduced the amount of exopolysaccharide produced,[40] suggesting that the bacteria need to control either how much EPS I they synthesize or when they synthesize it in order to invade nodules. These suppressors could (1) reduce the activity of a positively acting factor, (2) reduce the activity of an enzyme required for the synthesis of the exopolysaccharide, or (3) result from the derepression of a gene functionally analogous to the *psi* gene of *R. phaseoli.*[42,43]

R. MELILOTI CAN PRODUCE A SECOND EXOPOLYSACCHARIDE THAT CAN SUBSTITUTE FOR THE ROLE OF SUCCINOLYCAN IN NODULE INVASION

The symbiotic defects of *exo* mutants can be suppressed by the presence of a mutation, *expR101*, that causes overproduction of a second exopolysaccharide, EPS II.[44] Genetic analyses show that the products of a cluster of at least six *exp* genes located on the second symbiotic megaplasmid, as well as the product of the *exoB* gene, are required for EPS II synthesis. The presence of the *expR101* mutation causes increased transcription of the *exp* genes, resulting in overproduction of EPS II. As a consequence of this genetic analysis, we were able to construct strains that produced EPS I or EPS II exclusively or neither EPS. *Medicago sativa* plants inoculated with an EPS II producing strain formed nitrogen-fixing nodules. However, the EPS II producing strain formed empty, ineffective nodules on four other plant species which were effectively nodulated by the EPS I producing strain. Therefore, it appears that exopolysaccharides are involved in determining host range at the level of nodule invasion rather than at the level of primary recognition of the host.

The structure of EPS II has been determined[39,44] and is shown in FIGURE 1B. The independent discovery of this exopolysaccharide was reported by Zhan *et al.,*[45] and the partial structure they report is in agreement with that shown in FIGURE 1B. Both exopolysaccharides are acidic, contain glucose and galactose, and have acetyl and pyruvate (1-carboxyethylidene) modifications. However, the structures differ in many respects: (1) EPS II does not contain any succinate groups; (2) EPS II contains more galactose than does EPS I; (3) EPS II is unbranched; (4) EPS II has both α and β glycosidic linkages, whereas EPS I has only β glycosidic linkages; and (5) the pyruvate group in EPS I is linked to glucose, whereas in EPS II it is linked to galactose. However, it is interesting that each exopolysaccharide has a single Glc-β(1,3)-Gal linkage in its backbone and that these two rather diverse exopolysaccharides may share the common structural motif of O-6-acetylglucose- β(1,3)-galactose.

POSSIBLE ROLES FOR EXOPOLYSACCHARIDES IN NODULATION

Exopolysaccharides may have more than one function in nodulation. Inasmuch as *exoG* mutants, which make low molecular weight EPS I, but not high molecular weight EPS I, form effective nodules at reduced efficiency relative to wild-type mutants, but much better than mutants that make no exopolysaccharide, it is possible that both the high and low molecular weight forms of EPS have symbiotic functions. One of the most intriguing possibilities for the role of low molecular weight forms of the exopolysaccharides in nodulation is that they act as signals to the plant during the process of nodule invasion and development. Carbohydrates have previously been shown to function as signal molecules in plants.[46] Several lines of evidence suggest that oligosaccharide fragments of EPS may have a symbiotic function. First, *exoG* mutants, which produce mainly low molecular weight EPS I, can invade nodules, albeit at reduced efficiency.[35] Second, both we and John Leigh and his colleagues (personal communication) have obtained preliminary evidence that a low molecular weight fraction of EPS I can partially suppress the symbiotic deficiencies of *R. meliloti exo* mutants. A similar finding was reported previously by Djordjevic *et al.*[32] who found that *exo* mutants of *Rhizobium* sp. NGR234 and *R. trifolii* can form effective nodules if either high molecular weight EPS or oligosaccharide subunits of EPS are supplied exogenously. However some difference exists between the *R. meliloti*-alfalfa system and those studied by Djordjevic *et al.,*[32] because neither we nor others[33] have been able to suppress the symbiotic deficiencies of *R. meliloti exo* mutants by the addition of purified high molecular weight exopolysaccharide isolated from their *exo*[+] parent Rm1021.

In the course of these experiments, we observed that the symbiotic deficiencies of *exoH* mutants, which fail to succinylate their EPS I, could also be partially suppressed by the addition of a low molecular weight fraction of EPS I. This observation is consistent with the report of Leigh and Lee[47] that *exoH* mutants produce less low molecular weight EPS I than do the wild-type; perhaps the symbiotic defects of these mutants result from a failure to process the nonsuccinylated EPS I into an oligosaccharide signal molecule. Furthermore, this suppression of *exoH* mutants appeared to be somewhat more efficient than that of *exoA* mutants. This observation is consistent with the possibility that low and high molecular weight forms of the exopolysaccharide may have different symbiotic functions and suggests the additional possibility that there may be different structural requirements for these two types of roles.

Other possible roles for exopolysaccharides include serving as carriers for extracellular enzymes or signal molecules, forming part of the infection thread matrix, constraining descendents of attached bacteria to the immediate vicinity of the plant, or helping to evade or suppress plant defense responses.[44] With respect to this later possibility, it is interesting that $\beta(1,3)$-glucanases are among the major hydrolytic enzymes induced when plants are exposed to pathogens[48] and that each of the two *R. meliloti* exopolysaccharides has a single $\beta(1,3)$ linkage in its backbone. Furthermore, this linkage is known to be modified in the case of EPS II and likely to be modified in the case of EPS I.

ACKNOWLEDGMENTS

We thank the other members of the laboratory for their support and encouragement.

REFERENCES

1. GLAZEBROOK, J. & G. C. WALKER. 1991. Methods Enzymol. **204:** 398–418.
2. GRAVANIS, G., M. MILAS, M. RINAUDO & B. TINLAND. 1987. Carbohydr. Res. **160:** 259–265.
3. HISAMATSU, M., J. ABE, A. AMEMURA & T. HARADA. 1978. Carbohydr. Res. **66:** 289–294.
4. ZEVENHUIZEN, L. T. P. M. 1973. Carbohydr. Res. **26:** 409–419.
5. HISAMATSU, M., K. SANO, A. AMEMURA & T. HARADA. 1978. Carbohydr. Res. **61:** 89–96.
6. WILLIAMS, A. G. & C. J. LAWSON. 1979. *Patent Specification,* 1979. GB 1539064.
7. BJORNDAL, H., C. ERBING, B. LINDBERG, G. FAHRAEUS & H. LJUNGGREN. 1971. Acta Chem. Scandinav. **25:** 1281–1296.
8. HISAMATSU, M., J. ABE, A. AMEMURA & T. HARADA. 1980. Agric. Biol. Chem. **44:** 1049–1055.
9. JANSSON, P.-E., L. KENNE, G. LINDBERG, H. LJUNGGREN, J. LONNGREN, U. RUDEN & S. SVENSSON. 1977. J. Am. Chem. Soc. **99:** 11–14.
10. AMAN, P., M. MCNEIL, L.-E. FRANZEN, A. G. DARVILL & P. ALBERSHEIM. 1981. Carbohydr. Res. **95:** 263–282.
11. ZEVENHUIZEN, L. P. T. M. & A. R. VANNEERVAN. 1983. Carbohydr. Res. **118:** 127–134.
12. TOLMASKY, M. E., R. J. STANELONI, R. A. UGALDE & L. F. LELOIR. 1980. Arch. Biochem. & Biophys. **203:** 358–364.
13. TOLMASKY, M. E., R. J. STANELONI & L. F. LELOIR. 1982. J. Biol. Chem. **257:** 6751–6757.
14. BETLACH, M. R., M. A. CAPAGE, D. H. DOHERTY, R. A. HASSLER, N. M. HENDERSON, R. W. VANDERSLICE, J. D. MARRELLI & M. B. WARD. 1987. *In* Industrial Polysaccha-

rides: Genetic Engineering, Structure/Property Relations and Applications. M. Yalpani, ed.: 35–50. Elsevier Science Publishers B. V. Amsterdam.
15. LINTON, J. D., M. EVANS, D. S. JONES & D. N. GOULDNEY. 1987. J. Gen. Microbiol. 133: 2961–2969.
16. BAUER, W. D. 1981. Ann. Rev. Plant Physiol. 32: 407–499.
17. HALVERSON, L. J. & G. STACEY. 1986. Microbiol. Rev. 50: 193–225.
18. LONG, S. R. 1984. In Plant-Microbe Interactions. T. Kosuge & E. Nester, eds.: 265–306. Macmillan. New York.
19. VANCE, C. P., K. L. M. BOYLAN, S. STADE & D. A. SOMERS. 1985. Symbiosis 1: 69–84.
20. VERMA, D. P. S. & S. LONG. 1983. Int. Rev. Cytol. (Suppl. 14): 211–245.
21. DOWNIE, J. A. & A. W. B. JOHNSTON. 1986. Cell 47: 153–154.
22. LEROUGE, P., P. ROCHE, C. FAUCHER, F. MAILLET, G. TRUCHET, J. C. PROME & J. DENARIE. 1990. Nature (Lond.) 344: 781–784.
23. SUTHERLAND, I. W. 1979. Microbial Polysaccharides and Polysaccharases. Academic Press (London) Ltd. London.
24. FINAN, T. M., A. M. HIRSCH, J. A. LEIGH, E. JOHANSON, G. A. KULDAU, S. DEEGAN, G. C. WALKER & E. R. SIGNER. 1985. Cell 40: 869–877.
25. LEIGH, J. A., J. W. REED, J. F. HANKS, A. M. HIRSCH & G. C. WALKER. 1987. Cell 51: 579–587.
26. LEIGH, J. A., E. R. SIGNER & G. C. WALKER. 1985. Proc. Natl. Acad. Sci. USA 82: 6231–6235.
27. LANG-UNNASCH, N. & F. M. AUSUBEL. 1985. Plant Physiol. 77: 833–839.
28. NORRIS, J. H., L. A. MARCOL & A. M. HIRSCH. 1988. Plant Physiol. 88: 321–328.
29. CHAKRAVORTY, A. K., W. ZURKOWSKY, J. SHINE & B. G. ROLFE. 1982. J. Mol. Appl. Genet. 1: 585–596.
30. CHEN, H., M. BATLEY, J. REDMOND & B. G. ROLFE. 1985. J. Plant Physiol. 120: 331–349.
31. BORTHAKUR, D., C. E. BARBER, J. W. LAMB, M. J. DANIELS, J. A. DOWNIE & A. W. B. JOHNSTON. 1986. Mol. Gen. Genet. 203: 320–323.
32. DJORDJEVIC, S., H. CHEN, M. BATLEY, J. W. REDMOND & B. G. ROLFE. 1987. J. Bacteriol. 169: 53–60.
33. MULLER, P., M. HYNES, D. KAPP, K. NIEHAUS & A. PUHLER. 1988. Mol. Gen. Genet. 211: 17–26.
34. FINAN, T. M., B. KUNKEL, G. F. DE VOS & E. R. SIGNER. 1986. J. Bacteriol. 167: 66–72.
35. LONG, S., J. W. REED, J. W. HIMAWAN & G. C. WALKER. 1988. J. Bacteriol. 170: 4239–4248.
36. REUBER, T. L., S. LONG, & G. C. WALKER. 1991. J. Bacteriol. 173: 426–434.
37. MANOIL, C. & J. BECKWITH. 1985. Proc. Natl. Acad. Sci. USA 82: 8129–8133.
38. LONG, S., S. MCCUNE & G. C. WALKER. 1988. J. Bacteriol. 170: 4257–4265.
39. HER, G. R., J. GLAZEBROOK, G. C. WALKER & V. N. REINHOLD. 1990. Carbohydr. Res. 198: 305–312.
40. DOHERTY, D., J. A. LEIGH, J. GLAZEBROOK & G. C. WALKER. 1988. J. Bacteriol. 170: 4249–4256.
41. SUTHERLAND, I. W. 1982. Adv. Microbiol. Physiol. 23: 79–150.
42. BORTHAKUR, D., J. A. DOWNIE, A. W. B. JOHNSTON & J. W. LAMB. 1985. Mol. Gen. Genet. 200: 278–282.
43. BORTHAKUR, D. & A. W. B. JOHNSTON. 1987. Mol. Gen. Genet. 207: 149–154.
44. GLAZEBROOK J. & G. C. WALKER. 1989. Cell 56: 661–672.
45. ZHAN, H., S. B. LEVERY, C. C. LEE & J. A. LEIGH. 1989. Proc. Natl. Acad. Sci. USA 86: 3055–3059.
46. DARVILL, A. G., P. A. ALBERSHEIM, P. BUCHELI, S. DOARES, N. DOUBRAVA, S. EBERHARD, D. J. GOLLIN, M. G. HAHN, V. MARFA-RIERA, W. S. YORK & D. MOHNEN. 1989. In NATO ASI Series, Vol. H36, Signal Molecules in Plants and Plant-Microbe Interactions. B. J. J. Lutgenberg, ed.: 41–48. Springer-Verlag. Berlin Heidelberg.
47. LEIGH, J. A. & C. C. LEE. 1988. J. Bacteriol. 170: 3327–3332.
48. KOMBRINK E., M. SCHROEDER & K. HAHLBROCK. 1988. Proc. Natl. Acad. Sci. USA 85: 782–786.
49. LONG, S. R. 1989. Ann. Rev. Genet. 23: 483–506.

Construction and Use of a *Bacillus subtilis* Mutant Deficient in Multiple Protease Genes for the Expression of Eukaryotic Genes[a]

XIAO-SONG HE,[b] YUAN-TAY SHYU,[b] SHYROSZE
NATHOO,[c] SUI-LAM WONG,[c] AND ROY H. DOI[b,d]

[b]*Department of Biochemistry and Biophysics*
University of California
Davis, California 95616

[c]*Department of Biological Sciences*
The University of Calgary
Calgary, Alberta T2N 1N4, Canada

Bacillus subtilis has the potential for being an extremely useful host expression system, because it can secrete proteins into the growth medium.[1] This would facilitate, for instance, the purification of any eukaryotic protein that could be expressed and secreted in an efficient manner. However, one of the major problems with this system has been the large amounts of extracellular proteases that are produced by *B. subtilis* and that degrade any foreign proteins that are secreted simultaneously. Therefore, it is essential to obtain mutant strains that are missing the extracellular proteases. The other major problem with the production of eukaryotic proteins has been the relatively low efficiency of secretion of foreign proteins from *B. subtilis* and the subsequent formation of inclusion bodies. However, the potential also exists for synthesizing large amounts of foreign proteins intracellularly.

To overcome the protease degradation problem, we have been deleting the protease genes present in *B. subtilis* by cloning the protease gene, making deletions in the gene *in vitro* and inserting the deleted gene into the chromosome by gene conversion techniques.[2,3] In this paper we describe the construction of a *B. subtilis* strain in which four extracellular protease genes and one major intracellular gene have been deleted. In testing this strain with expression vectors containing the human tissue-type plasminogen activator (h-t-PA) gene or the rice α-amylase gene, we found that these gene products, although not secreted, were synthesized and accumulated intracellularly in larger quantities than in other *B. subtilis* strains.

BACTERIAL STRAINS AND MEDIA

The *B. subtilis* strains used in this research are listed in TABLE 1. Tryptose blood agar base (TBAB, Difco) or Spizizen's minimal medium[4] were used for transformation. Luria broth (LB) medium[5] supplemented with 0.5% glucose and 5 μg/ml

[a]This research was supported in part by Wyeth Laboratories Contract No. 860184 and Public Health Service grant GM19673 from the National Institutes of Health.
[d]To whom correspondence should be addressed.

69

kanamycin (LBGK) was used for expression of foreign gene products. The 2XSG medium[6] was used for growing cells for the intracellular protease assays.

RESULTS

Deletion of bpf Gene from B. subtilis DB403

The triple-protease deficient strain DB403 we constructed before[3] still contained about 1% of the extracellular protease activity of the wild-type *B. subtilis* strain. When grown on casein agar plates for extended periods, a small halo was seen around its colony. Recently the bacillopeptidase F (*bpf*) gene that codes for a serine peptidase was isolated and characterized.[7,8] To inactivate the *bpf* gene on the *B. subtilis* chromosome, a deletion was made from the cloned gene, which removes a 1.2-kb fragment including the ribosome binding site, signal sequence, propeptide sequence, and part of the mature protease coding sequence (FIG. 1A). This deleted *bpf* gene was recloned into pUB18 (a *B. subtilis* cloning vector constructed by

TABLE 1. *Bacillus subtilis* Strains Used in this Research

Strain	Genotype	Source
DB5 (1A5)	*trpC2 tre-12 metC3 glyB133*	BGSC[14]
DB104	*his nprR2 nprE18 aprEΔ3*	Kawamura and Doi[2]
DB403	*trpC2 npr apr epr*	Wang *et al.*[3]
KN2	*trpC2 lys-1 phe-1 metC3*	
	nprR2 nprE18 aprEΔ3 isp1	Nakamura *et al.*[13]
DB428	*trpC2 npr apr epr bpf*	pUB18-*Δbpf* gc DB403
DB429	*trpC2 metC3 npr apr epr bpf*	DB5 tf DB428
DB430	*trpC2 npr apr epr bpf isp1*	KN2 tf DB429
DB511 (BRE)	*trpC lys-3 recE4 amyE*	Nicholson and Chambliss[19]
DB551	*trpC2 npr apr epr bpf isp1 amyE*	DB511 tf DB430

ABBREVIATIONS: gc = gene conversion; tf = transformation.

replacing the *Eco*RI-*Pvu*II region of pUB110[9] with the multiple cloning site of pUC18,[10] Wang and Doi, unpublished results) to form the plasmid pUB18-*Δbpf,* which was transformed into competent cells of *B. subtilis* DB403 to induce conversion of the intact chromosomal *bpf* gene. Any transformant in which gene conversion occurred formed a colony with either reduced or enlarged halo size depending on the direction of the gene conversion. To find clones in which the mutant gene has been integrated into the chromosome, clones with reduced halo size were selected and the plasmid therein was cured. The frequency of gene conversion events was about 3% which was typical of the frequency observed previously.[2,11]

To verify that the *bpf* gene in the chromosome now contained the deletion, Southern mapping of the chromosomal DNA was conducted. As the results in FIGURE 1B indicate, the size of the *bpf*-containing fragment was reduced by about 1.2 kb by the presence of the deletion as expected. Furthermore, the activity of extracellular protease, as measured with cultures grown in liquid medium, was reduced significantly compared to that found in DB403 (FIG. 2A), which was known to produce about 1% of the extracellular proteolytic activity of the wild type.[3] No halos could be observed when this newly constructed strain was grown on casein agar

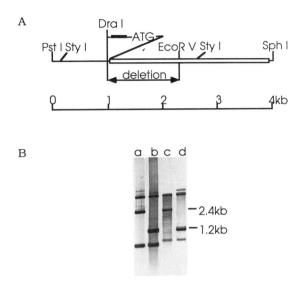

FIGURE 1. Deletion in the bacillopeptidase F (*bpf*) gene. (**A**) Restriction map showing the region deleted from the *bpf* gene. The ribosome binding site (*solid bar*) and the DNA sequence encoding the signal peptide pro-region as well as part of the mature protease were deleted *in vitro*. A plasmid pUB18-Δ*bpf* carrying the deleted *bpf* gene was transformed into competent *B. subtilis* DB403 cells to induce conversion of the wild-type chromosomal *bpf* to the mutant gene. (**B**) Verification of the deletion of the chromosomal *bpf* gene by Southern blotting. *Lane a,* intact *bpf* gene on plasmid; *lane b,* deleted *bpf* gene on plasmid; *lane c,* intact *bpf* gene on DB403 chromosome; *lane d,* deleted *bpf* gene on DB428. All the DNA samples were digested by *Sty*I. The 2.8-kb *Pst*I-*Sph*I fragment containing the deleted *bpf* gene was labeled and used as the probe.

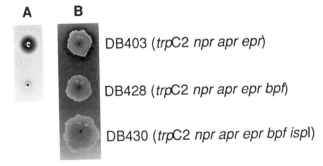

FIGURE 2. Extracellular protease levels of *B. subtilis* DB403, DB428, and DB430. (**A**) 25 ml of culture supernatants (in LB medium) harvested 3 hours after the onset of stationary phase were concentrated with 80% ammonium sulfate. The pellets were dissolved in 0.5 ml of 10 mM Tris-HCl, pH 7.8, 5 mM CaCl$_2$, and dialyzed against the same buffer. Aliquots of 15 μl were applied into the holes punched with a Pasteur pipette on an agar plate containing 1% gelatin and incubated at 37°C for 24 hours. (**B**) Cells were grown on 2XSG plates containing 0.5% skim milk for 24 hours at 37°C. The plates were left at room temperature for another 3 days before the photograph was taken.

plates for 4 days (FIG. 2B). This four extracellular protease-deficient strain has been designated *B. subtilis* DB428 (*trpC2 nprE aprE epr bpf*).

Inactivation of the Major Intracellular Protease Gene, isp1, from DB428

The *isp1* gene encoding the major intracellular serine protease ISP-1 of *B. subtilis* has been cloned, sequenced, and used to construct an *isp1⁻* strain by *in vitro* deletion and gene conversion.[12] This gene was found to be closely linked to *metC* (99.5% cotransduction by PBS1). To inactivate the ISP-1 of DB428, the *isp1⁻* gene of the *B. subtilis* KN2[13] was introduced into DB428 by the procedure shown in TABLE 1. First the *metC3⁻* gene from the chromosomal DNA of *B. subtilis* DB5 (1A5)[14] was introduced into DB428 by means of congression with the kanamycin-resistant plasmid pUB-δ*bpf*. Transformants were selected on kanamycin-supplemented TBAB plate and screened by tryptophan-supplemented Spizizen's minimal medium lacking methionine to find a Met⁻ transformant. This strain was designated DB429. Next the chromosomal DNA of KN2 was used to transform the competent cells of DB429. The Met⁺ transformants were selected on Spizizen's minimal medium supplemented with tryptophan. These transformants were further subjected to Southern mapping of their chromosomal DNA to select for the strains that had been transformed by the *isp1⁻* together with the *metC⁺*. The probe used was a mixture of two 20-mer oligonucleotides synthesized according to the published sequence of the *isp1* gene.[12] Out of the 10 Met⁺ transformants, 8 had the same band pattern as that of KN2 (not shown). One of them was designated DB430 (*trpC2 nprE aprE epr bpf isp1*).

DB430 was assayed for its intracellular protease activity, carried out in parallel with the *B. subtilis* strains DB104, KN2, and DB428 (FIG. 3). DB104 and KN2 are known to have the same deficient protease genes (*nprR2 nprE18 aprEΔ3*) except for the intracellular protease gene *ips1*, which is positive in DB104 but negative in KN2. The intracellular protease level of KN2 is lower than that of DB104 as expected. Compared to its parental strain DB428, the newly constructed strain DB430 showed a significantly lower level of intracellular protease activity. It is also apparent that the deletion in extracellular protease genes had lowered the intracellular protease activity level: the protease activities of DB428 and DB430 are significantly lower than those of DB104 and KN2, respectively. In fact, DB430 has the lowest activity of the four strains assayed.

DB430 was found to grow and sporulate normally. Although it always forms a larger colony on 2XSG plates containing casein after prolonged incubation (FIG. 2B), no difference was detected when its growth curve in liquid medium was compared with that of its parental strains (not shown).

Construction of Foreign Gene Expression Plasmids

The construction of foreign gene expression plasmids is shown in FIGURE 4. The *B. subtilis* terminator-probing vector pWT18 was chosen the expression vector, which contains a promoter fragment P1P2 functioning during the vegetative growth phase.[15] For the expression of human tissue-type plasminogen activator (h-t-PA) in *B. subtilis,* the t-PA coding sequence was obtained from the plasmid pET3a-tPA (He and Doi, unpublished results) as a 1.8-kb *Xba*I-*Eco*RV fragment, which contains the ribosome binding site of the *Escherichia coli* phage T7 gene 10[16] and an open reading frame encoding the mature tPA[17] plus two extra amino acid residues (Met and Gly) at its NH₂-terminus. This fragment was ligated with the *Xba*I/*Pvu*II-digested pWT18

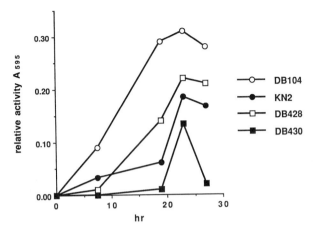

FIGURE 3. Intracellular protease activity levels of *B. subtilis* strains. *B. subtilis* strains DB104 (*his nprR2 nprE18 aprEΔ3*), KN2 (*trpC2 lys-1 phe-1 metC2 nprR2 nprE18 aprEΔ3 isp1*), DB428 (*trpC2 npr apr epr bpf*), and DB430 (*trpC2 npr apr epr bpf isp1*) were grown in 2XSG medium at 37°C. An aliquot of 10 ml was taken at each time point. Cells harvested by centrifugation were washed twice with 50 mM Tris-HCl, pH 7.5, 0.9% NaCl, resuspended in 0.5 ml of 50 mM Tris-HCl, pH 7.5, 1 mM CaCl₂ containing 2 mg/ml of lysozyme, incubated at 37°C for 10 minutes, disrupted by sonication, and followed by centrifugation to get clear cell extracts. For protease assays, 0.1 ml of the cell extract was mixed with 5 mg of Hide Powder Azure (Calbiochem Corp.) suspended in 0.9 ml of 50 mM Tris-HCl, pH 7.8, 1 mM CaCl₂, and incubated at 37°C for 110 minutes. The reaction mixture was centrifuged and A_{595} of the supernatant was taken as the relative protease activity level.

plasmid DNA and transformed into *B. subtilis* to get the expression plasmid pWT-tPA, in which the expression of t-PA gene is under the control of a *B. subtilis* promoter (P1P2) and a T7 ribosome binding site.

For the expression of rice α-amylase in *B. subtilis,* a similar expression plasmid, pYS1, was constructed by ligating a 1.9-kb *Xba*I-*Hin*dIII (filled-in) fragment encoding a rice α-amylase isozyme with the *Xba*I/*Pvu*II-digested pWT18 as just described

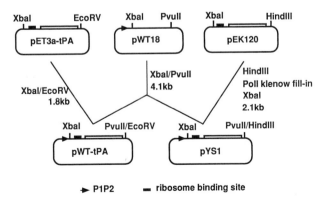

FIGURE 4. Construction of expression plasmids.

(FIG. 4). The rice α-amylase gene fragment was from the plasmid pEK120, which was previously constructed by subcloning the rice α-amylase gene[18] into the vector pET8c[16] in this laboratory (Shyu and Doi, unpublished results).

Expression of h-t-PA in B. subtilis

The expression plasmid pWT-tPA was introduced into B. subtilis strains DB403, DB428, and DB430 by transformation. All three strains, in addition to DB403 containing the plasmid pWP18 (a promoter probe vector containing a promoterless aprE gene, Wang & Doi, unpublished result) as a negative control, were grown in LBGK medium at 37°C until late log phase (Klett unit 300). Cells were harvested, washed, resuspended in buffer containing the protease inhibitors EDTA and PMSF, and disrupted by lysozyme treatment and sonication. The total cellular protein (T) was fractionated into soluble (S) and insoluble (I) fractions by centrifugation. The protein samples were applied to duplicated SDS-polyacrylamide gels. After electrophoresis, one gel was stained with Coomassie blue to show the total protein patterns in the different fractions, whereas the other was analyzed by immunoblotting with goat anti-human t-PA antibody to detect t-PA (FIG. 5). The results shows that t-PA was synthesized in all three B. subtilis strains containing the expression plasmid (lanes Tb, Tc, and Td). Most of the t-PA in the cells was in the insoluble fraction (lanes Ib, Ic, and Id) rather than the soluble fraction (lanes Sb, Sc, and Sd). In fact, it

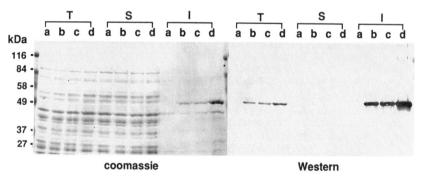

coomassie   **Western**

FIGURE 5. Expression of h-t-PA in different B. subtilis strains. Lane a, DB403(pWP18); lane b, DB403(pWT-tPA); lane c, DB428(pWT-tPA); lane d, DB430(pWT-tPA). T = total cellular protein; S = soluble protein; I = insoluble protein. Cells of different strains with or without h-t-PA expression plasmid pWT-tPA were harvested from 10 ml of late-log phase (Klett unit 300) cultures in LBGK medium, washed with SET buffer (20% sucrose, 50 mM Tris-HCl, pH 7.6, 50 mM EDTA), and resuspended with 0.9 ml of SET containing 0.5 mM of PMSF. Next 0.1 ml of lysozyme (20 mg/ml in SET) was added. The mixture was incubated at 37°C for 10 minutes and followed by sonication. An 0.1-ml portion was saved as the total cellular protein (T), whereas the rest was centrifuged at 15,000 rpm for 15 minutes. The supernatant was removed as the soluble protein fraction (S). The pellet was washed once with SET/PMSF and recovered as the insoluble fraction (I). Each fraction was mixed with sample application buffer,[25] boiled for 5 minutes, and centrifuged before analysis by 7.5% SDS-PAGE. Each lane of the T and S fractions corresponded to 50 μl of original culture. Each lane of I fraction corresponded to 1 ml of original culture. After electrophoresis, one gel was subjected to Coomassie staining. The other gel was subjected to Western blotting carried out as previously described.[26] Goat anti-h-t-PA (American Diagnostica Inc.) was used as the first antibody. HRP-conjugated rabbit anti-goat antibody (Bio-Rad) was used as the secondary antibody.

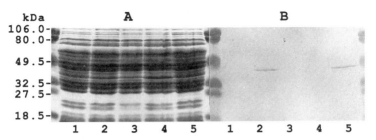

FIGURE 6. Expression of rice α-amylase in different *B. subtilis* strains. *B. subtilis* cells were cultured and harvested as described in FIGURE 5. Whole cell lysates were obtained by sonication. Each gel lane contained approximately 0.25 ml of original culture. (**A**) 10% SDS-PAGE was used to separate proteins from the whole cell lysates of *B. subtilis* DB428, DB430, DB511, and DB551 (*lanes* 1, 2, 4, and 5, respectively) containing rice amylase expression vector pYS1 and that of *B. subtilis* DB430(pWP18) (*lane* 3) as a negative control. (**B**) The proteins were electrotransferred from an SDS-polyacrylamide (10%) gel onto nitrocellulose paper. After blocking with 1% gelatin, the nitrocellulose sheets were incubated with rabbit anti-rice α-amylase antiserum, washed and incubated with AP-conjugated goat anti-rabbit IgG antiserum. 5-bromo-4-chloro-3-indolyl-phosphate (BCIP) was used as the substrate for color development.

is the most abundant component of the insoluble fraction. It is apparent that the yield of t-PA in DB430 (*ispl*⁻) is higher than that in DB403 and DB428 (*ispl*⁺), indicating that inactivation of the ISP-1 helps the accumulation of foreign gene products in the cells.

Expression of Rice α-Amylase Gene in B. subtilis

An α-amylase deficient *B. subtilis* mutant, DB551, was obtained by gene congression of DB430 with chromosomal DNA of DB511 (*B. subtilis* BRE[19]) and plasmid pWT18. The transformants were selected by TBAB plates containing 10 µg/ml kanamycin and 1% starch. The plates were incubated for 24 hours at 37°C and stained for amylase activity with iodine vapor. Halo-free colonies reflect amylase-deficient strains. One of them was designated DB551. No difference was found in general growth characteristics between DB430 and DB551 (data not shown). *B. subtilis* strains DB511, DB428, DB430, and DB551 were transformed by pYS1 and pWP18 to test for production of rice α-amylase in *B. subtilis*. FIGURE 6 clearly indicates that DB430 (lane 2) and DB551 (lane 5), which lack intracellular protease ISP-1, were superior to their parental strains DB428 (lane 1) and DB511 (lane 4), respectively, in the production of recombinant rice α-amylase.

DISCUSSION

We have constructed *B. subtilis* strain DB430 which is deficient in four extracellular genes (*nprE, aprE, eprE,* and *bpf*) and one major intracellular serine protease gene (*ispl*). This strain has been used as the expression host for two eukaryotic genes, namely, human tissue-type plasminogen activator (h-t-PA) gene and rice α-amylase gene. In both cases, DB430 gave a higher yield of the foreign gene product than did other *B. subtilis* strains tested. To our knowledge this is the first report showing the advantage of using an *isp1⁻* strain as the *B. subtilis* host for the expression of eukaryotic genes.

It is known that eukaryotic proteins are usually very sensitive to bacterial proteases. Newly constructed strain DB430, although deficient in five protease genes, still contains at least one minor intracellular protease gene, *isp2*,[20] and one minor extracellular protease gene, *mpr*.[21] Inasmuch as most of the *B. subtilis* protease genes, including *isp1*, are believed to be expressed during sporulation,[22] we placed the h-t-PA and rice α-amylase genes under the transcriptional control of promoter P1P2 of *B. subtilis* σ^A (σ^{43} is now referred to as σ^A) operon gene,[15] which functions during vegetative growth phase. The medium we used contains glucose which repressed sporulation,[23] and the cells were harvested before the onset of the stationary phase. We noted that even in this circumstance, inactivation of the *isp1* gene enhanced the yield of both h-t-PA and α-amylase. This suggests the possibility that there is already a low-level expression of *isp1* during the log phase.

The enhancement effect of the *isp1⁻* gene on the yield of h-t-PA is not as dramatic as is that of rice α-amylase, probably because the h-t-PA expressed in *B. subtilis* cells is present in insoluble inclusion bodies, which are more resistant to proteolysis.[24]

The newly constructed DB430 deficient in four extracellular protease genes should be a useful host for secretion of foreign gene products from *B. subtilis*. We also found that it has a lower activity level than do other *isp1⁻* strains when cell-free extracts were prepared and assayed for protease activity (FIG. 3). Since the cells had been washed thoroughly to remove extracellular proteases that might be present in the medium, the possible explanation is that there is a continual processing and secretion of extracellular proteases during the experimental procedures, or the preproenzyme from inside the cell was somehow processed and activated during the procedures. Therefore, the DB430 strain should be a suitable host not only for secretion of foreign gene products, but also for producing high levels of foreign and native proteins within the *B. subtilis* cells.

ACKNOWLEDGMENTS

We thank Drs. A. Nakamura and T. Beppu for providing *B. subtilis* KN2 and Dr. G. Chambliss for his gift of *B. subtilis* BRE.

REFERENCES

1. DOI, R. H., S.-L. WONG & F. KAWAMURA. 1986. Potential use of *Bacillus subtilis* for secretion and production of foreign proteins. Trends Biotechnol. **4:** 232–235.
2. KAWAMURA, F. & R. H. DOI. 1984. Construction of a *Bacillus subtilis* double mutant deficient in extracellular alkaline and neutral proteases. J. Bacteriol. **160:** 442–444.
3. WANG, L.-F., R. BRUCKNER & R. H. DOI. 1990. Construction of a *Bacillus subtilis* mutant deficient in three extracellular proteases. J. Gen. Appl. Microbiol. **35:** 489–492.
4. ANAGNOSTOPOULAS, C. & J. SPIZIZEN. 1961. Requirements for transformation in *Bacillus subtilis*. J. Bacteriol. **81:** 741–746.
5. MANIATIS, T., E. F. FRITSCH & J. SAMBROOK. 1982. Molecular Cloning: A Laboratory Manual. Cold Spring Harbor Laboratory. Cold Spring Harbor, NY.
6. LEIGHTON, T. J. & R. H. DOI. 1971. The stability of messenger ribonucleic acid during sporulation of *Bacillus subtilis*. J. Biol. Chem. **246:** 3189–3195.
7. WU, X.-C., S. NATHOO, A. S.-H. PANG, T. CARNE & S.-L. WONG. 1990. Cloning, genetic organization and characterization of a structural gene encoding bacillopeptidase F

from *Bacillus subtilis.* J. Biol. Chem. **265:** 6845–6850.

8. SLOMA, A., G. A. RUFO, JR., C. F. RUDOLPH, B. J. SULLIVAN, K. A. THERIAULT & J. PERO. 1990. Bacillopeptidase F of *Bacillus subtilis:* Purification of the protein and cloning of the gene. J. Bacteriol. **172:** 1470–1477.

9. MCKENZIE, T., T. HOSHINO, T. TANAKA & N. SUEOKA. 1986. The nucleotide sequence of pUB110: Some salient features in relation to replication and its regulation. Plasmid **15:** 93–106.

10. YANISCH-PERRON, C., J. VIEIRA & J. MESSING. 1985. Improved M13 phage cloning vectors and host strains: Nucleotide sequences of the M13mp18 and pUC19 vectors. Gene **33:** 103–119.

11. IGLESIAS, A. & T. TRAUTNER. 1983. Plasmid transformation in *Bacillus subtilis:* Symmetry of gene conversion in transformation with a hybrid plasmid containing chromosomal DNA. Mol. Gen. Genet. **189:** 73–76.

12. KOIDE, Y., A. NAKAMURA, T. UOZUMI & T. BEPPU. 1986. Cloning and sequencing of the major intracellular serine protease gene of *Bacillus subtilis.* J. Bacteriol. **167:** 110–116.

13. NAKAMURA, A., Y. KOIDE, F. KAWAMURA, S. HORINOUCHI, T. UOZUMI & T. BEPPU. 1990. Construction of a *Bacillus subtilis* strain deficient in three proteases. Agric. Biol. Chem. **54:** 1307–1309.

14. DEDONDER, R. A., J.-A. LEPESANT, J. LEPESANT-KEJZLAROVA, A. BILLAULT, M. STEIN-METZ & F. KUNST. 1977. Construction of a kit of reference strains for rapid genetic mapping in *Bacillus subtilis* 168. Appl. Environ. Microbiol. **33:** 989–993.

15. WANG, L.-F. & R. H. DOI. 1987. Promoter switching during development and the termination site of the σ^{43} operon of *Bacillus subtilis.* Mol. Gen. Genet. **207:** 114–119.

16. STUDIER, F. W., A. H. ROSENBERG, J. J. DUNN & J. W. DUBENDORFF. 1990. Use of T7 RNA polymerase to direct expression of cloned genes. Methods Enzymol. **185:** 60–89.

17. LEE, S. G., N. K. KALYAN, J. WILHELM, W.-T. HUM, R. RAPPAPORT, S.-M. CHENG, S. DHEER, C. URBANO, R. W. HARTZELL, M. RONCHETTI-BLUME, M. LEVNER & P. P. HUNG. 1988. Construction and expression of hybrid plasminogen activator prepared from tissue-type plasminogen activator and urokinase-type plasminogen activator genes. J. Biol. Chem. **263:** 2917–2924.

18. O'NEILL, S. D., M. H. KUMAGAI, A. MAJUMDAR, N. HUANG, T. D. SUTLIFF & R. L. RODRIGUEZ. 1990. The α-amylase gene in *Oryza sativa:* Characterization of cDNA clones and mRNA expression during seed germination. Mol. Gen. Genet. **221:** 235–244.

19. NICHOLSON, W. L. & G. H. CHAMBLISS. 1985. Isolation and characterization of a *cis*-acting mutant conferring catabolic repression resistance to α-amylase synthesis in *Bacillus subtilis.* J. Bacteriol. **161:** 875–881.

20. SRIVASTAVA, O. P. & A. I. ARONSON. 1981. Isolation and characterization of a unique protease from sporulating cells of *Bacillus subtilis.* Arch. Microbiol. **129:** 227–232.

21. SLOMA, A., C. F. RUDOLPH, G. A. RUFO, JR., B. J. SULLIVAN, K. A. THERIAULT, D. ALLY & J. PERO. 1990. Gene encoding a novel extracellular metalloprotease in *Bacillus subtilis.* J. Bacteriol. **172:** 1024–1029.

22. BURNELL, T. J., G. W. SHANKWEILER & J. H. HAGEMAN. 1986. Activation of intracellular serine proteinase in *Bacillus subtilis* cells during sporulation. J. Bacteriol. **165:** 139–145.

23. SCHAEFFER, P., J. MILLET & J. P. AUBERT. 1965. Catabolic repression of bacterial sporulation. Proc. Natl. Acad. Sci. USA **54:** 704–711.

24. KLEID, D. G., D. YANSURA, B. SMALL, D. DOWBENKO, D. M. MOORE, M. J. GRUBMAN, P. D. MCKERCHER, D. O. MORGAN, B. H. ROBERTSON & H. L. BACHRACH. 1981. Cloned viral protein vaccine for foot-and-mouth disease: Responses in cattle and swine. Science **214:** 1125–1129.

25. LAEMMLI, U. K. 1970. Cleavage of structural proteins during the assembly of the head of bacteriophage T4. Nature (London) **227:** 680–685.

26. HARLOW, E. & D. LANE. 1988. Antibodies: A Laboratory Manual. Cold Spring Harbor Laboratory. Cold Spring Harbor, NY.

Drug Development through the Genetic Engineering of Antibiotic-Producing Microorganisms[a]

C. RICHARD HUTCHINSON,[b] C. W. BORELL,[c]
M. J. DONOVAN,[d] F. KATO,[e] H. MOTAMEDI,[f]
H. NAKAYAMA,[g] S. L. OTTEN, R. L. RUBIN,[f]
S. L. STREICHER,[f] K. J. STUTZMAN-ENGWALL,[h]
R. G. SUMMERS, E. WENDT-PIENKOWSKI,
AND W. L. WESSEL

School of Pharmacy and Department of Bacteriology
University of Wisconsin
Madison, Wisconsin 53706

The search for microbial metabolites with potential for use as pharmaceutical agents and the development of commercial processes for their production have been a major endeavor for over 50 years. Interest in such work stems from the discovery of penicillin[1] and streptomycin[2] and the finding that these substances are potent therapeutic agents.

Initially, screening of culture broths for antibiotic activity against common pathogens or related Gram-negative and Gram-positive bacteria sufficed to uncover a large number of compounds, many of which became widely used antiinfectives (e.g., the erythromycins, cephalosporins, and tetracyclines). As the clinical use of antibiotics became widespread, however, resistant forms of the targeted pathogens became an increasing problem that is still with us. This necessitated the search for new antibiotics with a broader spectrum of activity or effectiveness against the resistant microbes, using screens that either had a greater sensitivity or were focused only on the problematic pathogens.[3] The chemical and biological modification of existing antibiotics also resulted in more effective analogs of naturally occurring drugs like the β-lactams.[4] All of these developments have offset the diminishing rate of discovery of new antibiotics since the early 1970s; yet, it is common today to screen 10,000 isolates before a truly promising new antibiotic is discovered.[5] While the search for new pharmaceutical products from microorganisms is being vigorously pursued by the use of sophisticated antibiotic screening methods plus techniques designed to look for other types of biological activities (e.g., anticancer, antiviral,

[a] This research was supported in part by grants from the National Institutes of Health (CA 35381), Xechem, Inc., and Calbiochem, Inc.

[b] Address for correspondence: C. Richard Hutchinson, School of Pharmacy, University of Wisconsin, 425 N. Charter St., Madison, WI 53706.

CURRENT ADDRESSES: [c] Difco Laboratories, Detroit, Michigan; [d] Department of Molecular Genetics, Smith Kline & French Laboratories, Swedeland, Pennsylvania; [e] Department of Microbiology, School of Pharmaceutical Sciences, Toho University, Chiba, Japan; [f] Department of Infectious Disease, Merck, Sharpe & Dohme Research Laboratories, Rahway, New Jersey; [g] Japan Research Institute, Sumitomo Pharmaceutical Co., Tokyo, Japan; [h] Fermentation Process Research, Pfizer Central Research, Groton, Connecticut.

immunosuppressant, and insecticidal),[3,5] a different approach to drug discovery and development would be enthusiastically welcomed.

A clue to a new approach comes from the realization that although the spectrum of isolable microbial metabolites is predicted by the existing genetic diversity of microorganisms, this can be altered almost without limitation by recombinant DNA methods. Therefore, if it is possible to genetically engineer a microorganism to produce new antibiotics or just new metabolites, then it should be feasible to design a new type of drug screening or development strategy tailored toward a particular therapeutic goal with a high potential for success. In the present article, we examine this issue from two viewpoints: the production of new metabolites and the enhanced production of known drugs by genetically engineered species of the *Streptomyces*, soil bacteria best known for their ability to produce antibiotics and other biologically active metabolites.[6]

PRODUCTION OF NEW METABOLITES BY GENETIC HYBRIDS

The utility of modifying antibiotics by fermentation methods has been known since 1948 when penicillin V, an acid-stable analog of benzylpenicillin, was produced by adding sodium phenoxyacetate to the *Penicillium chrysogenum* fermentation.[7] This so-called "directed fermentation" and the methods of "mutasynthesis" and "hybrid biosynthesis" developed later[8,9] rely on satisfactory precursor uptake and plasticity of the enzymes used in microbial secondary metabolism, that is, metabolism not essential for cellular growth. Because these two demands are seldom well met in practice, all three methods suffer from inefficient conversion of an added precursor to the desired metabolite. Formation of a precursor *in vivo* would circumvent the uptake problem and perhaps others, as recently demonstrated by the use of recombinant streptomycetes.

Developments in *Streptomyces* genetics since 1980, summarized recently by Hopwood and Chater,[10] led to the realization that the genes for secondary metabolite formation cloned from one organism could be expressed in other streptomycetes and related bacteria once they were suitably transformed. This was first demonstrated by cloning the genes for the synthesis of actinorhodin (1, FIG. 1) from *Streptomyces coelicolor* and by showing that these genes were expressed in *Streptomyces parvulus* on the basis of the production of 1 upon transformation of this host with a plasmid vector carrying the *act* gene cluster.[11] Further work with the *act* genes as well as with other gene clusters, such as those for the production of tetracenomycin C[12] and oxytetracycline,[13] have supported this fact. Expression in Gram-negative and most Gram-positive bacteria, however, is not a simple matter because of the high guanosine-cytosine (GC) content of *Streptomyces* genes, which results in a different codon usage pattern from that of other bacteria, and also the need for specialized transcriptional factors.

The formation of novel metabolites by recombinant streptomycetes is a major advance from the formation of known metabolites, although directed fermentation and mutasynthesis studies[8,9] had demonstrated that this flexibility exists in the secondary metabolic pathways. Nonetheless, certain barriers have to be surmounted: foreign DNA must be introduced stably into the host strain; foreign genes need to be expressed at the correct time; the preexisting and introduced enzymes must cooperate metabolically; and the host requires resistance to any new antibiotic produced. (Naturally occurring gene clusters for antibiotic production contain self-resistance genes for the latter reason.) Some of these prerequisites can be met by design, but others have to be solved empirically because it is not usually known in advance how

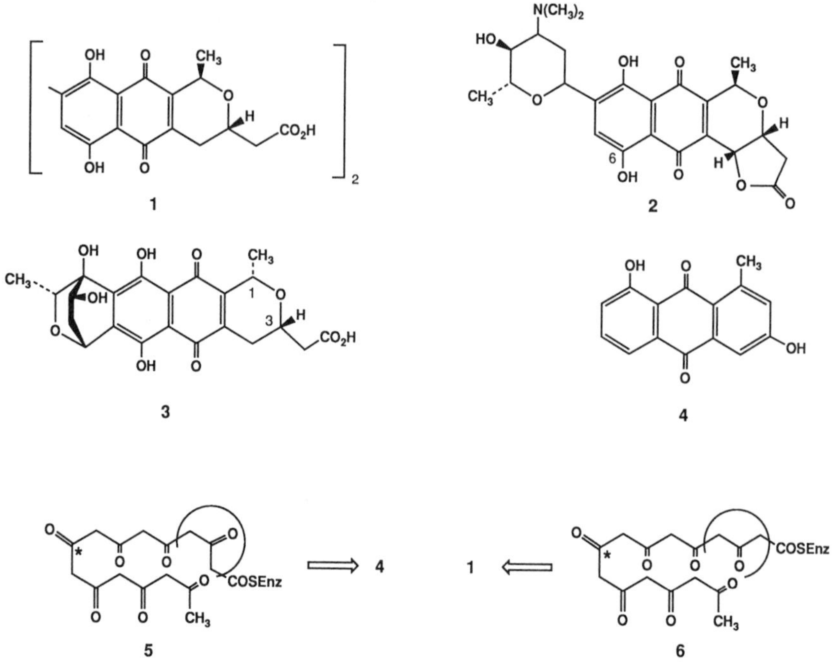

FIGURE 1. Normal and "hybrid" metabolites produced by streptomycetes on transformation with the *S. coelicolor act* genes. The brackets in **1** indicate that it is a dimeric form of the structure shown. The "*" in **5** and **6** indicates the carbonyl group that is reduced by the action of the *actIII* gene product, and the *thick arrows* represent the several biosynthetic steps necessary for the conversion of **5** to **4** and **6** to **1**.

well an enzyme will act on an unnatural substrate or how much resistance will be needed to ensure survival of a recombinant bacterium.

Pioneering biosyntheses of novel metabolites were reported in 1985 by David Hopwood and his collaborators who were examining the effect of introducing some or all of the *act* genes into strains of *Streptomyces* that make isochromane quinone metabolites which are closely related to actinorhodin.[14] New metabolites, mederrhodin A (**2**, FIG. 1) and dihydrogranatirhodin (**3**), were produced by interspecific transformants in large amounts, compared to the typical yields of the normal metabolites, but the *act* genes behaved differently in the two host strains. Mederrhodin A (which resulted from hydroxylation of the normal metabolite medermycin at position 6) was formed only when the *actV* gene was introduced; on the other hand, **1** and dihydrogranatirhodin (whose stereochemistry differs from that of the normal metabolite dihydrogranaticin at position 1 or 3) were produced only when all of the *act* genes were present. Similar studies by the Strohl and Floss groups have shown that the *actI, III,* and *VII* genes caused the production of aloesaponarin II (**4**, FIG. 1) when they were introduced into several other *Streptomyces* species[15] and that the *actIII* gene alone produced products lacking an aromatic hydroxyl group, which is characteristic of the reduction of a carbonyl group and loss of the resulting hydroxyl by dehydration.[15]

Interestingly, **4** is probably formed by way of the hypothetical compound **5**, the putative intermediate containing eight carbonyl groups (a "polyketide") and formed from an acetate and seven malonates, by the subset of *act* genes that is thought to encode a "polyketide synthase" (PKS), the set of enzymes that should normally form intermediate **6**, a conformer of **5**, and cyclize it to form aromatic rings. The only difference between structures **5** and **6** is orientation of the encircled bonds and their attached atoms. Does this mean that the "octaketide cyclase," which should normally act on **6** to give the cyclic backbone of **1**, is not highly specific, that is, lacks the fidelity expected of an enzyme? This may be true because **4** is a minor shunt metabolite of the actinorhodin pathway in *S. coelicolor*.[15] But it also may reflect differences in enzyme behavior in the new physiological backgrounds that lack the other enzymes of actinorhodin biosynthesis, which would normally (and probably rapidly) convert **6** to the normal intermediates of the actinorhodin pathway.

A better understanding of these findings and their implications for the production of new metabolites requires detailed information about the PKS genes and enzymes of the *Streptomyces*. A PKS is thought to share many of the properties of a bacterial fatty acid synthase (FAS).[16] As illustrated in FIGURE 2, the FAS produces long-chain fatty acids by the iteration of four basic reactions: condensation of acetate

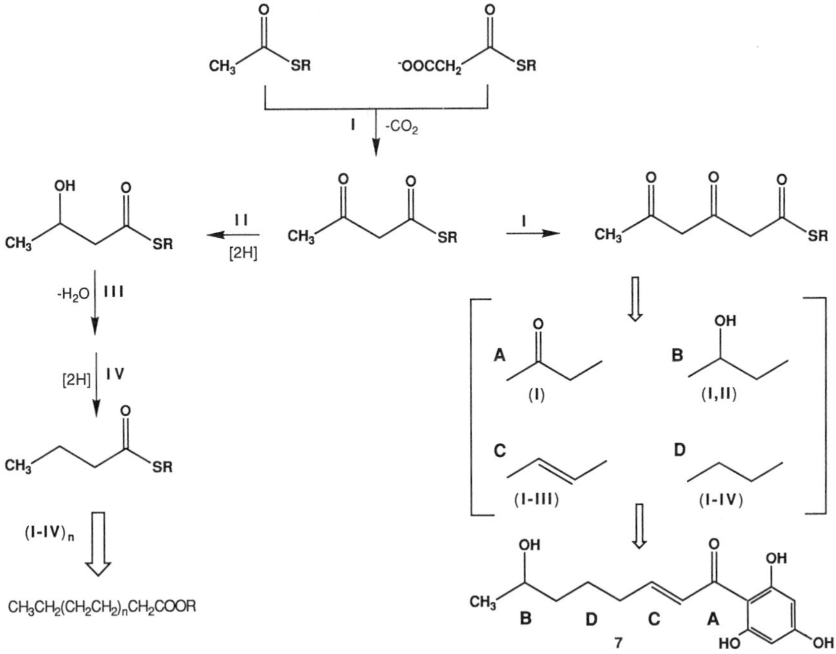

FIGURE 2. Comparison of the pathways for the biosynthesis of bacterial fatty acids and polyketides. Step **I** is catalyzed by a β-ketoacyl synthase, step **II** by the β-ketoacyl reductase, step **III** by the β-hydroxyacyl dehydratase, and step **IV** by the enoyl reductase. "−SR" indicates attachment of the intermediates to the ACP or β-ketoacyl synthase; "−CO$_2$" = loss of carbon dioxide; "[2H]" = reduction; and "−H$_2$O" = dehydration. The *thick arrow* on the left-hand side indicates the repetition of steps **I–IV** n times to provide a saturated fatty acid containing n + 4 carbon atoms. To form the hypothetical polyketide **7**, steps **I–IV** would be combined in different ways (indicated in parentheses below each substructure by the roman numerals representing the steps used) to form the substructures **A** to **D**, which appear in the final metabolite at the corresponding lettered positions. The aromatic ring of **7** is formed from three of the **A** substructures.

and malonate to give a β-ketobutyrate, reduction of the β-keto group, dehydration of the resulting β-hydroxy group, and reduction of the double bond so formed. Each of these intermediates remains attached to the FAS by a thioester bond. A type II FAS, characteristic of most bacteria and all plants, is a multienzyme complex consisting of a β-ketoacyl synthase (also known as the condensing enzyme), a β-ketoacyl reductase, a β-hydroxyacyl dehydratase, and an enoyl reductase plus the acyl carrier protein (ACP) that acts to shuttle the four intermediates during the assembly process. As the final fatty acid is removed from the ACP by transacylation to a water molecule, coenzyme A (CoA), or another acceptor, the FAS also requires the action

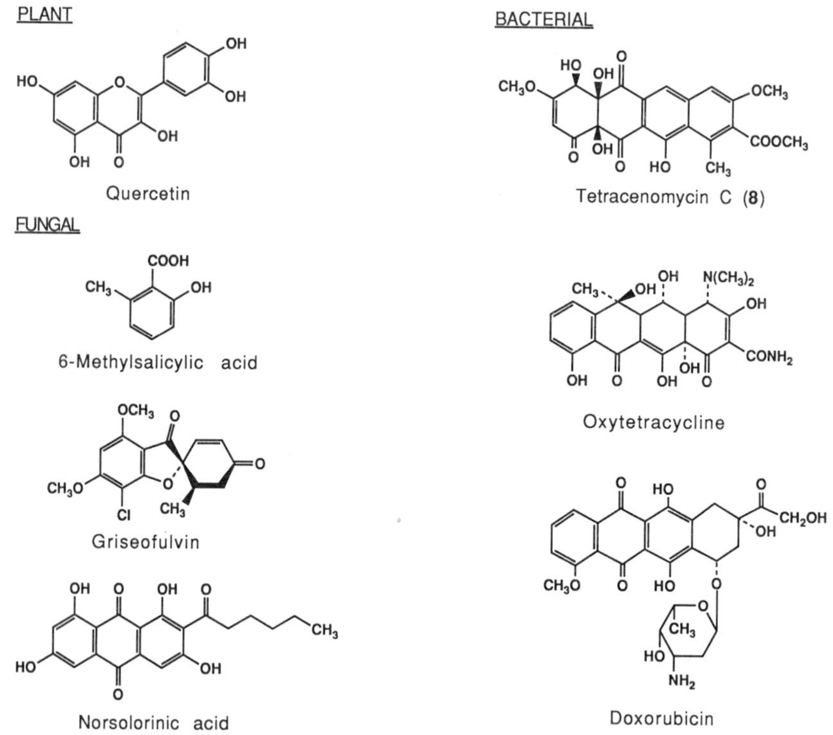

FIGURE 3. Typical "aromatic" polyketides that contain one or more aromatic rings and are found in plants, fungi, or bacteria.

of a thioesterase. The size of the fatty acid product is controlled by this enzyme, which is known to act on 16 carbon substrates much faster than on shorter chain fatty acid esters, and by the β-ketoacyl synthase, whose activity decreases markedly as the chain length approaches 16 carbons. During chain growth an acyltransferase of the FAS is used to attach malonylCoA to the ACP at the beginning of each cycle. Additional acyltransferases may be needed for attaching acetate to the ACP or β-ketoacyl synthase at the beginning of the process or for shifting the assembly intermediates from the ACP to the β-ketoacyl synthase during each cycle.

A PKS that forms typical aromatic polyketide metabolites (FIG. 3) differs from a type II FAS principally in the way it combines the four basic iterated steps (FIG. 2).

Several condensation steps often take place before reduction or dehydration events. These three reactions plus the reduction of α,β-unsaturated esters may be combined in almost any sequence in cases in which the final molecule contains structural elements resulting from reduction alone or reduction and dehydration processes. Polyketides of the hypothetical structure **7** could be formed in this way.

Another unique property of the PKS enzyme complexes is the cyclization events that form six-membered aromatic rings (FIG. 4). These are biological equivalents of intramolecular Claisen or aldol reactions and may involve spontaneous loss of the transiently formed β-hydroxyl groups, because this would be an energetically favorable process leading to aromatization. Cyclization and perhaps dehydration are thought to occur stepwise[15] and to require specific "polyketide cyclase" enzymes. This is illustrated in FIGURE 4a for the cyclization of two intermediates of the actinorhodin pathway that requires the products of the *actVII* and *actIV* genes. Such enzymes have to be able to choose between alternative cyclizations, because the same polyketide could form more than one type of tetracyclic compound, as shown in FIGURE 4b for the polyketide intermediate **9** that should be involved in the biosynthesis of tetracenomycin C.

To summarize the current thinking about the inherent properties of this type of PKS, they are enzyme complexes that must be able (1) to choose the correct starter unit (acetate, coumarate, propionate, or malonamide are used for the polyketides shown in FIG. 3), (2) to determine the number of condensation reactions between the starter and extender (usually malonate) units, (3) to fold the polyketide intermediate properly for its ensuing cyclization, and (4) to form the correct bonds leading to the (polycyclic) aromatic structure characteristic of a given PKS. Reduction of specific β-keto groups and dehydration of the resulting β-hydroxyls to form double bonds are further attributes. Acyltransferase and thioesterase activities analogous to those employed by a FAS may be a necessary part of these complexes or be provided by the FAS of the same organism. The intermediates of polyketide assembly are assumed to be bound to the ACP or β-ketoacyl synthase components via a thioester bond and to be released from the ACP by hydrolysis.

Evidence to support this concept for the PKS enzymes of streptomycetes comes from genetic rather than enzymatic studies at the present time. (The enzymology of a fungal[17,18] and a plant[19] PKS has been studied and provides useful analogies to what can be expected for the streptomycete enzymes.[16]) Mutations of the *S. coelicolor actI*[20] and *Streptomyces glaucescens tcmla*[21] PKS genes affect the earliest steps of actinorhodin and tetracenomycin C (**8**, FIG. 3) biosynthesis, preventing the formation of diffusible metabolites. Sequence analysis of the *actI*[22] (first carried out with the *Streptomyces violaceoruber gra* genes[23] for the biosynthesis of granaticin, a metabolite similar to actinorhodin [compare **1** and **3** in FIG. 1]) and *tcmla*[24] genes revealed the presence of three open reading frames (ORFs). The putative ORF1 proteins strongly resemble the *Escherichia coli fabB* gene product that forms the homodimeric β-ketoacyl synthase I of its FAS. There is excellent alignment of the region around the active site cysteine (to which the acyl intermediates of fatty acid biosynthesis are attached) among these proteins and the FAS-condensing enzymes from *E. coli,* yeast, and mammals.[24] The *actI, gra,* and *tcmla* ORF1 proteins also contain a conserved region—GHSxG—that is found at the corresponding position in the FABB protein, and in the acetyl/malonylCoA transferases of the bacterial, fungal, yeast, and mammalian FAS, and the thioesterase domain of mammalian FASs. Inasmuch as discrete acyltransferase or thioesterase genes have not been found in the *act, gra,* or *tcm* clusters, this region could represent the active site for such activities in these PKS enzymes (the serine would be the point of substrate attachment), or it could be the site at which acetate and malonate are bound prior to their transfer to the

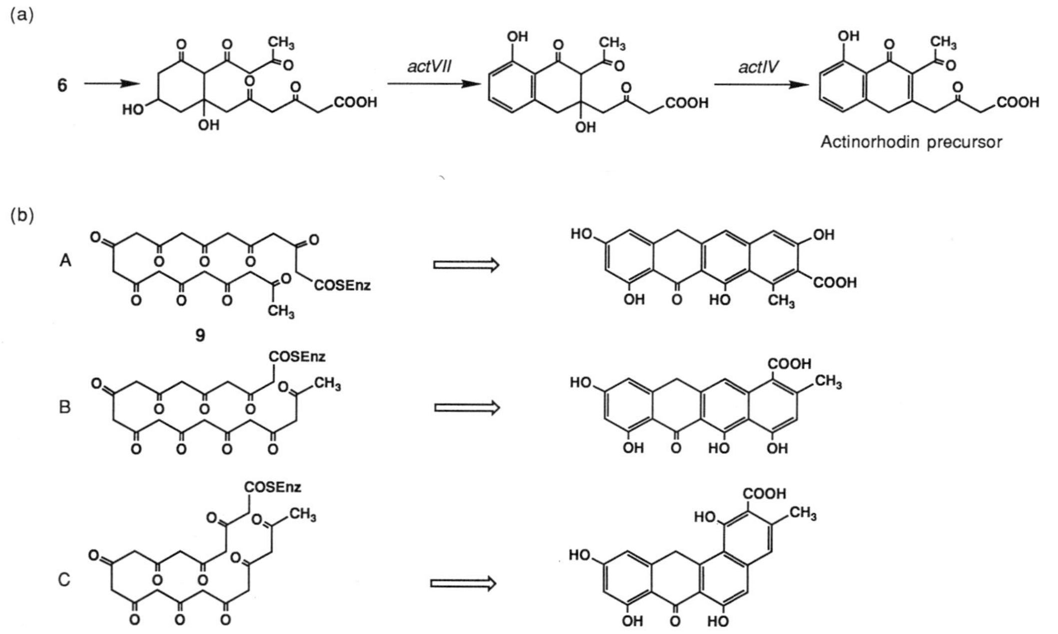

FIGURE 4. (a) The hypothesis for two of the cyclization steps in the early stages of actinorhodin biosynthesis. *S. coelicolor actVII* mutants accumulate the first compound in the sequence shown, which is believed to be formed by reduction and cyclization of the octaketide **6** (FIG. 1), and *actIV* mutants accumulate the second compound, which is formed from the first one by dehydration and cyclization. The apparent absence of spontaneous dehydration leading to aromatization of different six-membered rings in these two compounds may reflect the loss of a dehydratase activity of the PKS in the two mutants or it may be artifactual. **(b)** Some of the possible cyclization modes for the decapolyketide **9**, a putative intermediate of tetracenomycin C biosynthesis. The products of the folding of **9** into structures A, B, and C are shown to the right of *thick arrows* that represent the four cyclization events used in product formation. Each pair of decapolyketide and cyclized product is used in the biosynthesis of known bacterial polyketides.

cysteamine-SH of the ACP or the cysteine-SH of the β-ketoacyl synthase, as in bacterial and mammalian FAS.[25] The amino acid sequences of the putative ORF1 and ORF2 proteins are very similar and both *orfs* appear to be translationally coupled in each set of PKS genes. These facts are consistent with the formation of a heterodimeric β-ketoacyl synthase; however, the ORF2 proteins probably play a different role in polyketide assembly because they lack the active site cysteine that characterizes the ORF1 proteins. ACP genes are the third components of the PKS

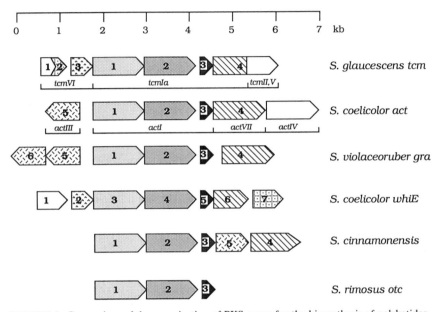

FIGURE 5. Comparison of the organization of PKS genes for the biosynthesis of polyketides from six *Streptomyces* spp. Each PKS *orf* is indicated by a thick wedge, pointed in the direction of gene transcription. The size of the *orfs* is indicated in kilobases (kb) of DNA at the top of the figure. Identical shading patterns of individual *orfs* indicate that their putative gene products have a significant similarity (lack of shading indicates no similarity), and in some cases, an equivalent function as discussed in the text. These functions are indicated by the location of the corresponding mutations underneath the *tcm* and *act* gene clusters. The unshaded, C-terminal part of *tcmIa orf4* strongly resembles an *N*-acetylserotonin *O*-methyltransferase of bovine pineal gland and complements the *tcmII* and −*V* mutations that block C-3 *O*-methylation of tetracenomycin C precursors. This *orf* thus appears to encode a bifunctional enzyme, which is also a property of the *actVII* product that presumably has cyclase and dehydratase activities. The PKS region of the oxytetracycline-producing *S. rimosus otc* gene cluster has not been sequenced completely. (Sequence data for the *S. cinnamonensis* and *S. rimosus* PKS genes were provided by Teresa Arrowsmith and David Sherman, respectively, of the John Innes Institute.)

regions in the three clusters, clearly recognizable by the strong similarity of their deduced products (size, isoelectric point, and amino acid sequence) to the ACPs of the bacterial and plant FAS, and the ACP-like regions of the multienzyme polypeptides of the yeast and mammalian FAS.[23,24]

The PKS genes of several clusters governing the formation of aromatic polyketides have now been cloned and sequenced.[22] A comparison of six of these regions (FIG. 5) shows the presence of considerable homology among the three types of *orfs* just described and among some of the genes believed to encode other components of the PKS. The *actIII* homologs must encode β-ketoacyl reductases: the products of such genes exhibit a strong resemblance to alcohol and ribitol dehydrogenases,[26] and transformation of various *Streptomyces* spp. with the *actIII* gene has the effect of reduction and dehydration of specific carbonyl groups in the putative polyketide intermediates.[15] The *actVII* homologs are thought to be polyketide cyclases, as *S. coelicolor actVII* mutants accumulate a product (FIG. 4a) resulting from the lack of cyclization and dehydration.[15] Similarly, because *S. glaucescens tcmVI* mutants accumulate tetracenomycin F_2 (FIG. 6), in which the last bond needed for the formation of the tetracyclic 5-deoxytetracenomycin F_1 has not been formed,[27] some of the *tcmVI* gene products (FIG. 5) also must be a polyketide cyclase. The other product encoded in this region may be an "oxidase," because tetracenomycin F_2 also lacks the carbonyl found at position 5 in tetracenomycin F_1 (FIG. 6). The order of the three *orfs* encoding the β-ketoacyl synthase and ACP components is invariant among the six cases shown in FIGURE 5. The fact that a putative cyclase *orf* follows these three

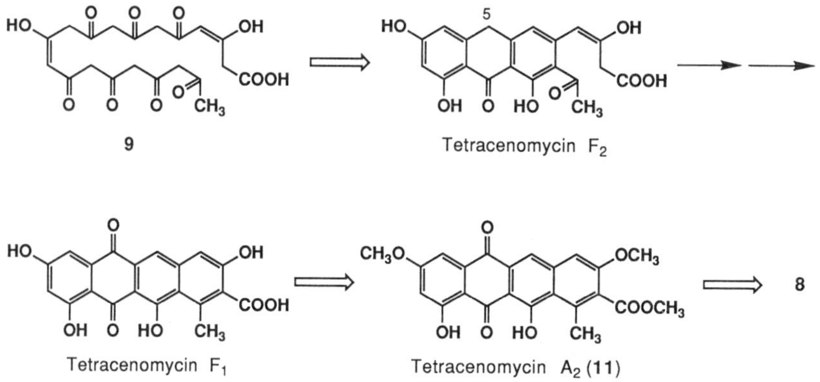

FIGURE 6. A summary of the hypothetical biosynthetic pathway to tetracenomycin C. Decapolyketide 9 is believed to be converted to 8 (FIG. 3) via the intermediates shown. *Thick arrows* indicate several steps and the *thin arrow*, a single step.

genes in four of five cases could mean that the products of these four *orfs* act in concert. Our finding that *tcmla orfs 1* to *4*, but not *orfs 1* to *3*, are sufficient to form tetracenomycin F_2 in a *S. glaucescens* Tcm C null mutant[21] and *Streptomyces lividans* when they are expressed under the control of a constitutive promoter[27] is consistent with this belief.

Considering the similarity of the gene products among these six examples of *Streptomyces* PKSs, it is not surprising to find that a PKS gene from one species can compensate for the defect in the corresponding gene from another species. This has been successfully demonstrated with the *gra orf4* gene and *actVII* mutation[22] and with a set of *tcmla orf1* and *orf3* homologs from the salinomycin-producing *Streptomyces albus* ATCC 21838 and some of the *tcmla* mutations.[29] Full restoration of actinorhodin or tetracenomycin C production was observed in both cases.

On the other hand, formation of new metabolites can also result from the substitution of polyketide synthase or cyclase (or reductase[15]) genes from one species for those from another species. Indeed, we have observed the formation of five new

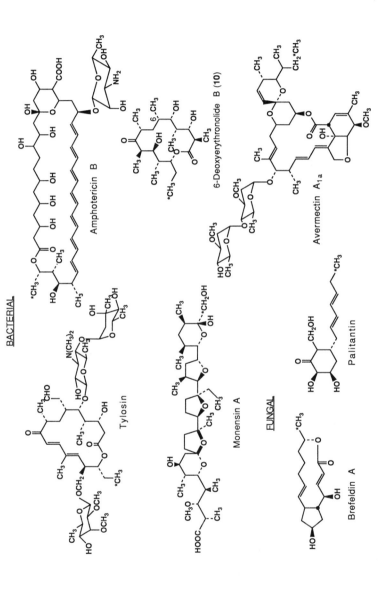

FIGURE 7. Typical "reduced" polyketides found in bacteria or fungi and produced by reduction and dehydration events during assembly of the carbon chain. The absence of aromatic rings and extensive reduction or reduction and dehydration of biosynthetic intermediates distinguishes these metabolites from those shown in FIGURE 3. The "*CH$_3$–" group in each structure is the point at which assembly of the polyketide intermediate begins. The bacterial metabolites are made from acetate, propionate, butyrate, and 2-methylbutyrate plus their α-carboxylated forms; the fungal compounds (brefeldin A and palitantin) are made from acetate and malonate only.

compounds, unrelated by chromatographic analysis to known intermediates of the tetracenomycin pathway,[30] in extracts of *tcmla* mutants transformed with a DNA fragment that contains *orfs 1* to *3* and *5* of the *S. cinnamonensis* PKS genes shown in FIGURE 5.[29] This suggests that further gene-mixing experiments should lead to the goal of producing new drugs by genetic engineering.

Investigations of bacterial PKSs for the formation of polyketides of a more reduced nature like tylosin and avermectin A_{1a} (FIG. 7) are in progress in other laboratories. Preliminary results from a study of the PKS that makes a 14-membered macrolide intermediate of the erythromycin pathway, called 6-deoxyerythronolide B (**10,** FIG. 7), show the presence of large, multienzyme polypeptides with active sites corresponding to systematic repetitions of the active site motifs found in the different domains of mammalian FAS.[31,32] Together with the knowledge that the assembly of such polyketides resembles fatty acid biosynthesis more closely than the process just outlined for the biosynthesis of aromatic polyketides,[33,34] these developments suggest that this type of PKS is analogous to the type I FAS enzymes of a few bacteria and fungi, yeast, and mammals.[16,25] Site-specific modification of such enzymes and suitable gene-mixing experiments in this area also show great promise in achieving the goal of producing new drugs. Leading examples have already appeared.[31,35,36]

ENHANCED PRODUCTION OF KNOWN METABOLITES BY RECOMBINANT *STREPTOMYCES* SPP.

There are several sensible approaches to the development of recombinant streptomycetes that overproduce secondary metabolites. One could: (1) introduce additional copies of a gene governing the slowest step of the biosynthetic pathway or replace the genetic elements controlling its expression with more efficient ones; (2) increase the number of the gene(s) conferring resistance to the metabolite on the reasonable assumption that the level of self-resistance could be a limiting factor in production, particularly if the mechanism of resistance involves metabolite excretion; (3) overexpress the gene(s) that regulates expression of the bulk of the structural genes (the ones that encode pathway enzymes) in the cluster, if there is evidence for positive control of the expression of these genes; or (4) increase the expression of the genes that control nutrient use or that provide the primary metabolites used in secondary metabolism.

We tested the effect of enhanced expression of the *tcmla orfs 1* to *3* genes on the production of tetracenomycin C by the wild-type *S. glaucescens* GLA.0 strain during work aimed at the overproduction of PKS gene products in *E. coli* and *S. glaucescens*. When *S. glaucescens* was transformed with pELE37, a high copy number plasmid bearing *tcmla orfs 1* to *3* under the control of *ermE**, a strong, constitutive promoter made by Bibb and Janssen[37] from the *ermE* erythromycin resistance gene of *Saccharopolyspora erythraea,* an approximately 28-fold overproduction of tetracenomycin A_2 (**11,** FIG. 6), the penultimate precursor of tetracenomycin C,[30] was observed.[28] A similar result was obtained when the vector carried only the *tcmal orf3* gene (the ACP),[28] although in this case we used the *mel* promoter of the pIJ702 vector[38] to drive expression of this gene. Metabolite production was not increased, however, when the vector carried only *tcmla orf1* and about one-half of *orf2*.[28] The simplest interpretation of these results is that the ACP is rate limiting in tetracenomycin A_2 formation. Another possibility is that excess ACP is able to increase the formation of the other pathway enzymes by some type of inducation. We favor the former explanation on the grounds that an ACP is not a catalytic protein but is used in stoichiometric amounts to provide malonylSR (FIG. 2). Hence, the activity of the β-ketoacyl synthase could be affected more by the relative amount of ACP in relation to the amounts of its two other components. That the production of tetracenomycin C was not simultaneously increased reflects a rate-limiting step in conversion of **11** to **8** (FIG. 6), which is consistent with the fact that **11** is the most abundant intermediate in the wild-type strain. It also could be due to a self-limiting resistance to **8**, because **11** and the earlier intermediates of tetracenomycin C biosynthesis do not have the growth inhibitory effects of tetracenomycin C. Further exploration of the biochemistry and physiology of the PKS enzymes in this and other streptomycetes must be done to understand the exact reason for our observations, but on the surface they suggest that modification of PKS gene expression could be a fruitful way to construct overproducing strains.

There is growing evidence that expression of many of the genes for secondary metabolism is often controlled positively by regulatory elements contained within or outside the gene clusters.[39] The earliest indication of this came from studies of

TABLE 1. Examples of Genes That Regulate Secondary Metabolism in *Streptomyces* spp.

Gene	Function
actII	Increases actinorhodin production 30- to 40-fold when introduced into *Streptomyces coelicolor;* positively regulates expression of the *S. coelicolor actI, IV, VII*, and *III* genes that encode the polyketide synthase of actinorhodin biosynthesis
redD	Increases undecylprodigiosin production approximately 30-fold when introduced into *S. coelicolor*[47]
strR	Increases streptomycin production five- to seven-fold when introduced into *Streptomyces griseus;* enhances three- to six-fold the amidinotransferase activity required for streptomycin production when introduced into *S. griseus* and two of its Str− mutants[48]
brpA	Controls expression of the resistance and at least six structural genes of bialaphos production in *Streptomyces hygroscopicus*[49]
dnrR₁	Increases ε-rhodomycinone production up to 78-fold when introduced into *Streptomyces peucetius*
dnrR₂	Increases daunorubicin production up to 10-fold when introduced into *S. peucetius;* restores daunorubicin resistance in *S. peucetius* H6101

mutations that appeared to block formation of most or all of the enzymes in the actinorhodin pathway, as deduced from the lack of cosynthesis of *actII* mutants with the other classes of *act* mutants known to secrete intermediates of actinorhodin biosynthesis.[20] When the *actII* locus was identified[40] and cloned in the wild-type *S. coelicolor* strain, it caused a 30- to 40-fold increase in actinorhodin production.[41] Subsequent work showed that the presence of the *actII* gene was required for transcription of the *actIII* gene[26] and for expression of the *actI* and *VII* genes.[42] Additional examples of the apparent positive control of secondary metabolite genes are listed in TABLE 1. A comprehensive review of this topic is available,[39] which also discusses the types of regulatory genes not found in the metabolite production gene clusters.

Work on the daunorubicin (dnr) production genes of *Streptomyces peucetius* showed the effect of introducing large portions of the *dnr* gene cluster, cloned in the pKC505 cosmid shuttle vector developed at Eli Lilly and Company,[43] into *S. peucetius*

29050 and strains derived from it.[44] The production of ε-rhodomycinone (12, FIG. 8) and daunorubicin (13) by these strains was monitored (12 is a key intermediate in the pathway to 13 [FIG. 8] and, in the wild-type strain, accumulates to a much higher level than does any other intermediate). Considerable increases in ε-rhodomycinone and daunorubicin production were exhibited by the transformants,[44] which could have been due to the presence of extra copies of the structural, regulatory, or resistance genes. That these effects might be due to positively acting regulatory genes was deduced from the properties of two small segments of DNA, which appeared to be responsible for most, if not all, of the metabolite overproduction observed in subcloning experiments.[45] We named the 2.0-kb BglII-BamHI DNA segment dnrR₁ and found that it greatly increases the production of only 12 (TABLE 1); the 1.9 BamHI DNA segment, dnrR₂, located about 12 kb away from dnrR₁, on the other

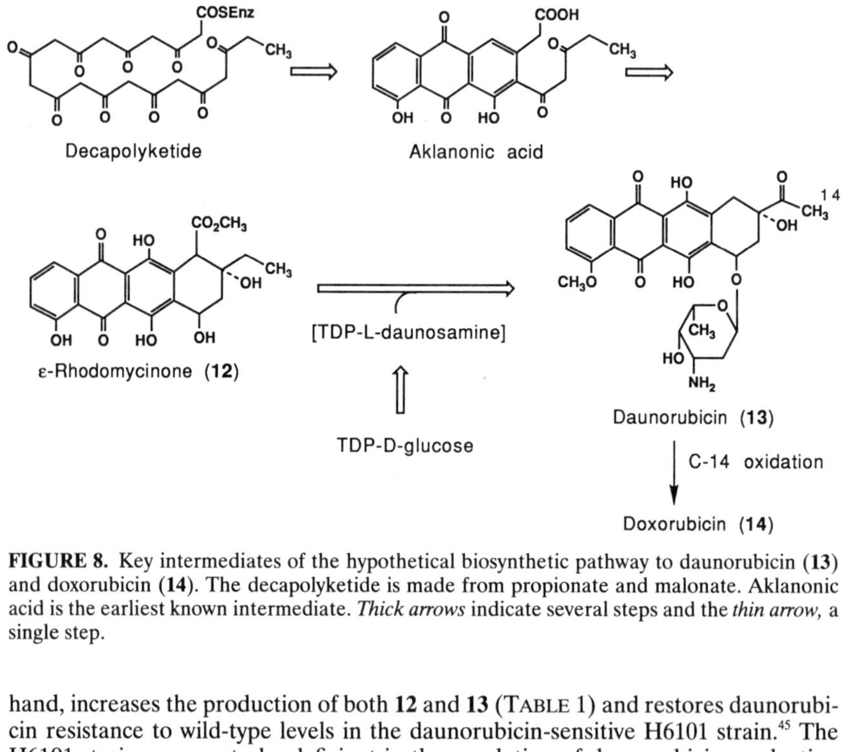

FIGURE 8. Key intermediates of the hypothetical biosynthetic pathway to daunorubicin (13) and doxorubicin (14). The decapolyketide is made from propionate and malonate. Aklanonic acid is the earliest known intermediate. *Thick arrows* indicate several steps and the *thin arrow*, a single step.

hand, increases the production of both 12 and 13 (TABLE 1) and restores daunorubicin resistance to wild-type levels in the daunorubicin-sensitive H6101 strain.[45] The H6101 strain appears to be deficient in the regulation of daunorubicin production because it does not make 13 or any of its precursors, does not convert any of these substances to 13, and is partially sensitive to daunorubicin. The fact that transformation of H6101 with dnrR₁ largely restored the production of 12, and transformation with dnrR₂ restored both 12 and 13, together with the fact that no other subclones from the dnr cluster had these effects, supports our belief that dnrR₁ and dnrR₂ might somehow be regulating the expression of several dnr genes. Sequence analysis of the dnrR₁ segment reveals an orf that would encode a product that strongly resembles the products of the S. coelicolor actII and redD genes (FIG. 9), two positive regulators of secondary metabolite production in S. coelicolor (TABLE 1). Moreover, S. lividans

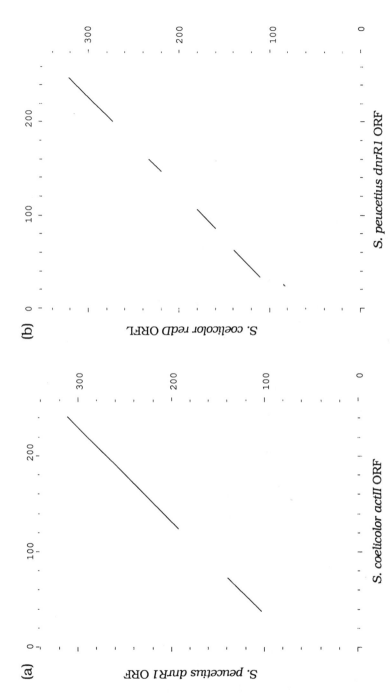

FIGURE 9. Comparison of the deduced products of the *S. peucetius dnrR₁* gene with (**a**) the *S. coelicolor actII*[46] and (**b**) the *redD orfI*[47] genes. The gene products were compared by COMPARE/DOTPLOT using a window of 30 and a stringency of 20.[50] The *solid diagonal lines* represent regions with similar amino acid residues for the two proteins in **a** and **b**.

(*dnrR₁*) transformants overproduced actinorhodin,[45] and an *S. coelicolor actII* mutant regained the actinorhodin production ability on transformation with the *dnrR₁* gene.[46] From these results, we predict that the *dnrR₁* gene will be found to regulate the expression of several *dnr* genes, probably those involved in the production of **12**. Why the presence of extra copies of this gene in the wild-type *S. peucetius* strain results in the overproduction of **12** is under study, but the behavior of *dnrR₁* and the genes listed in TABLE 1 have important implications[39] for the construction of recombinant streptomycetes that can overproduce therapeutically and commercially important drugs like **13** and its hydroxylated derivative, doxorubicin (**14**).

ACKNOWLEDGMENTS

We thank Teresa Arrowsmith, David Sherman, David Hopwood, and Paco Malpartida for permission to cite unpublished gene sequence data.

REFERENCES

1. FLEMING, A. 1929. Br. J. Exp. Pathol. **10:** 226–236.
2. SCHATZ, A., E. BUGIE & S. A. WAKSMAN. 1944. Proc. Soc. Exp. Biol. Med. **55:** 66–69.
3. NISBET, L. J. & J. W. WESTLEY. 1986. Ann. Rep. Med. Chem. **21:** 149–157.
4. MORIN, R. B. 1982. Chemistry and Biology of Betalactam Antibiotics. Vols. 1–3. Academic Press. New York.
5. OMURA, S. 1986. Microbiol. Rev. **50:** 259–279.
6. CRANDALL, L. W. & R. L. HAMILL. 1986. Antibiotics produced by *Streptomyces:* Major structural classes. *In* Antibiotic-Producing Streptomyces. S. W. Queener & L. E. Day, eds. Vol. 9: 355–402. Academic Press. New York.
7. BEHRENS, O. K., J. CORSE, J. P. EDWARDS, L. GARRISON, R. G. JONES, Q. F. SOPER, F. R. VAN ABEELE & C. W. WHITEHEAD. 1948. J. Biol. Chem. **175:** 793–809.
8. DAUM, S. J. & J. R. LEMKE. 1979. Ann. Rev. Microbiol. **33:** 241–265.
9. HUTCHINSON, C. R. 1988. Med. Res. Rev. **8:** 557–567.
10. HOPWOOD, D. A. & K. F. CHATER. 1989. Antibiotic biosynthesis in *Streptomyces*. *In* Genetics of Bacterial Diversity. D. A. Hopwood & K. F. Chater, eds.: 129–133. Academic Press. New York.
11. MALPARTIDA, F. & D. A. HOPWOOD. 1984. Nature **309:** 462–464.
12. MOTAMEDI, H. & C. R. HUTCHINSON. 1987. Proc. Natl. Acad. Sci. USA **84:** 4445–4449.
13. BINNIE, C., M. WARREN & M. J. BUTLER. 1989. J. Bacteriol. **171:** 887–895.
14. HOPWOOD, D. A., F. MALPARTIDA, H. M. KIESER, H. IKEDA, J. DUNCAN, I. FUJII, B. A. M. RUDD, H. G. FLOSS & S. OMURA. 1985. Nature **314:** 642–644.
15. BARTEL, P. L., C. B. ZHU, J. S. LAMPEL, D. C. DOSCH, N. C. CONNORS, W. R. STROHL, J. M. BEALE, JR. & H. G. FLOSS. 1990. J. Bacteriol. **172:** 4816–4826.
16. HOPWOOD, D. A. & D. H. SHERMAN. 1990. Ann. Rev. Genet. **24:** 37–66.
17. DIMROTH, P., H. WALTER & F. LYNEN. 1970. Eur. J. Biochem. **13:** 98–110.
18. DIMROTH, P., E. RINGELMAN & F. LYNEN. 1976. Eur. J. Biochem. **68:** 591–596.
19. HAHLBROCK, K. 1981. *In* The Biochemistry of Plants. E. E. Conn, ed. Vol. 7: 425–456. Academic Press. New York.
20. RUDD, B. A. M. & D. A. HOPWOOD. 1979. J. Gen. Microbiol. **114:** 35–43.
21. MOTAMEDI, H., E. WENDT-PIENKOWSKI & C. R. HUTCHINSON. 1986. J. Bacteriol. **167:** 575–580.
22. HOPWOOD, D. A., D. H. SHERMAN, C. KHOSLA, M. J. BIBB, T. J. SIMPSON, M. A. FERNANDEZ-MORENO, E. MARTINEZ & F. MALPARTIDA. In press.
23. SHERMAN, D. H., F. MALPARTIDA, M. J. BIBB, H. KIESER, M. J. BIBB & D. A. HOPWOOD. 1989. EMBO J. **8:** 2717–2725.
24. BIBB, M. J., S. BIRO, H. MOTAMEDI, J. F. COLLINS & C. R. HUTCHINSON. 1989. EMBO J. **8:** 2727–2736.

25. WAKIL, S. J. 1989. Biochemistry **28:** 4523–4530.
26. HALLAM, S. E., F. MALPARTIDA & D. A. HOPWOOD. 1988. Gene **74:** 305–320.
27. NAKAYAMA, H., H. MOTAMEDI, R. G. SUMMERS, E. WENDT-PIENKOWSKI, W. L. WESSEL & C. R. HUTCHINSON. Unpublished work.
28. SUMMERS, R. G., E. WENDT-PIENKOWSKI, W. L. WESSEL & C. R. HUTCHINSON. Unpublished work.
29. KATO, F., E. WENDT-PIENKOWSKI, T. ARROWSMITH, M. J. DONOVAN, F. MALPARTIDA, D. A. HOPWOOD & C. R. HUTCHINSON. Unpublished work.
30. YUE, S., H. MOTAMEDI, E. WENDT-PIENKOWSKI & C. R. HUTCHINSON. 1986. J. Bacteriol. **167:** 581–586.
31. DONADIO, S., M. J. STAVER, J. B. MCALPINE, S. J. SWANSON & L. KATZ. 1991. Science **252:** 675–679.
32. CORTES, J., S. F. HAYDOCK, G. A. ROBERTS, D. J. BEVITT & P. F. LEADLAY. 1990. Nature **348:** 176–178.
33. YUE, S., J. S. DUNCAN, Y. YAMAMOTO & C. R. HUTCHINSON. 1987. J. Am. Chem. Soc. **109:** 1253–1255.
34. CANE, D. E. & C.-C. YANG. 1987. J. Am. Chem. Soc. **109:** 1255–1257.
35. MCALPINE, J. B., J. S. TUAN, D. P. BROWN, K. D. BREBNER, D. N. WHITTERN, A. BUKO & L. KATZ. 1987. J. Antibiotics **40:** 1115–1122.
36. EPP, J. K., M. L. B. HUBER, J. R. TURNER, T. GOODSON & B. E. SCHONER. 1989. Gene. **85:** 293–301.
37. BIBB, M. J. & G. R. JANSSEN. 1987. Unusual features of transcription and translation of antibiotic resistance genes in two antibiotic-producing *Streptomyces* species. *In* Genetics of Industrial Microorganisms: Proceedings of the Fifth International Symposium. M. Alacevic, D. Hranueli & Z. Toman, eds. Part B: 309–318. Pliva. Zagreb, Yugoslavia.
38. KATZ, E., C. J. THOMPSON & D. A. HOPWOOD. 1983. J. Gen. Microbiol. **129:** 2703–2714.
39. CHATER, K. F. 1990. Bio/technology **8:** 115–121.
40. MALPARTIDA, F. & D. A. HOPWOOD. 1986. Molec. Gen. Genet. **205:** 66–73.
41. HOPWOOD, D. A., F. MALPARTIDA & K. F. CHATER. 1986. Gene cloning to analyse the organization and expression of antibiotic biosynthesis genes in *Streptomyces*. *In* Regulation of Secondary Metabolite Formation. H. Kleinkauf, H. von Dohren, H. Dornauer & G. Nesemann, eds. 13–22. VCH. Weinheim, Germany.
42. STROHL, W. R. Personal communication.
43. RICHARDSON, M. A., S. KUHSTOSS, P. SOLENBERG, N. A. SCHAUS & R. N. RAO. 1987. Gene. **61:** 231–241.
44. OTTEN, S. L., K. J. STUTZMAN-ENGWALL & C. R. HUTCHINSON. 1990. J. Bacteriol. **172:** 3427–3434.
45. STUTZMAN-ENGWALL, K. J., S. L. OTTEN & C. R. HUTCHINSON. 1992. J. Bacteriol. **174,** in press.
46. MALPARTIDA, F. Personal communication.
47. NARVA, K. E. & J. S. FEITELSON. 1990. J. Bacteriol. **172:** 326–333.
48. OHNUKI, T., T. IMANAKA & S. AIBA. 1985. J. Bacteriol. **164:** 85–94.
49. ANZAI, H., T. MURAKAMI, S. IMAI, A. SATOH, K. NAGAOKA & C. J. THOMPSON. 1987. J. Bacteriol. **169:** 3482–3488.
50. DEVEREUX, J., P. HAEBERLI & O. SMITHIES. 1984. Nucl. Acids Res. **12:** 387–395.

Cloning of an NADH-Dependent Butanol Dehydrogenase Gene from *Clostridium acetobutylicum*[a]

DANIEL J. PETERSEN,[b] RICHARD W. WELCH,[b]
KARL A. WALTER,[c] LEE D. MERMELSTEIN,[c]
ELEFTHERIOS T. PAPOUTSAKIS,[c,d]
FREDERICK B. RUDOLPH,[b] AND GEORGE N. BENNETT[b]

[b]*Rice University*
Department of Biochemistry and Cell Biology
Houston, Texas 77251

[c]*Northwestern University*
Department of Chemical Engineering
Evanston, Illinois 60208

The production of 1-butanol by the anaerobic bacterium *Clostridium acetobutylicum* has gained renewed interest recently, as the potential generation of "metabolically engineered" strains of this organism, with superior yield and selectivity, could provide an economically feasible alternative to current petrochemical derived production of this solvent. The enzyme butanol dehydrogenase (BDH) catalyzes the final step in butanol synthesis, the conversion of butyraldehyde to butanol. This activity is induced at the onset of solvent synthesis and thus is implicated in the control of butanol formation. *C. acetobutylicum* has both NADH- and NADPH-dependent butanol dehydrogenase activities. The cloning of an NADPH-dependent alcohol dehydrogenase (ADH) was recently reported by Youngleson *et al.*[1] There is substantial evidence, however, to suggest that BDH activity dependent on the NADH cofactor is of greater importance for butanol production in the batch fermentation.[2-4]

Recent work by our group has resulted in the purification to near homogeneity of two distinct NADH-dependent isozymes, BDH I and BDH II, from *C. acetobutylicum* ATCC 824.[4] The sequence of the 25 NH_2-terminal amino acids of BDH II was determined from purified protein blotted to polyvinylidene difluoride (FIG. 1A). The resulting sequence was reverse translated to give a highly degenerate set of potential DNA sequences. From the A + T bias of the organism and codon usage data from previously sequenced genes,[5-7] estimations were made for the "wobble" base of each codon whenever possible. Oligonucleotide probes were designed and synthesized from stretches (of sufficient length) of the resulting sequences that would give minimum degeneracy. Whenever the DNA sequence could not be presumed by the technique just described, inosine residues were incorporated into the oligonucleotides.

End-labeled oligonucleotide probes were purified and tested for specificity of hybridization to restriction enzyme-digested *C. acetobutylicum* chromosomal DNA. Initial 22- and 23-mer oligonucleotide probes designed from the extreme NH_2-

[a]This work was supported by Grant BCS-8912094 from the National Science Foundation. D.J.P. was supported by a National Science Foundation Predoctoral Fellowship.
[d]To whom correspondence should be sent.

1	2	3	4	5	6	7	8	9	10	11	12	13	14	15	16	17	18	19	20	21
Met	-Val	-Asp	-Phe	-Glu	-Tyr	-Ser	-Ile	-Pro	-Thr	-Cys	-Ile	-Phe	-Phe	-Gly	-Lys	-Asp	-Lys	-Ile	-Asn	-Val -
ATG	GTI	GAT	TTT	GAA	TAT	ACI	ATI	CCI	ACI	TGT	ATI	TTT	TTT	GGI	AAA	GAT	AAA	ATI	AAT	GT
			C										C	C		G		G		

22	23	24	25
Leu	-Gly	-Cys	-Glu

FIGURE 1A. Sequences of mixed oligonucleotide probes, BDH-3, specific for the BDH II gene. The amino acid positions, amino acid names (as determined by Edman degradation sequencing), and the probe sequences are indicated from top to bottom.

terminal or central region of the determined protein sequence, respectively, failed to give specific hybridization signals. A larger 62-mer oligonucleotide probe named BDH-3 (FIG. 1A), however, gave strong unique hybridization signals. No hybridization was observed to purified λ phage arms or *Escherichia coli* chromosomal DNA.

BDH-3 was used to screen a λEMBL3 phage, *C. acetobutylicum* genomic library. The library was diluted to obtain 1,000–2,000 plaques per plate following infection of *E. coli* NM519 cells. Of approximately 10,000 plaques screened, four exhibited homology to BDH-3. Following plaque purification, DNA was isolated from one of these phage, and the entire genomic insert was subcloned as a *Sal* I fragment into appropriately cleaved pUC19 to give a recombinant plasmid designated pDP51.

Various restriction enzymes were used to generate a physical map of pDP51 (FIG. 1B). The 7.9-kb insert of pDP51 was demonstrated to be of clostridial origin and exhibit no homology to the previously cloned NADPH-dependent ADH by a Southern blot experiment (data not shown). Electrophoresed blotted restriction enzyme digests of *C. acetobutylicum* chromosomal DNA or the cloned NADPH-dependent ADH were probed with [^{32}P]-radiolabeled pDP51. In all cases the fragments exhibiting homology corresponded in size to those expected from the physical map of pDP51.

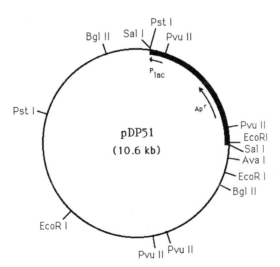

FIGURE 1B. Physical map of pDP51. Restriction endonuclease sites within the 7.9-kilobase (kb) *Sal*I insert and the pUC19 vector (*bold line*) are shown. The direction of transcription from the β-galactosidase promoter (P_{lac}) and β-lactamase gene (Ap^r) is indicated by *arrows*.

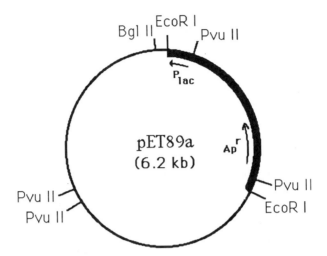

FIGURE 1C. Physical map of pET89a. Restriction endonuclease sites within the 3.5-kb *Eco*RI insert and the pUC19 vector are shown with other designations as in **B**. The 3.5-kb insert is in the opposite orientation in pET89b.

Restriction enzyme digests of pDP51 were probed with [^{32}P]-end labeled BDH-3 oligonucleotide in order to identify fragments for subcloning that contain the NH$_2$-terminal region of the BDH II gene. The hybridization signals indicated that this region was located within the central *Eco*RI fragment of the insert. This fragment was subcloned in both orientations into appropriately cleaved pUC19 to generate pET89a (FIG. 1C) and pET89b (not shown).

The results of assays for NADH-dependent BDH activity were performed on crude extracts of *E. coli* HB101 harboring the various subclones and the vector pUC19 alone, using the method of Welch.[4] The results, presented in TABLE 1, suggest that a clostridial BDH is expressed from plasmids pDP51, pET89a, and pET98b. Furthermore, the fact that pET89a and pET89b contain the same 3.5-kb *Eco*RI insert in the opposite orientation with respect to the *lac* promoter suggests that the BDH gene is transcribed from a promoter of clostridial origin recognized by the *E. coli* RNA polymerase.

Analysis of labeled proteins produced by the maxicell technique showed that pDP51 encodes a polypeptide that corresponds with the molecular sizes of BDH I and BDH II. Inasmuch as these BDH isozymes have highly homologous NH$_2$-termini, it could not immediately be ruled out that the BDH II based oligonucleotide

TABLE 1. Enzyme Activities of pDP51 and Its Derivatives

Plasmid	Insert Size (kb)	Activity (U/mg)[a]
pUC19	0	0.029
pDP51	7.9	0.267
pET89a	3.5	0.255
pET89b	3.5	0.290

[a]U = micromole NADH oxidized per minute. Assays were conducted in *E. coli* HB101.

had hybridized to the gene corresponding to BDH I. Hence, to determine definitively which BDH was encoded within pDP51, the NADH-dependent BDH activity was purified from *E. coli* HB101 (pDP51) using the protocol developed for clostridial extracts.[8] The sequence of the 25 NH$_2$-terminal amino acids of the resulting 42-kD protein was determined and found to be identical to that previously determined for BDH II, confirming that BDH II was encoded within pDP51-1.

SUMMARY

The acetone-butanol fermentation of *C. acetobutylicum* is characterized by the unique shift from acid to solvent production. The mechanism of the solventogenic switch involves the induction of several enzymes, including NADH-dependent butanol dehydrogenase (BDH) at the onset of solventogenesis. This enzyme is responsible for the final conversion of butyraldehyde to butanol, and is distinct from the NADPH-dependent alcohol dehydrogenase (ADH) also present in the organism. To characterize the genetic control of this gene, we have cloned and expressed it in *E. coli*.

A λEMBL3 phage library of *C. acetobutylicum* DNA was screened via plaque hybridization using a [^{32}P]-radiolabeled, 32-fold degenerate, 62-mer oligonucleotide probe. The probe was designed by reverse translation of the NH$_2$-terminal amino acid sequence of purified BDH II. Southern blot experiments indicate that the phage insert was of clostridial origin and had no homology with the previously cloned NADPH-dependent ADH. Subcloning of DNA from purified positive plaques has localized the gene to a 3.5-kb *Eco*RI fragment from which the enzyme is well expressed. The sequence of the 25 NH$_2$-terminal amino acids for the cloned enzyme purified from *E. coli* was determined and found to be identical to that for the clostridial NADH-dependent BDH II. Maxicell analysis of [^{35}S]-radiolabeled plasmid-encoded proteins identified a species encoded by the clostridial insert with the expected M_r of 42 kD.

REFERENCES

1. YOUNGLESON, J. S., J. D. SANTANGELO, D. T. JONES & D. R. WOODS. 1988. Cloning and expression of a *Clostridium acetobutylicum* alcohol dehydrogenase gene in *Escherichia coli*. Appl. Environ. Microbiol. **54:** 676–682.
2. HUESEMANN, M. H. W. & E. T. PAPOUTSAKIS. 1989. Enzymes limiting butanol and acetone formation in continuous and batch fermentations of *Clostridium acetobutylicum*. Appl. Microbiol. Biotechnol. **31:** 435–444.
3. DURRE, P., A. KUHN, M. GOTTWALD & G. GOTTSCHALK. 1987. Enzymatic investigations on butanol dehydrogenase and butyraldehyde dehydrogenase in extracts of *Clostridium acetobutylicum*. Appl. Microbiol. Biotechnol. **26:** 268–272.
4. WELCH, R. W. 1990. Purification and studies of two butanol (ethanol) dehydrogenases and the effects of rifampicin and chloramphenicol on other enzymes important in the production of butyrate and butanol in *Clostridium acetobutylicum* ATCC 824. Ph.D. Thesis. Rice University.
5. JANSSEN, P. J., W. A. JONES, D. T. JONES & D. R. WOODS. 1988. Molecular analysis and regulation of the *glnA* gene of the Gram-positive anaerobe *Clostridium acetobutylicum*. J. Bacteriol. **170:** 400–408.
6. YOUNGLESON, J. S., W. A. JONES, D. T. JONES & D. R. WOODS. 1989. Molecular analysis and nucleotide sequence of the *adh1* gene encoding an NADPH-dependent butanol dehydrogenase in the Gram-positive anaerobe *Clostridium acetobutylicum*. Gene **78:** 355–364.

7. ZAPPE, H., W. A. JONES, D. T. JONES & D. R. WOODS. Structure of an Endo-β-1,4-glucanase gene from *Clostridium acetobutylicum* P262 showing homology with endoglucanase genes from *Bacillus* spp. Appl. Environ. Microbiol. **54:** 1289–1292.
8. WELCH, R. W., F. B. RUDOLPH & E. T. PAPOUTSAKIS. 1989. Purification and characterization of the NADH-dependent butanol dehydrogenase from *Clostridium acetobutylicum* (ATCC) 824. Arch. Biochem. Biophys. **273:** 309–318.

Transformation of a Methylotrophic Bacterium, *Methylobacterium extorquens,* with a Broad-Host-Range Plasmid by Electroporation

SHUNSAKU UEDA,[a] SEIJI MATSUMOTO,
SHOICHI SHIMIZU, AND TSUNEO YAMANE

Department of Food Science and Technology
School of Agriculture
Nagoya University
Chikusa-ku, Nagoya 464-01 Japan

Methylotrophic bacteria that are able to grow on one-carbon compounds such as methanol and monomethylamine as sole carbon sources have received considerable attention as commercial sources such as single cell protein, amino acids, pyrroloqui-noline quinone, and poly-β-hydroxybutyrate. The use of methylotrophs could be enhanced by the development of an applicable recombinant DNA technology. In genetic studies with methylotrophic bacteria as host organisms, a conjugative trans-fer system was used exclusively for introducing DNA into the cells.[1] However, this system was technically cumbersome, time consuming, and limited in the number and type of plasmids that could be used. The technique of electroporation has been used to transfer DNA into mammalian cells,[2] plant protoplasts,[3] and yeast cells.[4] Recently, this technique was used to transform a variety of prokaryotic cells with plasmid DNAs.[5-7] However, this transformation technique has not been reported with methylotrophic bacteria. To develop such a transformation system, we used a pink-pigmented facultative methylotrophic bacterium, *Methylobacterium extorquens,* and a broad-host-range plasmid, pLA2917, as a model system. We observed that up to 8×10^3 transformants per microgram of DNA could be obtained using conditions for electroporation of 10 pulses and 300 μs in duration at a field strength of 10 kV per centimeter.

MATERIALS AND METHODS

Bacterial Strains and Plasmids

The methylotrophic bacterium *M. extorquens,* used in this study, was formerly called *Protaminobacter ruber* and is the same strain as *P. ruber* ATCC8457. Cultures were grown aerobically at 30°C in an inorganic salt medium[8] containing 1% (v/v) methanol (MIS medium). *Escherichia coli* strain HB101[9] was grown with shaking at 37°C in an LB medium.[10] Solid media were prepared by supplementing the liquid media with 1.5% (w/v) agar. When kanamycin (Km) was used, it was added to the

[a]Address for correspondence: Dr. Shunsaku Ueda, Department of Food Science and Technology, School of Agriculture, Nagoya University, Chikusa-ku, Nagoya 464-01, Japan.

media to give a final concentration of 50 μg/ml. The broad-host-range plasmid pLA2917 (21 kb)[11] was maintained in *E. coli* HB101.

DNA Manipulations

Plasmid DNA was prepared from *E. coli* as described.[12] Plasmid DNA was isolated from *M. extorquens* using the method of Kim and Lidstrom.[13] Plasmid preparations were treated with ribonuclease A (EC3.1.27.5, Boehringer Mannheim Biochemicals, Indianapolis, Indiana). Restriction endonucleases (Boehringer Mannheim Biochemicals) were used as recommended by suppliers. Plasmids and restricted DNA fragments were analyzed by 0.5% (w/v) agarose gel electrophoresis. DNA fragments were recovered from agarose gel using DEAE-cellulose membrane (NA45, Schleicher & Schuell Inc., Keene, New Hampshire). Southern hybridization was carried out as described by Southern.[14] Analysis of DNA transferred onto nitrocellulose membrane was performed using an ECL gene detection system (Amersham International, Amersham, United Kingdom).

Electroporation Apparatus

Electroporation experiments were performed using a somatic hybridizer, model SSH-2 (Shimadzu Co., Kyoto, Japan). Square-shaped pulses up to 700 V are delivered by SSH-2. Maximum time duration to be set was optionally modified from 100 to 500 μs. A type SSH-C11 electroporation chamber was used with parallel electrodes of 0.5-mm spacing. The maximum field strength of 14 kV/cm is available with this chamber.

Transformation of M. extorquens by Electroporation

All operations were carried out at 4°C unless otherwise specified. *M. extorquens* was grown in MIS medium to the middle-logarithmic phase (1.4×10^9/ml) and harvested by centrifugation at $6,000 \times g$ for 10 minutes. The cells were washed with an electroporation buffer (10 mM Tris-HCl; 2 mM $MgCl_2 \cdot 6H_2O$; 10% [w/v] sucrose, pH 7.5) and resuspended in the same buffer at a cell concentration of 7.0×10^{10}/ml. The cell suspension and the solution of pLA2917 (70 μg/ml) were mixed at a ratio of 9:1 (v/v) in an Eppendorf tube. The mixture (10 μl) was then transferred into a space between the electrodes of the chamber and held there for 3 minutes. After the electric pulse, a 5-μl aliquot of the mixture was transferred to an Eppendorf tube containing 200 μl of the MIS medium, and the cell suspension was incubated for 2 hours at 30°C to allow expression of the antibiotic-resistant genes of pLA2917 prior to plating. Km-resistant transformants of *M. extorquens* were scored on the selective plates after incubating for 4–5 days at 30°C.

RESULTS

Transformation of M. extorquens with pLA2917 by Electroporation

Inasmuch as plasmid pLA2917 replicates in the cells of *M. extorquens*, we examined the transformation of this bacterium with pLA2917 by electroporation.

Information is lacking on the electric conditions required for electroporation-mediated transformation with methylotrophs including *M. extorquens*. Therefore, we chose the following electric conditions: field strength, 14 kV/cm; pulse duration, 300 μs; number of pulses, 10 times. Under these conditions kanamycin-resistant (Km^r) transformants of *M. extorquens* appeared at a frequency of 3.0×10^2/μg DNA. These transformants contained plasmid DNA with the same molecular size as that of pLA2917. Furthermore, when a 3.4-kb *Hind*III-*Sal*I fragment of pLA2917 was used as a probe in Southern blot analysis, the plasmid DNA isolated from Km^r *M. extorquens* strongly hybridized with the probe (FIG. 1). This result indicates that pLA2917 was introduced into the cells of *M. extorquens* by electroporation.

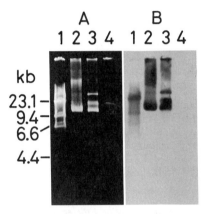

FIGURE 1. Agarose gel electrophoresis (**A**) and Southern blot hybridization analysis (**B**) of a plasmid DNA from Km^r-transformant of *M. extorquens*. *Lane 1, Hind*III-digested lambda DNA molecular weight marker (kb); *lane 2,* pLA2917 from *E. coli* HB101; *lane 3,* cleared lysate from a Km^r-transformant of *M. extorquens; lane 4,* cleared lysate from *M. extorquens*. The probes used were lambda DNA digested with *Hind*III for *lane 1* and the 3.4-kb *Hind*III-*Sal*I fragment of pLA2917 for *lanes 2 to 4,* respectively.

Effects of the Electric Conditions on Transformation Efficiency

A high efficiency of transformation by electroporation was generally obtained with the electric conditions that kill 50–70% of the viable cells.[15] At a fixed duration of 300 μs and 10 pulses, a field strength of 10 kV/cm gave a 50% survival rate of *M. extorquens* (FIG. 2). FIGURE 2 also shows the effect of field strength on transformation efficiency. The field strength was changed from 0 to 14 kV/cm at a duration of 300 μs and 10 pulses. No transformant was obtained at a field strength below 5 kV/cm, but transformation efficiency increased from 1.0×10^2 to 2.8×10^3/μg DNA when the field strength was increased from 6 to 10 kV/cm. At a field strength of 14 kV/cm, efficiency decreased to 36% of the maximum value ($\simeq 1.0 \times 10^3$/μg DNA). These results indicate that the electric field strength was effective in the electric breakdown of the cell membrane of *M. extorquens* and influences the transformation efficiency. To examine the effect of pulse duration on transformation efficiency, pulse durations of 100–500 μs were used at a fixed field strength of 10 kV/cm and 10 pulses. As shown in FIGURE 3, the transformation efficiency increased with increasing pulse duration from 100–300 μs. Maximum efficiency of 8×10^3/μg DNA was

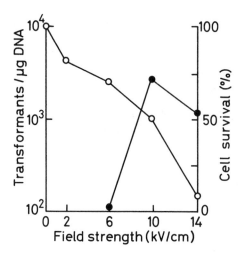

FIGURE 2. Effect of field strength on cell survival of *M. extorquens* (○) and transformation efficiency (●). Duration and number of pulses were 300 μs and 10 times, respectively.

obtained at a pulse duration of 300 μs. The effect of the number of electric field pulses on the transformation efficiency was investigated at a field strength of 10 kV/cm and a pulse duration of 300 μs. As shown in FIGURE 4, the transformation efficiency increased as the number of pulses increased to 6. However, increasing the number of pulses beyond 6 did not increase the transformation efficiency. The maximum efficiency of 1.7×10^3/μg DNA was obtained at 10-times pulsation.

Dependency of the Transformation Efficiency on DNA Concentration

FIGURE 5 shows the relation between the DNA concentration used and the number of transformants obtained. The yield of transformants was proportional to the concentration of plasmid DNA over the range employed (7 ng/ml to 7 μg/ml). Conversely, transformation efficiency decreased by increasing the amount of DNA,

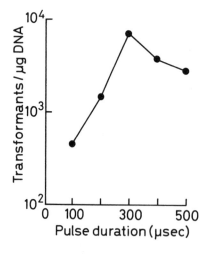

FIGURE 3. Effect of pulse duration on transformation efficiency. Field strength and number of pulses were 10 kV/cm and 10 times, respectively.

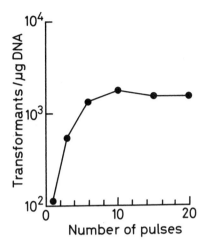

FIGURE 4. Effect of number of pulses on transformation efficiency. Field strength and duration were 10 kV/cm and 300 μs, respectively.

indicating that plasmid DNA was present in saturating amounts for the cell density used.

DISCUSSION

This work describes the use of electric pulse-mediated transformation for the pink-pigmented methylotrophic bacterium *M. extorquens*. Although the transfer of plasmids into a variety of bacteria has been carried out by conjugation or transformation, most methylotrophic strains do not possess a natural transformation system and have not been reported to be transformable by artificial methods, with a few exceptions.[16,17] In these studies, not only was the transformation frequency very low, but also the results could not be reproduced. Therefore, plasmid DNA has been introduced into methylotrophs by conjugation using the broad-host-range plasmids of the *E. coli* incompatibility groups of IncP and IncQ.[11,18] However, this technique is

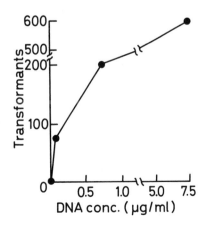

FIGURE 5. Effect of DNA concentration on transformation of *M. extorquens*. Field strength, duration, and number of pulses were 10 kV/cm, 300 μs, and 10 times, respectively.

limited to special plasmids that can provide conjugation or mobilization functions. Electroporation is another way to transfer plasmids into methylotrophs as well as several genera of bacteria.[19] We used electroporation to transform the methylotrophic bacterium *M. extorquens* with the 21-kb plasmid pLA2917. Transformation efficiency was affected by the electric conditions of field strength, pulse duration, and pulse number. We obtained the transformation efficiencies from 2×10^3 to $8 \times 10^3/\mu g$ of pLA2917 DNA with 10 pulses at a field strength of 10 kV/cm and a pulse duration of 300 μs. Two types of pulse shapes, exponential or square-wave form, generated by electroporation apparatuses are available. With either type, wide-ranging transformation efficiencies (10^1–$10^6/\mu g$ DNA) have been reported with different bacterial strains and different plasmids.[19-21] These results indicate that the optimal conditions may depend not only on the bacterial strain but also on the plasmid used. As a result, conditions should be determined carefully in each case. The electroporation apparatus (SSH-2) used in this study generates a pulse of square wave. Although it is difficult to compare the transformation efficiencies obtained by electroporation, the transformation frequency of 2–8 $\times 10^3/\mu g$ of pLA2917 DNA for *M. extorquens* found in this study should provide for a practical use in its genetic manipulation. Unfortunately, it was not possible to examine the effect of plasmid size on transformation efficiency because of the absence of other plasmids that are able to replicate in *M. extorquens*. When smaller plasmids become available to use with *M. extorquens*, we expect a higher transformation frequency by electroporation than that found with the 21-kb plasmid pLA2917.

SUMMARY

Electroporation was used to transform the methylotrophic bacterium *Methylobacterium extorquens* with broad-host-range plasmid pLA2917, which contains a gene specifying resistance to kanamycin. Plasmid DNA was introduced into *M. extorquens* in the presence of an electric pulse, and kanamycin-resistant transformants were obtained. These transformants harbored plasmid DNA that was identical to plasmid pLA2917. We examined several factors independently and found up to 8 $\times 10^3$ transformants per microgram of DNA using 10 pulses with a duration of 300 μs at a field strength of 10 kV/cm.

REFERENCES

1. DE VRIES, G. E., U. KUES & U. STAHL. 1990. Physiology and genetics of methylotrophic bacteria. FEMS Microbiol. Rev. **75:** 57–102.
2. CHU, G., H. HAYAKAWA & P. BERG. 1987. Electroporation for the efficient transfection of mammalian cells with DNA. Nucleic Acids Res. **15:** 1311–1326.
3. FROMM, M. E., L. P. TAYLOR & V. WALBOT. 1986. Stable transformation of maize after gene transfer by electroporation. Nature (Lond.) **319:** 791–793.
4. HASHIMOTO, H., H. MORIKAWA, Y. YAMADA & A. KIMURA. 1985. A novel method for transformation of intact yeast cells by electroporation of plasmid DNA. Appl. Microbiol. Biotechnol. **21:** 336–339.
5. CRAIG, F. F., J. G. COOTE, R. PARTON, J. H. FREER & N. J. L. GILMOUR. 1989. A plasmid which can be transferred between *Escherichia coli* and *Pasteurella haemolytica* by electroporation and conjugation. J. Gen. Microbiol. **135:** 2885–2890.
6. FIEDLER, S. & R. WIRTH. 1988. Transformation of bacteria with plasmid DNA by electroporation. Analyt. Biochem. **170:** 38–44.

7. POTTER, H. 1988. Electroporation in biology: Methods, applications, and instrumentation. Analyt. Biochem. **174**: 361–373.
8. SATO, K., S. UEDA & S. SHIMIZU. 1977. Form of vitamin B_{12} and its role in a methanol-utilizing bacterium, *Protaminobacter ruber.* Appl. Envir. Microbiol. **33**: 515–521.
9. BOYER, H. W. & D. ROULLAND-DUSSOIX. 1969. A complementation analysis of the restriction and modification of DNA in *Escherichia coli.* J. Molec. Biol. **41**: 459–472.
10. MILLER, J. H. 1972. Experiments in Molecular Genetics. Cold Spring Harbor Laboratory. Cold Spring Harbor, NY.
11. ALLEN, L. N. & R. S. HANSON. 1985. Construction of broad-host-range cosmid cloning vectors: Identification of genes necessary for growth of *Methylobacterium organophilum* on methanol. J. Bacteriol. **161**: 955–962.
12. BIRNBOIM, H. C. & J. DOLY. 1979. A rapid alkaline extraction procedure for screening recombinant plasmid DNA. Nucleic Acids Res. **7**: 1513–1523.
13. KIM, Y. M. & M. E. LIDSTROM. 1989. Plasmid analysis in pink facultative methylotrophic bacteria using a modified acetone-alkaline hydrolysis method. FEMS Microbiol. Lett. **60**: 125–130.
14. SOUTHERN, E. M. 1975. Detection of specific sequences among DNA fragments separated by gel electrophoresis. J. Molec. Biol. **98**: 503–517.
15. SAMBROOK, J., E. F. FRITSCH & T. MANIATIS. 1989. Molecular cloning: A Laboratory Manual, 2nd ed,: 75. Cold Spring Harbor Laboratory. Cold Spring Harbor, NY.
16. HABER, C. L., L. N. ALLEN, S. ZHAO & R. S. HANSON. 1983. Methylotrophic bacteria: Biochemical diversity and genetics. Science **221**: 1147–1153.
17. SPENCE, D. W. & G. C. BARR. 1981. A method for transformation of *Paracoccus denitrificans.* FEMS Microbiol. Lett. **12**: 159–161.
18. WINDASS, J. D., M. J. WORSEY, E. M. PIOLI, D. PIOLI, P. T. BARTH, K. T. ATHERTON, E. C. DART, D. BYROM, K. POWELL & P. J. SENIOR. 1980. Improved conversion of methanol to single-cell protein by *Methylophilus methylotrophus. Nature 287:* 396–401.
19. WIRTH, R., A. FRIESENEGGER & S. FIEDLER. 1989. Transformation of various species of gram-negative bacteria belonging to 11 different genera by electroporation. Molec. Gen. Genet. **216**: 175–177.
20. KUSAOKE, H., Y. HAYASHI, Y. KADOWAKI & H. KIMOTO. 1989. Optimum conditions for electric pulse-mediated gene transfer to *Bacillus subtilis* cells. Agric. Biological Chem. **59**: 2441–2446.
21. LUNCHANSKY, J. B., P. M. MURIANA & T. R. KLAENHAMMER. 1988. Application of electroporation for transfer of plasmid DNA to *Lactobacillus, Lactococcus, Leuconostoc, Listeria, Pediococcus, Bacillus, Staphylococcus, Enterococcus* and *Propionibacterium.* Molec. Microbiol. **2**: 637–646.

Cloning and Expression of an HIV-1 Specific Single-Chain F_v Region Fused to *Escherichia coli* Alkaline Phosphatase[a]

JOHANN KOHL, FLORIAN RÜKER,[b]
GOTTFRIED HIMMLER, EBRAHIM RAZAZZI,
AND HERMANN KATINGER

Institut für Angewandte Mikrobiologie
Universität für Bodenkultur
Nußdorfer Lände 11
A-1190 Vienna, Austria

The variable region of an antibody is that part of the molecule that is responsible and sufficient for antigen binding. Although Fab' and (Fab')$_2$ fragments can be produced from whole antibodies by enzymatic cleavage using either papain or pepsin, no proteolytic enzyme is available that can be used to readily produce F_v fragments. There are only a few reports that describe the successful enzymatic production of F_v fragments.[1,2] It was shown recently at different laboratories that genes coding for the Fab' fragment[3] or just the variable regions of an antibody can be successfully expressed in *Escherichia coli* to yield antigen-binding proteins with binding characteristics similar or identical to those of the original antibody. In the expression of variable regions alone, two approaches have been followed, namely, to covalently link[4,5] or not to link[6] the two variable domains. Plückthun and colleagues[7] have expressed identical antibody F_v regions in *E. coli* in different ways, either as a monomer by using a linker or unfused as a heterodimer. It was reported that no substantial differences in the behavior of the two recombinant antigen-binding proteins could be observed. The packing of the two variable domains by hydrophobic interaction between amino acids exposed at the surface of the molecules seems to be sufficiently strong to allow the formation of a biologically active heterodimer in *E. coli*. However, it can be argued that an equimolar expression rate and the correct assembly might be favored when the two domains are linked to each other by means of a synthetic linker peptide region engineered in between the two genes which allows the expression of the F_v fragment as a single polypeptide chain.

The small size of F_v fragments and the absence of the constant regions of the antibody suggest many potential advantages, such as better uptake in different organs, faster clearance from serum, and less background in imaging applications. Immunogenicity should also be reduced. Practical applications such as the expression in *E. coli* of engineered F_v fragments fused to toxin molecules to yield immunotoxins have already been reported.[8]

We have engineered the genes of human monoclonal antibody 3D6 which binds specifically to HIV-1 gp41[9-12] in order to produce a single-chain F_v fragment. We have expressed this engineered gene as a fusion protein together with *E. coli* alkaline

[a]The work presented here was supported in part by the Fonds zur Förderung der wissenschaftlichen Forschung, Project Nos. P6540 and P7556.
[b]Corresponding author.

phosphatase (EcPhoA)[13,14] and purified it by affinity chromatography using a synthetic peptide representing the epitope of antibody 3D6. The epitope has been mapped before.[11] We could demonstrate that this protein, called APsc3D6, is bifunctional, showing both phosphatase activity and specific antigen binding.

EXPERIMENT

The construction of a cDNA library from the hybridoma cell line 3D6 and the isolation of full-length clones coding for the heavy and light chain of the anti-HIV-1 gp41 antibody expressed by this cell line are described elsewhere (Rüker *et al.*, this volume, and ref. 15).

Restriction sites were introduced in the cDNAs by oligonucleotide-directed site-specific mutagenesis (*in vitro* mutagenesis system, Amersham, UK) at the 5' and 3' ends of the two variable regions as well as a start codon at the 5' end of the heavy chain variable region and a stop codon at the 3' end of the light chain variable region. See FIGURE 1 for details. The domains coding for the variable regions of the antibody were then isolated using the introduced restriction sites and were ligated in the presence of a synthetic double-stranded oligonucleotide coding for the linker peptide sequence (FIG. 2). The amino acid sequence of the linker peptide was taken from ref. 4. After this engineered gene was cloned and sequenced, the sequence was corrected by site-directed mutagenesis at the boundaries between the variable regions and the linker, to give the correct amino acid residues as indicated in FIGURE 2. This final single-chain F$_v$ fragment gene was again verified by sequencing.

We previously engineered the gene coding for *E. coli* alkaline phosphatase (EcPhoA) in order to facilitate the construction of translational fusions. The mutations that were introduced into the EcPhoA gene have been described in detail.[14] The vector pAPMUT2 contains the EcPhoA gene with an engineered *Eco*RV site 5' of the stop codon. Into this *Eco*RV site, the single-chain F$_v$ fragment gene was introduced in the correct reading frame, as shown in FIGURE 3. In this process, three additional codons (Leu, Glu, Ser) stemming from the polylinker of pUC19 were introduced between the two coding regions. The resulting fusion gene was then cloned in the prokaryotic expression vector pKK223-3[16] and transformed into *E. coli* JM105. The structure of the final expression vector called pJK5APsc is shown schematically in FIGURE 4.

Bacteria were grown in LB medium to a density of 1 OD$_{600}$ in the presence of ampicillin. IPTG (1 mM) was then added for the induction of the *tac* promoter, and the cells were incubated at 30°C for 4 hours. The APsc3D6 fusion protein that had accumulated in the periplasmic space of the bacteria was released by osmotic shock, according to ref. 17.

Proteins in the periplasmic extract were separated by hydrophobic interaction chromatography on phenylsepharose fast flow (Pharmacia, Sweden). Elution was performed with a linear ammonium sulfate gradient (60–0%). Fractions were collected, assayed for phosphatase activity, and positive fractions were pooled. The samples were subsequently desalted by gel filtration on Sephadex G25 (Pharmacia) and resuspended in equilibration buffer (20 mM Tris 2, pH 7.4/650 mM NaCl) for the affinity chromatography step. Affinity chromatography was performed by passing the samples over a column containing immobilized HIV-1 peptide.[11] After washing, samples were eluted with a 3 M solution of KSCN.

Samples of the periplasmic extract and of the purified material after the affinity chromatography step were run on SDS-PAGE gradient gels (8–25%, Pharmacia Phast system) in the absence of mercaptoethanol. Two identical gels were run; one

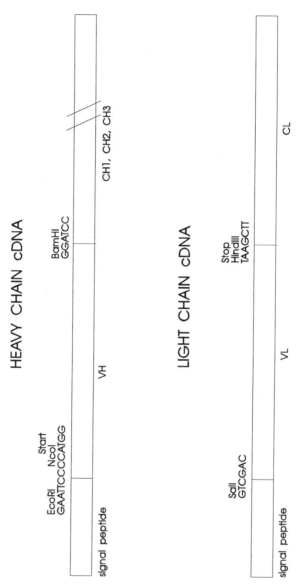

FIGURE 1. Mutations at the borders of the variable region of heavy and light chains, respectively.

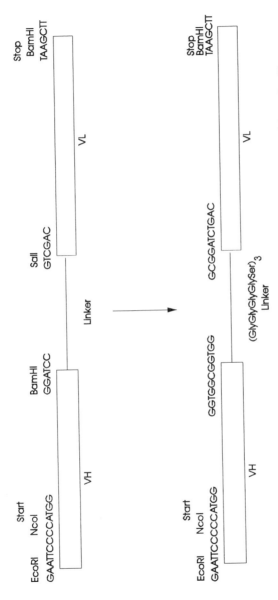

FIGURE 2. Assembly of the variable regions of the heavy and light chains, followed by correction of the nucleotide sequence.

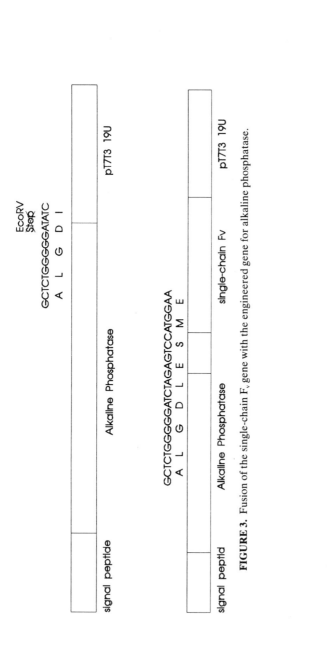

FIGURE 3. Fusion of the single-chain F$_v$ gene with the engineered gene for alkaline phosphatase.

FIGURE 4. Structure of the expression vector pJK5APsc.

was developed by silver staining, and the other was incubated with nitroblue-tetrazolium in diethanolamine buffer, pH 9.6, in order to detect phosphatase activity.

To check for specific antigen binding of the APsc3D6 protein, the affinity-purified sample was assayed on Western blots of HIV-1 proteins (Novopath immuno-blot assay, BioRad) and developed with *p*-nitrophenyl-phosphate. As a positive control, 3D6 antibody produced by the original hybridoma cell culture was applied to the test strips and then incubated with secondary anti-human-IgG antibody-alkaline phosphatase conjugate. Phosphatase activity was visualized with *p*-nitrophenyl-phosphate. As a negative control, protein from the periplasmic extract of *E. coli* transfected with a vector containing the unfused gene for alkaline phosphatase was used.

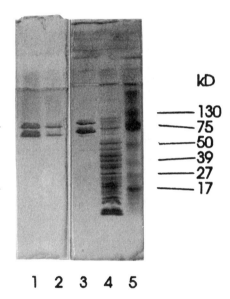

FIGURE 5. Purification of the APsc3D6 protein from *E. coli* JM105 harboring pJK5APsc. Two identical SDS-gradient gels (8–25%, Pharmacia Phast system) are shown, one of which is developed with nitroblue-tetra-zolium in order to detect phosphatase activity (*lanes 1 and 2*); the other one is developed by silver staining (*lanes 3, 4, and 5*). *Lanes 1 and 3,* affinity-purified APsc3D6 protein; *lanes 2 and 4,* periplasmic fraction; *lane 5,* protein size markers (BioRad).

RESULTS

The cloning scheme just shown has allowed us to construct a semisynthetic gene coding for a single-chain F$_v$ fragment derived from the human monoclonal antibody 3D6. In-frame fusion with an engineered gene for *E. coli* alkaline phosphatase has yielded a gene coding for the active fusion protein APsc3D6.

Due to the signal peptide encoded by the phosphatase gene, the recombinant fusion protein was directed to the periplasmic space of *E. coli,* allowing purification of the protein from the periplasmic extract of osmotically shocked bacteria.

Results of the purification can be seen in FIGURE 5. The expected molecular weight of the fusion protein is 75.5 kD, corresponding to the lower band that exhibits phosphatase activity. Because of the absence of β-mercaptoethanol during electrophoresis, alkaline phosphatase could reconstitute dimers in order to be active. The upper band that can be seen on the gel might represent a dimerized form of the fusion protein.

The purified preparation of the phosphatase—single-chain fusion protein was tested by HIV-1 specific Novopath immunoblot assay (BioRad). Specific binding activity identical to that of the original antibody was observed and is shown in FIGURE 6, where the main reacting bands of HIV-1, gp41 and gp160, as well as a cross-reaction with gp120 can be seen.

DISCUSSION

We have produced an active fusion protein consisting of the single-chain F_v fragment of a human monoclonal anti-HIV-1 antibody and *E. coli* alkaline phosphatase. The construction principle of the F_v fragment was based on data published for a mouse monoclonal antibody. This shows that the underlying principle is a general one and is probably applicable to any type of antibody.

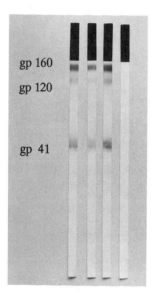

gp 160

gp 120

gp 41

FIGURE 6. Testing specific binding activity of the APsc3D6 protein by Novopath immunoblot assay (BioRad). From left to right: *lane 1,* affinity purified APsc3D6 protein; *lane 2,* periplasmic fraction containing APsc3D6 protein; *lane 3,* human monoclonal antibody 3D6; *lane 4,* negative control, periplasmic extract of *E. coli* harboring a plasmid only expressing *E. coli* alkaline phosphatase.

Expression of the fusion protein with the signal peptide of alkaline phosphatase at the 5′ end allowed us to purify active protein from the periplasmic extract of the bacteria. For purification, we used hydrophobic interaction chromatography followed by gel filtration and affinity chromatography. As a ligand, we used a synthetic peptide representing the epitope recognized by the antibody 3D6.[11]

The fusion protein shown here has proved to be a valuable tool in the diagnostics of AIDS, eliminating the need for using a secondary antibody in immunoassays and Western blots. Other applications, such as its use as an affinity ligand for purification of HIV-1 antigens, are presently being investigated.

SUMMARY

We have constructed a single-chain F_v fragment representing the variable domain of the human monoclonal antibody 3D6, binding specifically to HIV-1 gp41. This gene was fused to the coding region of *E. coli* alkaline phosphatase (EcPhoA) and expressed in *E. coli*. The EcPhoA signal peptide was used to direct the recombinant fusion protein to the periplasmic space of the bacteria, from where it was purified by hydrophobic interaction chromatography and gel filtration followed by antigen-affinity chromatography using a synthetic HIV-1 peptide as ligand. The purified fusion protein was bifunctional, showing both phosphatase activity as well as antigen-binding specificity identical to that of the original antibody.

ACKNOWLEDGMENTS

We thank Heidi Riegler for excellent technical help, Helen Shuttleworth for the vector pAP85 containing the EcPhoA gene, and Immuno AG for kindly providing the peptide column.

REFERENCES

1. INBAR, D., J. HOCHMANN & D. GIVOL. 1972. Localisation of antibody combining sites within the variable portions of heavy and light chains. Proc. Natl. Acad. Sci. USA **69:** 2659–2662.
2. HOCHMAN, J., D. INBAR & D. GIVOL. 1973. An active antibody fragment (F$_v$) composed of the variable portions of heavy and light chains. Biochemistry **12:** 1130–1135.
3. BETTER, M., P. CHANG, R. R. ROBINSON & A. H. HORWITZ. 1988. *Escherichia coli* secretion of an active chimeric antibody fragment. Science **240:** 1041–1043.
4. HUSTON, J. S., D. LEVINSON, M. MUDGETT-HUNTER, M.-S. TAI, J. NOVOTNY, M. N. MARGOLIES, R. J. RIDGE, R. E. BRUCCOLERI, E. HABER, R. CREA & H. OPPERMANN. 1988. Protein engineering of antibody binding sites: Recovery of specific activity in an anti-digoxin single-chain F$_v$ analogue in *Escherichia coli*. Proc. Natl. Acad. Sci. USA **85:** 5879–5883.
5. BIRD, R. E., K. D. HARDMAN, J. W. JACOBSON, S. JOHNSON, B. M. KAUFMAN, S.-M. LEE, T. LEE, S. H. POPE, G. S. RIORDAN & M. WHITLOW. 1988. Single-chain antigen-binding proteins. Science **242:** 423–426.
6. SKERRA, A. & A. PLÜCKTHUN. 1988. Assembly of a functional immunoglobulin Fv fragment in *Escherichia coli*. Science **240:** 1038–1041.
7. GLOCKSHUBER, R., M. MALIA, I. PFITZINGER & A. PLÜCKTHUN. 1990. A comparison of strategies to stabilize immunoglobulin F$_v$-fragments. Biochemistry **29:** 1362–1367.
8. CHAUDHARY, V. K., C. QUEEN, R. P. JUNGHANS, T. A. WALDMANN, D. J. FITZGERALD & I. PASTAN. 1989. A recombinant immunotoxin consisting of two antibody variable domains fused to Pseudomonas exotoxin. Nature (Lond.) **339:** 394–397.
9. GRUNOW, R., S. JAHN, T. PORSTMANN, S. T. KIESSIG, H. STEINKELLNER, F. STEINDL, D. MATTANOVICH, L. GÜRTLER, F. DEINHARDT, H. KATINGER & R. VON BAEHR. 1988. The high efficiency, human B cell immortalizing heteromyeloma CB-F7. J. Immunol. Methods **106:** 257.
10. DÖPEL, S. H., T. PORSTMANN, R. GRUNOW, A. JUNGBAUER & R. V. BAEHR. 1989. Application of a human monoclonal antibody in a rapid competitive anti-HIV ELISA. J. Immunol. Methods **116:** 229–233.
11. DÖPEL, S.-H., T. PORSTMANN, P. HENKLEIN & R. VON BAEHR. 1989. Fine mapping of an immunodominant region of the transmembrane protein of the human immunodeficiency virus (HIV-1) by different peptides and their use in anti-HIV ELISA. J. Virol. Methods **25:** 167–178.
12. JUNGBAUER, A., C. TAUER, E. WENISCH, F. STEINDL, M. PURTSCHER, M. REITER, F. UNTERLUGGAUER, A. BUCHACHER, K. UHL & H. KATINGER. 1989. Pilot scale production of a human monoclonal antibody against human immunodeficiency virus HIV-1. J. Biochem. Biophys. Methods **19:** 223–240.

13. SHUTTLEWORTH, H., J. TAYLOR & N. MINTON. 1986. Sequence of the gene for alkaline phosphatase from *Escherichia coli* JM83. Nucl. Acids Res. **14:** 8689.
14. KOHL, J., F. RÜKER, G. HIMMLER, D. MATTANOVICH & H. KATINGER. 1990. Engineered gene for *Escherichia coli* alkaline phosphatase for the construction of translational fusions. Nucl. Acids Res. **18:** 1069.
15. FELGENHAUER, M., J. KOHL & F. RÜKER. 1990. Nucleotide sequences of the cDNAs encoding the V-regions of H- and L-chains of a human monoclonal antibody specific to HIV-1-gp41. Nucl. Acids Res. **18:** 4927.
16. BROSIUS, J. & A. HOLY. 1984. Regulation of ribosomal RNA promoters with a synthetic *lac* operator. Proc. Natl. Acad. Sci. USA **81:** 6929–6933.
17. NOSSAL, N. G. & L. A. HEPPEL. 1966. The release of enzymes by osmotic shock from *Escherichia coli* in exponential phase. J. Biol. Chem. **241:** 3055–3062.

Comparison of the F_v Fragments of Different Phosphorylcholine Binding Antibodies Expressed in *Escherichia coli*

ANDREAS PLÜCKTHUN AND ILSE PFITZINGER

Genzentrum der Universität München
c/o Max-Planck-Institut für Biochemie
Am Klopferspitz
D-8033 Martinsried, Germany

The production of antigen-binding fragments of antibodies in *Escherichia coli* greatly facilitates further research on antibodies, their engineering, and numerous applications in biotechnology and medicine. We previously developed a system with which fully functional F_v fragments (the heterodimer of the V_H and V_L domain) and F_{ab} fragments (the heterodimer of the whole light chain $[V_L C_L]$ and the first two domains of the heavy chain $[V_H C_H]$) can be expressed in *E. coli*.[1,2] The purification can be carried out in a single step by hapten affinity chromatography.

Our strategy consists of the simultaneous expression and secretion of both chains of the antibody fragment into the periplasm of *E. coli*. Considerations in the development of this approach were that both chains must fold simultaneously and in each other's presence, and that the folding and assembly must take place in the oxidizing environment of the periplasm. The cytoplasm, where protein synthesis occurs, has a reducing environment, and we[3,4] have shown that neither the F_v nor the F_{ab} fragment can be obtained in functional form without the correct formation of the disulfide bonds in the variable domains. Thus, the transport of the two chains to the periplasm is a prerequisite for obtaining fully functional antigen-binding fragments.

Previous experiments were carried out with the antibody McPC603 (or M603), an IgA of the mouse that binds phosphorylcholine. Its particularly attractive features include the known three-dimensional structure[5] and substantial previous work on its binding properties.[6-10] In this article, we describe experiments comparing this antibody with the related antibody TEPC15 (or T15), both expressed as F_v fragments in *E. coli*. These antibodies, together with MOPC167 (or M167), are the three "prototypes" of the murine immune response after challenge with phosphorylcholine. A mutational analysis of M167 will be described elsewhere (Schweder and Plückthun, in preparation).

The F_v fragment is the smallest conceivable fragment that has the complete binding pocket of an antibody. We previously showed that the intrinsic binding constant of the F_v fragment is almost identical to that of the whole antibody.[1,11] Therefore, the presence of the constant domains has no influence on the structure of the antigen-binding site. However, the association energy between the V_L and the V_H domain is not very high, and the F_v fragment dissociates into these domains at high dilution.[11] This leads to complicated binding behavior, which can be deconvoluted into an intrinsic binding constant *identical* to the whole antibody and a finite association constant of the two domains.[11] The association constant between V_L and V_H depends on the precise structure of the antibody and therefore indirectly on its specificity, because the hypervariable loops contribute several important interactions

115

for the association of the two domains to form the F_v fragment. Furthermore, the antigen itself contributes to the stabilization of the F_v fragment.[11]

This dissociation of the F_v fragment into V_H and V_L can be prevented by either of three strategies[11]: first, chemical cross-linking of the two domains; second, the introduction of disulfide bonds; or third, the construction of a peptide linker to give a so-called single-chain F_v fragment.[11-13] We could show that such a single-chain F_v fragment can still be transported and that it folds correctly inside the periplasm.[11] It can also be purified by hapten affinity chromatography, and the measured hapten binding affinity is very similar to that of the whole antibody. Interestingly, however, the amount of functional fragment is not increased by the peptide linker. Thus, the mutual finding and association of the two domains in the periplasm of *E. coli* do not seem to constitute a problem.

Another strategy to stabilize the heterodimer is to express the F_{ab} fragment,[2,4,14,15] which contains constant domains C_L and C_H1 in addition to the variable domains. The recombinant F_{ab} fragment gave exactly the same binding constant as did the whole antibody or the F_{ab} fragment obtained by proteolysis of the whole antibody.[15] We also found that the proteolytic F_{ab} fragment differs from the recombinant one by glycosylation in the C_H1 domain.[15] This identity of binding constants between the recombinant and the proteolytic material clearly demonstrates that this glycosylation has no influence on binding. Because the hapten binding constants of the F_{ab} fragment and the whole antibody are also identical, the presence of the constant domains C_H2 and C_H3 (and any glycosylation therein[16]) has no influence on antigen binding either.

We did observe, however, that the yield of functional F_{ab} fragment is consistently smaller than that of the functional F_v fragment expressed in *E. coli*. A detailed analysis of this phenomenon[4] showed that the problem is not one of expression or secretion, but one of folding and assembly.

Other investigators have attempted to cut down the size of the F_v fragment even further by using only V_H as a binding domain.[17] Our investigations of single domains have shown, however, that the V_H domain is difficult to handle because of limited solubility (Glockshuber and Plückthun, unpublished experiments). The V_L domain, on the other hand, is very soluble and has a tendency to dimerize. We have not obtained any evidence for binding the phosphorylcholine antigen by either domain alone. The recombinant V_L domain of M603 produced in *E. coli* could recently be crystallized, and its crystal structure was solved to a resolution of 2.0 Å. It has now been refined to an R factor of 14%.[18,19]

We have now extended this expression strategy to the family of the other phosphorylcholine binding antibodies and corresponding mutants. This allows the use of protein-engineering experiments combined with structural and mechanistic studies to compare their binding, stability, folding, and catalytic properties. We report in this paper the gene synthesis and expression of the F_v fragment of the antibody T15 and its comparison with M603. The detailed examination of well-characterized antibodies is the prerequisite for the development of efficient screening and selection procedures in *E. coli*, and the family of phosphorylcholine binding antibodies is well suited for such experiments.

MATERIALS AND METHODS

Recombinant DNA Techniques and Protein Expression

Recombinant DNA techniques were based on those of Maniatis *et al.*[20] The antibody fragments were expressed in *E. coli* JM83, using a vector[21] with an f1 phage origin[22] and a resident repressor gene, a transcriptional terminator, which can therefore be used in many strains. In some cases, the *omp*T⁻ strain UT4400 (ref. 23) was also used. Site-directed mutagenesis was carried out according to Kunkel *et al.*[24] and Geisselsoder *et al.*[25]

Gene Synthesis

The oligonucleotides were synthesized with an Applied Biosystems Model 380A synthesizer. They were then purified by polyacrylamide gel electrophoresis, phosphorylated with polynucleotide kinase, hybridized, and ligated with T4 ligase using the same methodology as that in the synthesis of the genes of M603.[14]

The operon of the T15 antibody[26] was assembled as follows. A mutagenesis with four oligonucleotides of the expression plasmid encoding M603 (ref. 21) was carried out. Two of them introduced the required mutations in V_H, one introduced a silent change in the signal of V_L, creating a unique *Sty*I site in the plasmid, and one converted the COOH-terminal sequence of the V_L gene from M603 to T15. The piece between the *Sty*I and the *Bsp*MII site was assembled from six oligonucleotides (three for each strand) and first subcloned into pUC19 by having extended the synthetic region to make it compatible with polylinker cloning sites. From this pUC19 derivative, the relevant piece was obtained as a *Sty*I-*Bsp*MII fragment and ligated to the previously mutagenized expression plasmid, cut with the same enzymes. The complete sequence of the operon was verified in the expression plasmid by DNA sequencing (FIG. 1).

Protein Purification

Both recombinant antibody fragments were purified by phosphorylcholine affinity chromatography essentially as described previously[1] except that the bacterial growth was performed at 20°C and the cells were induced for 3 hours before the harvest.[11] The cells were then disrupted in a French pressure cell, and the soluble part of the lysate was directly applied onto the affinity column.

Cross-Linking

The cross-linking of the F_v fragment with glutaraldehyde was carried out as described previously.[11] The cross-linked F_v fragment was then repurified by affinity chromatography.

Hapten Binding

Hapten binding was followed by recording changes in protein fluorescence as described previously.[11]

RESULTS

The response of the immune system to phosphorylcholine-containing antigens has been studied intensively.[27] The natural immunogen is the surface polysaccharide

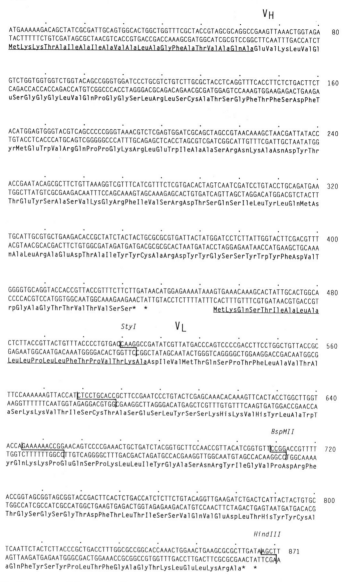

FIGURE 1. Sequence of the synthetic operon of T15. The signal sequences are *underlined* and stop codons denoted with a *star*. The relevant restriction sites as well as the junctions of the synthetic oligonucleotides are labeled. The beginning of the mature parts of V_H and V_L is also labeled.

of *Streptococcus pneumoniae* and several other microorganisms, which carries phosphorylcholine esterified on a sugar residue. The murine immune response results in the production of antibodies, all derived from the same V_H gene, and this family of antibodies is named after the antibody carrying the genomic sequence T15. In contrast, three different V_L genes are predominantly employed by the mouse. The resulting antibodies can be represented by myeloma protein "prototypes" McPC603 (or M603), TEPC15 (or T15), and MOPC167 (or M167).[27] These myeloma proteins are of the IgA class and are the best studied examples of the phosphorylcholine binding antibodies. We have thus chosen these particular proteins as model systems. The same three types of antibodies are also elicited if the immunization is carried out

```
Heavy chains
                                             CDR 1                         CDR 2
M603    EVKLVESGGGLVQPGGSLRLSCATSGFTFSDFYMEWVRQPPGKRLEWIAASRNKGNKYTT
T15     EVKLVESGGGLVQPGGSLRLSCATSGFTFSDFYMEWVRQPPGKRLEWIAASRNKANDYTT
M167    EVKVVESGGGLVQPGGSLRLSCATSGFTFSDFYMEWVRQTPGKRLEWIAASRSKAHDYRT
        ***.***********************************.************.*...* *
                                                      CDR 3
M603    EYSASVKGRFIVSRDTSQSILYLQMNALRAEDTAIYYCARNY-YGST-W-YFDVWGAGTT
T15     EYSASVKGRFIVSRDTSQSILYLQMNALRAEDTAIYYCARDY-YGSSYW-YFDVWGAGTT
M167    EYSASVKGRFIVSRDTSQSVLYLQMNALRAEDTATYYCTRDADYGNSYFGYFDVWGAGTT
        *******************.**************.***.*.  **.. . **********

M603    VTVSS
T15     VTVSS
M167    VTVSS
        *****

Light chains
                                            CDR 1                         CDR 2
M603    DIVMTQSPSSLSVSAGERVTMSCKSSQSLLNSGNQKNFLAWYQQKPGQPPKLLIYGASTR
T15     DIVMTQSPTFLAVTASKKVTISCTASESLYSSKHKVHYLAWYQKKPEQSPKLLIYGASNR
M167    DIVITQDELSNPVTSGESVSISCRSSKSLLY-KDGKTYLNWFLQRPGQSPQLLIYLMSTR
        ***.**.    .*.....*..** .*.**    .   .*.*. ..*.*.*.****  *.*
                                                 CDR 3
M603    ESGVPDRFTGSGSGTDFTLTISSVQAEDLAVYYCQNDHSYPLTFGAGTKLELKRA
T15     YIGVPDRFTGSGSGTDFTLTISSVQVEDLTHYYCAQFYSYPLTFGAGTKLELKRA
M167    ASGVSDRFSGSGSGTDFTLEISRVKAEDVGVYYCQQLVEYPLTFGAGTKLELKRA
          **.***.**********.**.*..**.. ***..  .****************
```

FIGURE 2. Comparison of the sequences of the V_H and V_L domains of M603, M167, and T15. In the sequences, deletions are marked with a *dash.* In the line below the three sequences, a *star* refers to an identity across all three sequences and a *period* refers to a conservative exchange. *Underlined residues* were not determined in the original publications, but obtained from homologies to similar antibodies. The hypervariable regions according to the definition of Kabat *et al.*[26] are *boxed* and *labeled.*

with phosphorylcholine-derivatized protein antigen, but IgM and IgG are usually obtained in this case.

To be able to carry out comparative studies on structure, folding, binding, and antibody-mediated catalysis,[28-30] we wished to express the F_v fragments of all three antibodies (M603, T15, and M167) in *E. coli.* We had previously developed the *E. coli* expression technology with the antibody M603.[1,2]

An alignment of the variable domains V_L and V_H for all three antibodies is shown in FIGURE 2. We used leucine at position L112 (numbered according to the crystal structure of M603; also numbered 112 in FIG. 2) in the original synthesis of the light chain gene of M603[14] for homology reasons, although isoleucine was reported from more recent sequencing experiments (Rudikoff, unpublished work; quoted in ref. 5). It is very unlikely that this substitution is of any significance as both residues are commonly found in antibodies at this position. The light chain sequence for T15 was not completely determined (summarized in ref. 26), but all residues of the sequence that were determined are identical to the sequence of the antibody S107 (ref. 26), and the missing residues were thus taken from the sequence of S107. The sequences shown in FIGURE 2 then correspond to the domains used in the gene synthesis.

The heavy chains of the three myeloma proteins are rather similar, because they are probably derived from the same V_H gene (reviewed in ref. 27). We could therefore take advantage of the existing synthetic V_H gene of M603[14] and obtain the T15 V_H gene by site-directed mutagenesis with several oligonucleotides simultaneously (see the Materials and Methods section). This mutagenesis was carried out directly in the expression vector for the F_v fragment that is described elsewhere.[21] On this vector, the two genes encoding V_H and V_L are arranged in a synthetic operon under the control of one regulatable promoter.[1,21]

The light chains, on the other hand, are sufficiently different so that such a mutagenesis strategy would not be advantageous. We therefore decided to completely synthesize the light chain of T15, analogous to the previous synthesis of the genes for M603. The synthetic strategy took advantage of restriction sites present in the synthetic operon of M603,[14] so that a synthetic cassette of T15 could replace the corresponding piece of M603. After complete assembly of the genes, they are arranged analogously to the genes of M603. These plasmids functionally express the F_v fragments of M603 and T15 (FIG. 3).

The phosphorylcholine binding F_v fragments of M603 and T15 can be purified by hapten affinity chromatography in a single step, directly demonstrating the functionality of the F_v fragments produced in $E.$ $coli$. Interestingly, however, the amounts of functional fragment isolated were lower for T15 (about 10–20%), although vectors, translation imitation regions, signal sequences, strains, and experimental conditions were identical. Codon use in the synthetic genes is essentially identical in the two operons, and rare codons[31] have largely been avoided in both constructs. No differences in possible hairpins are apparent in the two genes. The expression of the plasmid encoding T15 was also examined in a host strain deficient in the protease

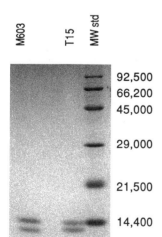

FIGURE 3. SDS PAGE of the F_v fragments of M603 and T15 expressed in $E.$ $coli$ and purified by hapten affinity chromatography. A 15% polyacrylamide gel stained with Coomassie brilliant blue is shown. The *upper band* corresponds to V_H (122 amino acids for M603 and 123 amino acids for T15); the *lower one* corresponds to V_L (115 amino acids for both M603 and T15).

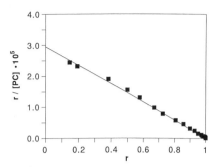

FIGURE 4. Scatchard plot of the binding of phosphorylcholine to the recombinant F_v fragment of the antibody T15. r denotes the fraction of the antibody with a bound ligand and [PC] the molar concentration of free hapten.

OmpT,[23,32] but no influence on the level of expression of functional T15 F_v fragment was found compared to that of an $ompT^+$ host (data not shown). Most likely, proteolytic sensitivity to an as yet unidentified protease causes this difference in expression level, and this effect is even more pronounced for M167 (Pfitzinger, Schweder, and Plückthun, unpublished observations).

We then examined the binding constant of the F_v fragment of T15. The dissociation of the F_v fragment at low concentration requires cross-linking of the V_H and V_L domains for fluorescence measurements[11] or high protein concentrations for equilibrium dialysis measurements[1] to obtain linear Scatchard plots with binding constants identical to the F_{ab} fragment or the whole antibody. Both experimental approaches have been shown to be in quantitative agreement.[11] This demonstrates that modification of surface lysines by glutaraldehyde with subsequent reduction to obtain the cross-linked F_v fragment, which preserves the positive charge, does not affect the binding of the hapten. We therefore employed fluorescence measurements of the glutaraldehyde cross-linked F_v fragments to determine hapten binding constants (FIG. 4).

The association constant obtained for the F_v fragment of T15 ($3.3 \times 10^5 \text{ M}^{-1}$) is in good agreement with binding constants previously reported for the whole antibody[6,8,33] ($2.3 \times 10^5 \text{ M}^{-1}$ to $5.5 \times 10^5 \text{ M}^{-1}$). Antibody T15 thus binds PC slightly better than does antibody M603, and this trend is seen in both the F_v fragments and the whole antibody, validating the recombinant methodology.

DISCUSSION

The three antibodies, M603, M167, and T15, are "prototypes" of the three classes of antibodies that are generated when a mouse is challenged with the antigen phosphorylcholine. Their heavy chains are derived from the same V_H gene, but each one uses a light chain derived from a different V_L gene.[27] This particular system of related antibodies now allows the investigation of several mechanistic questions of folding, binding, and antibody-mediated catalysis by the use of protein engineering.

The first aspect of interest is the structure of these three antibodies. Only the structure of the F_{ab} fragment of McPC603 has been experimentally determined,[5] but for the structure of the other two, only models have so far been proposed.[34] The main problem in modeling these related structures is that some loops forming part of the binding pocket are not equal in length (FIG. 5). This difference occurs mainly in the CDR3 loop of the heavy chain, which is partially derived from the D gene. It may also differ in length because of the so-called N-region diversity.[27] Thus, a large number of degrees of freedom is introduced into modeling the active site by the uncertainty of the structure of these loops. The availability of the recombinant F_v fragments will now allow a renewed attempt at structure determination by both NMR and X-ray crystallography. This is of particular importance for the critical evaluation of the various modeling attempts and for obtaining feedback about the quality of different theoretical approaches.

The second aspect that may be analyzed with these recombinant antibodies is the functional importance of different somatic mutations that have accumulated and constitute differences to the germ-line genes. The recombinant F_v fragments can now

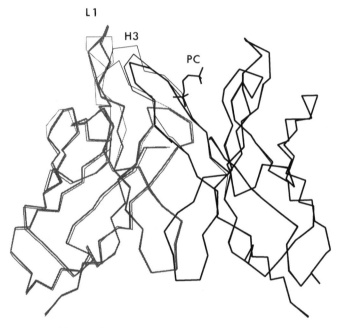

FIGURE 5. Superposition of initial model structures of the F_v fragments of T15 and M167 onto the experimental structure of M603. Only the α-carbon backbone and the position of the hapten phosphorylcholine (from M603) are shown. The heavy chains are shown with *black lines* (on the right), the *light chains* with gray lines (on the left). The *thick lines* denote the experimental structure of M603, the *medium lines* the model of T15, and the *thin lines* the model of M167. The two loops, in which the length of the antibody loops differ, are marked with L1 (CDR 1 of the light chain) and H3 (CDR 3 of the heavy chain). PC is phosphorylcholine.

be used to delineate which of these mutations were selected to secure chain association or antigen binding, and which of these mutations are due to random drift without particular benefit to the function of the molecule. This question also requires an examination of the question, which combinations of heavy and light chains are able to associate *in vivo* and function in phosphorylcholine binding.[35]

The third aspect that can now be scrutinized is that of catalysis. Previous results with the antibodies M167 and T15 obtained from mouse ascites[28,29] and with the recombinant F_v fragment of M603 and some of its mutants (ref. 30; Stadlmüller and Plückthun, manuscript in preparation) have shown that these antibodies have catalytic activity in hydrolyzing carbonate esters of choline. The activities reported for M167 and T15 are higher than those for M603, but the mechanistic and structural reasons have not been elucidated. These mechanistic questions can now be addressed experimentally by protein engineering of the F_V fragments.

SUMMARY

The development of general methods to express functional antibody fragments in *E. coli*[1] greatly facilitates the engineering of antibodies. Some of the essential features of the technology are summarized. As a model system, phosphorylcholine binding antibodies are used. The immune response against this antigen results in three classes of antibodies, exemplified by the myeloma proteins McPC603, TEPC15, and MOPC167. F_V fragments of these antibodies can now be conveniently prepared in *E. coli* to aid in understanding the structural logic of this well-characterized immune response.

REFERENCES

1. SKERRA, A. & A. PLÜCKTHUN. 1988. Assembly of a functional immunoglobulin F_V fragment in *Escherichia coli*. Science **240**: 1038–1041.
2. PLÜCKTHUN, A. & A. SKERRA. 1989. Expression of functional antibody F_V and F_{ab} fragments in *E. coli*. Meth. Enzymol. **178**: 497–515.
3. GLOCKSHUBER, R., T. SCHMIDT & A. PLÜCKTHUN. 1991. The disulfide bonds in antibody variable domains: Effects on stability, folding *in vitro* and functional expression in *E. coli*. Biochemistry, in press.
4. SKERRA, A. & A. PLÜCKTHUN. 1991. Secretion and *in vivo* folding of the Fab fragment of the antibody McPC603 in *Escherichia coli*: Influence of disulfides and *cis*-prolines. Protein Engin., in press.
5. SATOW, Y., G. H. COHEN, E. A. PADLAN & D. R. DAVIES. 1986. Phosphocholine binding immunoglobulin F_{ab} McPC603: An X-ray diffraction study at 2.7 Å. J. Mol. Biol. **190**: 593–604.
6. LEON, M. A. & N. M. YOUNG. 1971. Specificity for phosphorylcholine of six murine myeloma proteins reactive with pneumococcus C polysaccharide and β-lipoprotein. Biochemistry **10**: 1424–1429.
7. YOUNG, N. M. & M. A. LEON. 1977. The binding of analogs of phosphorylcholine by the murine myeloma proteins McPC603, MOPC167 and S107. Immunochemistry **14**: 757–761.
8. METZGER, H., B. CHESEBRO, N. M. HADLER, J. LEE & N. OTCHIN. 1971. Modification of immunoglobulin combining sites. *In* Progress in Immunology. Proceedings of the 1st International Congress of Immunology. Amos, B., ed.: 253–267. Academic Press. New York.
9. GOETZE, A. M. & J. H. RICHARDS. 1977. Structure-function relations in phosphorylcholine-binding mouse myeloma proteins. Proc. Natl. Acad. Sci. USA **74**: 2109–2112.
10. GOETZE, A. M. & J. H. RICHARDS. 1978. Molecular studies of subspecificity differences among phosphorylcholine-binding mouse myeloma antibodies using ^{31}P nuclear magnetic resonance. Biochemistry **17**: 1733–1739.
11. GLOCKSHUBER, R., M. MALIA, I. PFITZINGER & A. PLÜCKTHUN. 1990. A comparison of strategies to stabilize immunoglobulin F_V fragments. Biochemistry **29**: 1362–1367.
12. BIRD, R. E., K. D. HARDMAN, J. W. JACOBSON, S. JOHNSON, B. M. KAUFMAN, S. M. LEE, T. LEE, S. H. POPE, G. S. RIORDAN & M. WHITLOW. 1988. Single-chain antigen-binding proteins. Science **242**: 423–426.
13. HUSTON, J. S., D. LEVINSON, M. MUDGETT-HUNTER, M. S. TAI, J. NOVOTNY, M. N. MARGOLIES, R. J. RIDGE, R. E. BRUCCOLERI, E. HABER, R. CREA & H. OPPERMANN. 1988. Protein engineering of antibody binding sites: Recovery of specific activity in an anti-digoxin single-chain F_V analogue produced in *Escherichia coli*. Proc. Natl. Acad. Sci. USA **85**: 5879–5883.

14. PLÜCKTHUN, A., R. GLOCKSHUBER, I. PFITZINGER, A. SKERRA & J. STADLMÜLLER. 1987. Engineering of antibodies with a known three-dimensional structure. Cold Spring Harbor Symp. Quant. Biol. **52:** 105–112.

15. SKERRA, A., R. GLOCKSHUBER & A. PLÜCKTHUN. 1990. Structural features of the F_{ab} fragment of the antibody McPC603 not evident from the crystal structure. FEBS Lett. **271:** 203–206.

16. NISONOFF, A., J. E. HOPPER & S. B. SPRING. 1975. The Antibody Molecule. Academic Press. New York.

17. WARD, E. S., D. GÜSSOW, A. D. GRIFFITHS, P. T. JONES & G. WINTER. 1989. Binding activities of a repertoire of single immunoglobulin variable domains secreted from *Escherichia coli.* Nature **341:** 544–546.

18. GLOCKSHUBER, R., B. STEIPE, R. HUBER & A. PLÜCKTHUN. 1990. Crystallisation and preliminary X-ray studies of the V_L domain of the antibody McPC603 produced in *Escherichia coli.* J. Mol. Biol. **213:** 613–615.

19. STEIPE, B., A. PLÜCKTHUN & R. HUBER. 1991. Refined crystal structure of a recombinant immunoglobulin domain and a CDR1-grafted mutant. J. Mol. Biol., in press.

20. MANIATIS, T., E. F. FRITSCH & J. SAMBROOK. 1982. Molecular Cloning: A Laboratory Manual. Cold Spring Harbor Laboratory Press. Cold Spring Harbor, NY.

21. SKERRA, A., I. PFITZINGER & A. PLÜCKTHUN. 1990. The functional expression of antibody F_v fragments in *Escherichia coli:* Improved vectors and a generally applicable purification strategy. Biotechnology **9:** 273–278.

22. VIEIRA, J. & J. MESSING. 1987. Production of single-stranded plasmid DNA. Methods Enzymol. **153:** 3–11.

23. EARHARD, C. F., M. LUNDRIGAN, C. L. PICKETT & J. R. PIERCE. 1979. *Escherichia coli* K-12 mutants that lack major outer membrane protein a. FEMS Microbiol. Lett. **6:** 277–280.

24. KUNKEL, T. A., J. D. ROBERTS & R. A. ZAKOUR. 1987. Rapid and efficient site-specific mutagenesis without phenotypic selection. Methods Enzymol. **154:** 367–382.

25. GEISSELSODER, J., F. WITNEY & P. YUCKENBERG. 1987. Efficient site-directed *in vitro* mutagenesis. Biotechniques **5:** 786–791.

26. KABAT, E. A., T. T. WU, M. REID-MILLER, H. M. PERRY & K. GOTTESMAN. 1987. Sequences of proteins of immunological interest. Public Health Service, Natl. Institutes of Health, Bethesda, Maryland.

27. PERLMUTTER, R. M., S. T. CREWS, R. DOUGLAS, G. SORENSEN, N. JOHNSON, N. NIVERA, P. J. GEARHART & L. HOOD. 1984. The generation of diversity in phosphorylcholine-binding antibodies. Adv. Immunol. **35:** 1–37.

28. POLLACK, S. J., J. W. JACOBS & P. G. SCHULTZ. 1986. Selective chemical catalysis by an antibody. Science **234:** 1570–1573.

29. POLLACK, S. J. & P. G. SCHULTZ. 1987. Antibody catalysis by transition state stabilization. Cold Spring Harbor Symp. Quant. Biol. **52:** 97–104.

30. PLÜCKTHUN, A., R. GLOCKSHUBER, A. SKERRA & J. STADLMÜLLER. 1989. Properties of F_v and F_{ab} fragments of the antibody McPC603 expressed in *E. coli.* Behring Inst. Mitt. **87:** 48–55.

31. GROSJEAN, H. & W. FIERS. 1982. Preferential codon usage in procaryotic genes: The optimal codon-anticodon interaction energy and the selective codon usage in efficiently expressed genes. Gene **18:** 199–209.

32. GRODBERG, J. & J. J. DUNN. 1988. OmpT encodes the *Escherichia coli* outer membrane protease that cleaves T7 RNA polymerase during purification. J. Bacteriol. **170:** 1245–1253.

33. POLLET, R. & H. EDELHOCH. 1973. The binding properties of anti-phosphorylcholine mouse myeloma proteins as measured by protein fluorescence. J. Biol. Chem. **248:** 5443–5447.

34. PADLAN, E. A., D. R. DAVIES, S. RUDIKOFF & M. POTTER. 1976. Structural basis for the specificity of phosphorylcholine-binding immunoglobulins. Immunochemistry **13:** 945–949.

35. HAMEL, P. A., M. H. KLEIN & K. J. DORRINGTON. 1986. The role of the V_L and V_H segments in the preferential reassociation of immunoglobulin subunits. Mol. Immunol. **23:** 503–510.

Construction of Expression Vectors for Gene Fusions on the Model of β-Galactosidase-Human Fibroblast β-Interferon for the Purpose of Immunoenzyme Assay

A. N. MARKARYAN,[a] S. V. MASHKO,[b] L. V. KUKEL,[a]
A. L. LAPIDUS,[b] A. N. BACH,[a] AND A. M. EGOROV[c]

[a]Institute of Biochemistry
USSR Academy of Sciences
117071 Moscow, USSR

[b]Institute of Genetics and
Selection of Industrial Microorganisms
113545 Moscow, USSR

[c]M. V. Lomonosov Moscow State University
Chemistry Faculty
Chemical Enzymology Department
Moscow, USSR

Gene fusion has become widely used in modern recombinant DNA technology. Important applications have been found notably in the study of gene expression and gene regulation.[1,2] Fusion proteins prepared by ligating two structural genes are also applied in enzyme technology as in the construction of bifunctional enzymes in which the enzymatic moieties operate in sequence.[3,4] Hybrid proteins can also be used to elicit antibody formation.[5] Gene fusion also may find application in affinity chromatography, because purification of the desired protein can be facilitated by ligating an "affinity tail" to the gene product.[6,7] One promising area of fusion application is the construction of antigen-marker enzyme conjugates for ELISA purposes.[8,9] Variants of ELISA require the covalent coupling of an enzyme to antigen by chemical means. As highly purified enzymes and antigens are required, the essential reagents are rendered relatively expensive. Moreover, chemical coupling can generate artifacts. On the other hand, in-frame fusions of peptide antigens with marker enzymes such as β-galactosidase in *Escherichia coli* can easily be constructed. The *lac*Z gene has been used extensively in gene fusion experiments.[10] Enzymatically active chimeric β-galactosidase can easily be constructed because 26 NH_2-terminal residues[11,12] and two COOH-terminal residues[13] can be removed without impairing its enzymatic activity. It was reported earlier that DNA fragments inserted at the 3' end of the *lac*Z gene direct the synthesis of more stable hybrid proteins than do recombinants with the DNA inserted at the 5' end.[14] Therefore, construction of high efficiency β-galactosidase expression vectors with a polylinker engineered into the 3' end of the *lac*Z gene is of great importance. However, the high level synthesis of recombinant β-galactosidases in *E. coli* leads to their inactivation and the formation of insoluble particles.[15] For practical applications (e.g., enzyme fusion for ELISA), we need vectors that produce active, stable, and soluble forms of β-galactosidase. One aim of the present

125

study was to choose an optimal vector for fusion preparation on the basis of expressed enzyme properties.

In addition, some hybrid β-galactosidases are lethal to the cell when synthesized in large amounts, as are some membrane-bound or membrane-transported protein fusions.[16] We cannot predict whether new or hybrid proteins will adversely affect the growth of the cell, especially when expressed at high levels. These results emphasize the necessity of controlling gene expression when constructing gene fusion. Frequently applied transcription-controlling signals in expression vectors have different P/O sequences of the lac operon. Data reported earlier[17] show that the efficiency of repression for the lac system is essentially lower than that for the P_L/O_L sequence of phage λ in which the high *in vivo* activity of P_L can be repressed by a factor greater than 10. For this reason, leakage of a toxic product could be lethal to the cells.

To overcome this disadvantage, we decided in the present work to construct the expression vectors under the control of phage λ P_R promoter, with an efficiency close to that of P_L.[18] The resulting plasmids also carry the gene of repressor cIts857, whose presence allows the P_R-directed transcription of the hybrid gene in a thermoinducible fashion. Tighter control of transcription, needed especially for the expression of toxic products, was tested in *lac*Z gene fusion with human fibroblast β-interferon (hIFN-β). The toxicity of hIFN-β in bacterial cells was established earlier.[19]

MATERIAL AND METHODS

The following *E. coli* strains were used as recipients: C 600 (F^-, *thi, thr, leu B6, sup E44, lacY1, tonA21*) and TG1 (Δ(*lac-proAB*), *thi, strA, supE44, endA, sbcB, hsdR^-*, [F'traD36, *proAB^+, lacI^Q, lacZΔM15*]).

Bacteria were grown at 28°C in LB medium (1% Bacto Bacto yeast extract/0.5% NaCL) with 100 μg ampicillin per milliliter.

The following plasmids were used in constructions: pORF1,[20] pUR291,[13] pUC18,[21] and pPR124B, pPR37, TC7, and pINF27 (obtained in the laboratory of Prof. S. V. Mashko).[22] The methods of enzymatic treatment of DNA, plasmid isolation, electrophoretic analysis of DNA in agarose, and polyacrylamide gels have been described elsewhere.[23]

β-Galactosidase was assayed by hydrolysis of *o*-nitrophenyl-β-D-galactopyranoside.[24] Protein was measured by the Bradford method using bovine serum albumin as standard.[25]

β-Galactosidase synthesis was measured after temperature induction. Fresh overnight cultures were diluted (1:200) and grown at 28°C to OD_{600} = 0.6. β-Galactosidase synthesis was induced by heating the broth at 45°C for 15 minutes and further incubating it at 42°C for 1 and 2 hours. β-galactosidase activity was measured by lysing aliquots of the cultures with chloroform and SDS.[24] Enzyme activity was converted from units per bacterium to micromoler per minute per milligram using the conversion factor given by Miller.[24] Lysate supernatants were prepared by sonic disintegration and centrifuging the cell debris at 20,000 rpm for 15 minutes.

SDS-PAGE[26] was used to analyze proteins. The level of protein production was determined by scanning the Coomassie blue-stained gel on a CDS-200 computing densitometer (Beckman). The results represent an average of two experiments.

Heat stability measurements were carried out at 37°, 42°, and 54°C with lysate supernatants in 0.1 M sodium phosphate buffer, pH 7.2, containing 1 mM $MgSO_4$ and 0.1 M β-mercaptoethanol.

Western blotting was performed as described in ref. 27.

The ELISA procedure was as follows: Flat bottom microtiter plates (Dynatech) were coated for 12 hours at 4°C with 50 μl of solution containing 0.04 mg/ml protein A from *Staphylococcus aureus* in coating buffer (0.05 M NaHCO₃, pH 8.3). The plate was shaken out and washed with 100 μl wash solution (0.9% NaCl, 0.05% Tween 20), and the remaining sites on the plastic surface were blocked with 100 μl blocking solution (1% BSA, 50 mM NaH₂ PO₄, 0.15 M NaCl, pH 7.2). After repeating the washing procedure once, a 50-μl solution of anti-hIFN-β antibodies (Celltech) in blocking buffer (0.004 mg/ml) was added to each well. After a 1-hour incubation at 37°C, the plate was shaken and hit against a paper towel to remove the last remaining liquid. The lysate supernatant with β-galactosidase activity was then diluted 1:1 (50 μl) with 50 μl of 2 × blocking solution. All samples were run in duplicate, each well receiving 50 μl. After a 1-hour incubation, the plate was washed four times. Bound fusion protein was determined by adding 50 μl of substrate solution (1 mg/ml o-nitrophenyl-β-D-galactopyranoside). After a 30-minute incubation at 37°C, the absorbance was read at 405 nm using a Titrek multiscan photometer (Flow Laboratories).

RESULTS AND DISCUSSION

Three major types of construction are used to optimize protein production in *E. coli:* (1) hybrid ribosome-binding sites (plasmid placZ); (2) genes encoding hybrid proteins (plasmid pPRCZ)[28]; and (3) two-cistron mRNAs with partially overlapping genes (plasmid pONZ).[29,30] The latter method was successfully tested in experiments on optimization of expression of the genes encoding human or animal interferons, human interleukin-β1, and *E. coli* β-galactosidase.[30] It is known that efficient transcription from the strong P_R promoter inhibits plasmid replication.[15] This is why two phage fd Rho-independent transcription terminators were inserted downstream of the *lac*Z gene in all prepared plasmids. As the initial vector in our further constructions, we used pPR124B, carrying *c*Its857 gene, the P_R and P_{RM} promoters, and two tandemly positioned phage fd Rho-independent transcription terminators.

Construction of placZ (Fig. 1). The first approach consisted of linking the structural part of the foreign gene (in our case *lac*Z) to the prokaryotic regulatory elements, providing efficient transcription and translation. Plasmid pORF1 bearing the complete structural gene of β-galactosidase without its NH₂-terminal eight nonessential codons was used as a source of *lac*Z gene. Plasmid pLZ was derived from pORF1 in which a *Bal*I restriction enzyme site downstream of the *lac*Z gene was substituted by *Bgl*II.

The plasmid placZ was produced in the following way: the *Bam*HI-*Bgl*II DNA fragment of pLZ bearing the *lac*Z gene was cloned into the *Bam*HI-*Bgl*II site of pPR124B downstream of the phage λ P_R promoter and in front of transcription terminator signals. The ligated DNA mixture was used to transform competent cells for *lac*Z⁻ strain TG1. As the parent plasmid pLZ did not express β-galactosidase, Lac⁺ transformants harboring recombinant plasmids carrying the *lac*Z gene were identified by plating the transformed cells on indicator plates supplemented with ampicillin and X-gal under the derepression conditions (42°C). The plasmid placZ structure was also verified by restriction enzyme analysis.

Construction of pPRCZ (Fig. 1). In the second approach, a foreign DNA fragment was inserted into the coding part of a bacterial gene so that the reading frames of both genes were adjusted to one another. The hybrid plasmid carrying cro-*lacZ* fusions was constructed by joining DNA fragment *Bgl*II-*Pst*1 from plasmid pPR37 carrying the P_R promoter and the first 20 codons of the *cro* gene to the *lac*Z gene

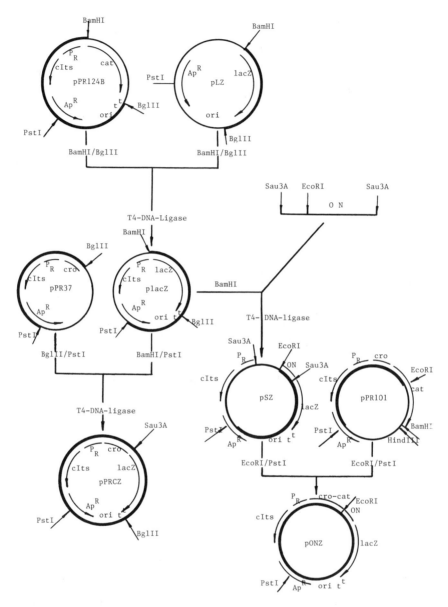

FIGURE 1. Schematic representation of the construction of plasmids placZ, pPRCZ, and pONZ.

*Bam*HI-*Pst*1 carried by plasmid *plac*Z. The *cro* and *lac*Z genes were fused by ligation, during which *Bam*HI and *Bgl*II sites were lost.

Construction of pONZ (Fig. 1). For this approach, the coding region of a foreign gene was attached to ATG codon present in the vector, to form hybrid operon transcribed from the phage P_R promoter.

FIGURE 2. Structure of oligos (ON). Fragment contains the natural trpE-trpD intercistronic region of *E. coli* trp operon and is flanked with ss cohesive ends produced by treating DNA with *Eco*RI + *Sau*3A. The stop (TGA and ATG) codons are in boldface.

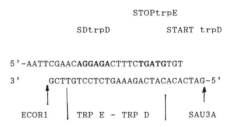

Initially, plasmid pSZ was obtained by cloning *Sau*3A- fragment from plasmid pINF27 into the *Bam*HI site of placZ. The *Eco*RI-*Pst*I DNA fragment of pSZ bearing the oligo (ON) and *lac*Z gene was then inserted into the *Eco*RI-*Pst*I site of pPR101 carrying P_R promoter, 20 NH_2-terminal codons of the *cro* gene and the coding part of the *cat* gene. The ON structure is shown in FIGURE 2. The resulting plasmid pONZ thus carries the *lac*Z gene which is the distal cistron of this hybrid operon. The efficiently translated cro'-cat'-trpE hybrid cistron is proximal to the P_R promoter. The differences in the structure of all three constructed plasmids are outlined in TABLE 1.

Analysis of β-galactosidases directed by placZ, pPRCZ, and pONZ. TABLE 2 compares the amounts of recombinant β-galactosidases produced by the thermoinducible plasmids placZ, pPRCZ, and pONZ. Exponentially growing cultures of strain TG1 harboring these plasmids were induced at 42°C for 60 and 120 minutes, and β-galactosidase activity was measured in cell cultures. Quantitations of the amounts of protein synthesized were based on SDS-PAGE, shown in FIGURE 3. For comparison, enzyme and protein levels were measured in the cultures of TG1 strain harboring plasmid pUR291 induced with IPTG. The results obtained show that the plasmid pPRCZ, which represents the fusion of *lac*Z gene with 20 codons of the *cro* gene, provides the highest level of protein synthesis (15% of total cellular proteins). The pONZ directs the synthesis of the protein, with a yield (12.5%) close to that of plasmid pPRCZ. In contrast, the most proximal fusion, placZ, produces less protein (9%). It is not surprising that the highest yield of synthesized protein was achieved with pPRCZ. Such plasmids in which the *cro* and *lac*Z genes are fused in-register produce elevated levels of cro-β-galactosidase, amounting to 30% of the total cellular protein.[15] The same effect on the level of protein synthesis was observed for cro-chloramphenicol acetyltransferase gene carrying on their NH_2-terminal part no

TABLE 1. Differences in NH_2-Terminal Structure of Recombinant β-Galactosidases Directed by placZ, pPRCZ, and pONZ

Plasmid	Region between P_R and *lac*Z
placZ	9 5'-PR-SDcro-ATGcroGATCCCGTC... *lac*Z
pPRCZ	20 9 5'-PR-SDcro-ATGcro ... N ... AAAGATCCCGTC... cro *lac*Z
pONZ	9 5'-PR- ... SDtrpD ... TGATGTGTGATCCCGTC... *lac*Z

TABLE 2. Synthesis of Recombinant β-Galactosidases

	Cell Cultures		Lysate Supernatants		Debris
Plasmid	Protein % Total	Specific Activity (μmol/min/mg)	Protein % Total	Specific Activity (μmol/min/mg)	Protein % Total
placZ	9.0	13.0	2.3	54.3	6.7
pPRCZ	15.4	3.25	5.6	25.7	9.8
pONZ	12.4	13.3	3.1	48.4	9.3
pUR291	4.9	—	—	—	—

less than nine codons of the *cro* gene.[22] This part of the *cro* gene is thought to determine a thermodynamically favorable and therefore an optimal mRNA secondary structure recognized by ribosomes.

Numerous genes that encode prokaryotic or eukaryotic proteins have been cloned in *E. coli*. Early observations[31,32] with such systems indicated that the high level expression of these genes results in the formation of insoluble proteinaceous aggregates. When the supernatants of lysates from strains with pPRCZ, placZ, and pONZ were analyzed by SDS-PAGE, less fusion protein was found than that in the total cell extracts. The nonextractable protein remained in the cell debris, indicating that recombinant enzymes precipitate intracellularly, as was previously reported for β-galactosidase fusion proteins.[31] A lower specific β-galactosidase activity was measured in cell cultures with plasmid pPRCZ than with placZ or pONZ (TABLE 2). From these data we conclude that a large portion of overproduced protein is enzymatically inactive because the cro-β-galactosidase has reduced the specific activity. This assumption was supported by the specific activity of highly purified cro-β-galactosidase which corresponds to 35% of that of wild-type β-galactosidase.

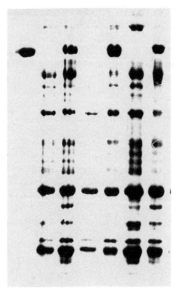

FIGURE 3. SDS-PAGE of recombinant β-galactosidases. Samples of cultures (*E. coli* TG1) were run on 8% SDS-PAGE. *Lane 1:* marker protein: β-galactosidase (116 kD); *lanes 2, 3:* cultures with placZ before and after induction (1 hour at 42°C); *lanes 4, 5:* cultures with pPRCZ before and after induction (1 hour at 42°C); *lanes 6, 7:* cultures with pONZ before and after induction (1 hour at 42°C).

To select the optimal vector for further fusion experiments, we compared solubility, specific activity, and stability of recombinant β-galactosidases produced by pPRCZ, placZ, and pONZ. TABLE 2 shows that the most active and soluble proteins were expressed by plasmids pONZ and placZ. To compare the stability of the wild-type enzyme and the recombinant β-galactosidases against thermal denaturation, the lysate supernatants of the cells were heated to 37°C, 42°C, and 54°C (FIG. 4). The thermoinactivation curves show that the hybrid cro-β-galactosidase directed by pPRCZ was more sensitive to high temperatures, especially 42°C. Plasmids placZ and pONZ producing truncated enzymes (without eight NH_2-terminal amino acids) show stability close to that of the wild-type enzyme.

Results based on protein properties therefore show that both plasmids pONZ and placZ could be applied for fusion preparations. In this work we used plasmid pONZ to obtain fusions with hIFN-β.

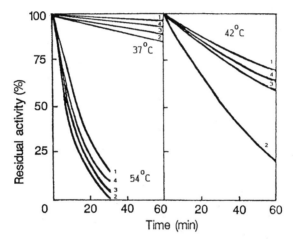

FIGURE 4. Heat stability of recombinant β-galactosidases. Lysate supernatants were incubated at 37°, 42°, and 54°C for the indicated length of time and residual activities were determined. 1 = native β-galactosidase from *E. coli* strain ML308; 2, 3, 4 = recombinant β-galactosidases synthesized by pPRCZ, placZ, and pONZ, respectively.

Construction of PGI1. The strategy used for the construction of plasmid PGII with an *in-frame* fusion between the structural genes of β-galactosidase and hIFN-β is outlined in FIGURE 5.

Initially, the *Cla*I-*Hind*III DNA fragment of PUR 291 bearing the truncated *lacZ* gene without the first 280 NH_2-terminal amino acids of β-galactosidase was inserted in pPR124B. After transformation, ampicillin-resistant colonies were selected, and the restriction enzyme analysis of the recombinant plasmids supported the plasmid pPZ structure.

The *Eco*RI DNA fragment (767 bp) of plasmid TC7 carrying the complete gene of hIFN-β was cloned into the *Eco*RI site of PUC 18. The plasmid pUCTC7 was then digested with Sau3A, and removed DNA fragment Sau3A-Sau3A (540 bp) bearing the hIFN-β gene was introduced into the *Bam*HI site of pPZ (at the 3'-region of the *lacZ* gene). In resulting plasmid pPZTC7, the reading frame of the fused hIFN-β and truncated *lac*Z genes is in phase, encoding a polypeptide of 918 residues. The linker between the two structural genes codes 10 residues (FIG. 6).

In the final step, the ClaI-ClaI DNA fragment of plasmid pONZ, carrying the efficiently translated *cro'-cat'-trpE* fused cistron proximal to the P_R promoter and the

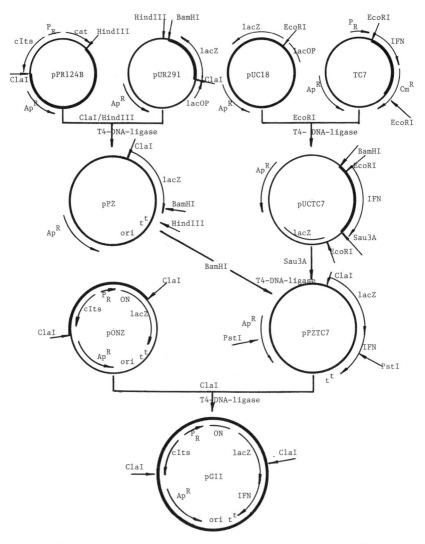

FIGURE 5. Schematic representation of the construction of plasmid pGI1.

DNA fragment encoding the NH$_2$-terminal part of the *lac*Z gene protein product (lacking its first eight amino acid codons), was inserted into the *Cla*I site of plasmid *pPZTC7*, and the resulting plasmid pGI1 was thus obtained.

Analysis of hybrid polypeptide directed by PGI1. The hybrid gene product of

```
     1020 1021  .    .    .    .    .    .    .    .    1    2
lacZ-TGG-TGT-CGG-GGA-TCC-CCG-GGT-ACC-GAG-CTC-GAA-TTC-ATG-AGC-IFN
     Trp-Cys-... ... ... ... ... ... ... ... ...-Met-Ser-...
```

FIGURE 6. Structure of linker region between *lacZ* and hIFN-β in pGI1.

plasmid PGI1 has an approximate molecular weight of 134 kD on SDS-PAGE. This band is not detected in repressed cells (FIG. 7A, lane 1) but is visible in extracts of cells after temperature induction (FIG. 7A, lanes 2 and 3). PGI1 produced a fusion protein amounting to 5% of the total cellular protein. In contrast, PONZ produced two times more protein than did PGI1. Inasmuch as both plasmids have identical promoter and ribosome binding site sequences, we suppose that intracellular proteolysis occurs, which diminishes the level of synthesized fusion protein. As judged from the results of SDS-PAGE, proteolysis was detected near the site at which the two polypeptides were joined (FIG. 7A, lanes 2 and 3).

Immunochemical studies of hybrid polypeptide directed by PGI1. To prove that plasmid PGI1 directs the synthesis of the desired hybrid polypeptide, we performed immunochemical studies. FIGURE 7A shows that after temperature induction, the supposed protein band with an apparent molecular weight of 134 kD appeared. To demonstrate that this protein band has the antigenicity of hIFN-β, we carried out Western blotting. Crude extracts possessing β-galactosidase activity were prepared and applied to electrophoresis to determine if a fusion protein is present and still recognized by anti-IFN antibodies. After electrophoresis, protein bands were visualized with Coomassie staining or blotted onto a nitrocellulose membrane and incubated with polyclonal anti-hIFN-β antibodies. Antibody binding was visualized by the enzymatic activity of peroxidase conjugated to an anti-rabbit IgG. The results (FIG. 7B) demonstrated that a high-molecular weight fusion protein encoded by PGI1 reacts (is denatured) with anti-hIFN-β antibodies.

The decisive question is whether the native fusion protein is recognized by anti-IFN antibodies. To answer this question, an ELISA procedure was performed as described in the Material and Methods section. The native fusion protein was at

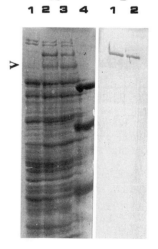

FIGURE 7. SDS-PAGE and subsequent Western blot. Proteins were separated on 8% SDS-polyacrylamide gels before blotting. **(A)** *Lane 1:* cell-free extract from pGI1 before induction; *lanes 2, 3:* the same extract after 1 and 2 hours of induction at 42°C, respectively; *lane 4:* marker enzymes (94, 67, and 43 kD). **(B)** Reaction of anti-hIFN-β antibodies with the fusion protein in *lanes 1 and 2* after 1 and 2 hours of induction at 42°C, respectively. *Arrow* indicates the position of β-galactosidases (116 kD).

first bound to the solid phase on which protein A from *S. aureus* and anti-IFN antibodies were coated subsequently. The bound fusion protein was then determined by measuring β-galactosidase activity with chromogenic substrate (FIG. 8).

Our results show that the fusion protein expressed by pGI1 retains β-galactosidase activity and antigenicity (capacity to bind anti-IFN antibodies).

CONCLUSIONS

Three different β-galactosidase expression vectors under the control of a strong P_R promoter were constructed for application in ELISA. From the expressed protein properties, plasmids pONZ and placZ were recommended for COOH-terminal fusion preparations. The fusion of the hIFN-β *in frame* to the β-galactosidase gene (plasmid pONZ) is described. A fusion protein was demonstrated in crude extracts of *E. coli* by Western blots using polyclonal anti-IFN-antibodies. The native fusion protein needed for ELISA was demonstrated by binding the hybrid protein with solid

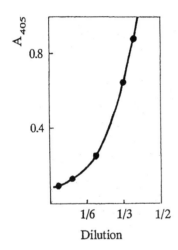

FIGURE 8. Dependence of OD_{405} on the dilution of lysate supernatant from pGI1 in the ELISA.

phase on which protein A and anti-IFN-antibodies had been consequently immobilized. Despite hIFN-β toxicity, when P_R is derepressed, a substantial amount of β-galactosidase-hIFN-β is produced.

ACKNOWLEDGMENTS

We would like to thank Dr. Lara Izotova for her assistance in performing the Western blot and ELISA procedures and Dr. Marina Lebedeva for a generous gift of plasmid TC7 bearing the gene of hIFN-β.

REFERENCES

1. CASADABAN, M. J., A. MARTINEZ-ARIAS, S. K. SHAPIRA & J. CHOU. 1983. Methods Enzymol. **100:** 293–308.

2. YOUNGMAN, P., P. ZUBER, J. B. PERKINS, K. SANDMAN, M. IGO & R. LOSICK. 1985. Science **228:** 285–291.
3. BULOW, L., P. LJUNGERANTZ & K. MOSBACH. 1985. Bio/Technology **3:** 821–823.
4. BULOW, L. 1987. Eur. J. Biochem. **168:** 443–448.
5. SCHUMAN, H., T. SILHAVY & J. BECKWITH. 1980. J. Biol. Chem. **255:** 168–174.
6. GERMINO, J. & D. BASTIA. 1984. Proc. Natl. Acad. Sci. USA **81:** 4692–4698.
7. NILSSON, B., L. ABRAHMSEN & M. UHLEN. 1985. EMBO J. **4:** 1075–1080.
8. PETERHANS, A., M. MECKLENBURG, F. MEUSSDOERFFER & K. MOSBACH. 1987. Anal. Biochem. **163:** 470–475.
9. LINDBLADH, C., M. PERSSON, L. BULOW, S. STAHL & K. MOSBACH. 1987. Biochem. Biophys. Res. Commun. **149:** 607–614.
10. SILHAVY, T. S., M. L. BERMAN & L. W. ENQUIST. Experiments in Gene Fusions. Cold Spring Harbor Laboratory Press. Cold Spring Harbor, NY.
11. MULLER-HILL, B. & J. KANIA. 1974. Nature **249:** 561–563.
12. BRAKE, A. I., A. V. FOWLER, I. ZABIN, J. KANIA & B. MULLER-HILL. 1978. Proc. Natl. Acad. Sci. USA **75:** 4824–4827.
13. RUTHER, U. & B. MULLER-HILL. 1983. EMBO. J. **2:** 1791–1794.
14. STANLEY, K. K. 1983. Nucleic Acids Res. **11:** 4077–4092.
15. ZABEAU, M. & K. K. STANLEY. 1982. EMBO. J. **1:** 1217–1224.
16. HALL, M. & T. SILHAVY. 1981. Ann. Rev. Genet. **15:** 91.
17. LANZER, M. & H. BUJARD. 1988. Proc. Natl. Acad. Sci. USA **85:** 8973–8977.
18. TRUKHAN, M. E., R. L. GOROVITZ, M. I. LEBEDEVA, A. L. LAPIDUS & S. V. MASHKO. 1988. Mol. Biol. (in Russian) **22:** 1033–1043.
19. REMAUT E., P. STANSSENS & W. FIERS. 1983. Nucl. Acids Res. **11:** 4677–4688.
20. WEINSTOCK, G. M., C. RHYS, M. L. BERMAN, B. HAMPAR, D. JACKSON, T. J. SILHAVY, J. WEISEMANN & M. ZWEIG. 1983. Proc. Natl. Acad. Sci. USA **80:** 4432–4436.
21. VIEIRA, J. & J. MESSING. 1982. Gene **19:** 259–268.
22. MASHKO, S. 1987. Optimization of expression of the heterologous genes in *E. coli*. Doctoral Thesis. Institute of Genetics and Selection of Industrial Microorganisms. Moscow.
23. MANIATIS, T., E. FRITSH & J. SAMBROOK. 1982. Molecular Cloning. A Laboratory Manual. Cold Spring Harbor Laboratory Press. Cold Spring Harbor, NY.
24. MILLER, J. H. 1972. Experiments in Molecular Genetics. Cold Spring Harbor Laboratory Press. Cold Spring Harbor, NY.
25. BRADFORD, M. 1976. Anal. Biochem. **72:** 248–254.
26. LAEMMLI, U. K. 1970. Nature (London) **227:** 680–685.
27. TOWBIN, H., T. STAEHELIN & J. GORDON. 1979. Proc. Natl. Acad. Sci. USA **76:** 4350–4354.
28. HARRIS, T. J. 1983. *In* Genetic Engineering. J. Setlow & A. Hollander, eds. Vol. 4: 127–185. Academic Press. New York.
29. MASHKO, S. V., A. L. LAPIDUS, M. I. LEBEDEVA, S. V. PODKOVYROV, T. G. PLOTNIKOVA, Y. I. KOZLOV, B. A. REBENTISH, S. V. KOSTROV, A. S. RYZHAVSKAYA, A. Y. STRONGIN, E. D. SVERDLOV & V. G. DEBABOV. 1984. Dokl. Akad. Nauk SSSR (in Russian) **278:** 1491–1496.
30. MASHKO, S. V., V. P. VEIKO, M. I. LEBEDEVA, A. V. MOCHULSKY, I. I. SCHECHTER, M. E. TRUKHAN, K. I. RATMANIVA, B. A. REBENTISH, V. E. KALUZHSKY & V. G. DEBABOV. 1990. Gene **88:** 121–126.
31. GOEDDEL, D. V., D. G. KLEID, F. BOLIVAR, H. L. HEYNEKER, D. G. YANSURA, R. CREA, T. HIROSE, A. KRASZEWSKI, K. ITAKURA & A. D. RIGGS. 1979. Proc. Natl. Acad. Sci. USA **76:** 106–110.
32. WEIS, J. H., L. W. ENQUIST, J. S. SALSTROM & R. J. WATSON. 1983. Nature **302:** 72–74.

Hybrid Genes Expressing Fusion Peptides

A Strategy for Bacterial Production of Recombinant Human Interleukin-3 with a Defined NH$_2$-Terminus

FRANZ KRICEK, CHRISTINE RUF, HEINZ ASCHAUER,
PETER STUCHLIK, CHRISTINE GRAF,
JAN-MARCUS SEIFERT, FRED-JOCHEN WERNER,
AND GERALD EHN

Sandoz Forschungsinstitut
Brunner Strasse 59
A-1235 Vienna, Austria

For evaluation of the clinical potential of the hematopoietic growth factor interleukin-3 (IL-3), the production of recombinant human IL-3 (rhIL-3) in appropriate host cells is necessary. Because of an AUG start codon at the 5' end of recombinant genes, however, many recombinant proteins purified from *Escherichia coli* extracts consist of a mixture of molecules starting either with an NH$_2$-terminal methionine or with the authentic amino acid of the mature natural protein. For clinical use, homogeneous preparations consisting of molecules with a defined NH$_2$-terminus are desirable. In this report, the cloning of an rhIL-3 gene fused at its 5' end to various DNA sequences that allow the fused genes to be efficiently expressed in *E. coli* and the production of NH$_2$-terminally homogeneous rhIL-3 by enzymatic cleavage are described.

RESULTS

Construction of Gene Fusion Plasmids

The DNA sequence coding for hIL-3, except 26 base pairs of its 5' end, was cut out from a plasmid harbored by *E. coli* strain GI609/hIL-3 (obtained from Genetics Institute, Inc., Cambridge, Massachusetts) and fused to sequences encoding NH$_2$-terminal parts of human granulocyte-macrophage colony-stimulating factor (rhGM-CSF) or monocyte-derived neutrophil-activating factor (NAF;NAP-1;IL-8) which have been found to be efficiently expressed in *E. coli* under the control of the tryptophan promoter.[1] The linkers restore the complete coding sequence and maintain the correct reading frame. Moreover, they code for the peptide sequence AspAspAspAspLys, thus allowing specific cleavage of the fusion protein by the proteolytic enzyme enterokinase to remove the fusion part of the molecule during the purification process.[2] As shown in FIGURE 1a, plasmids pSFI 804, pSFI 805, and pSFI 809 represent such constructs. In plasmid pSFI 809, only the enterokinase-specific sequence was fused to IL-3, thus possibly facilitating cleavage because of the similarity of the fusion protein with the natural substrate of enterokinase.[2]

136

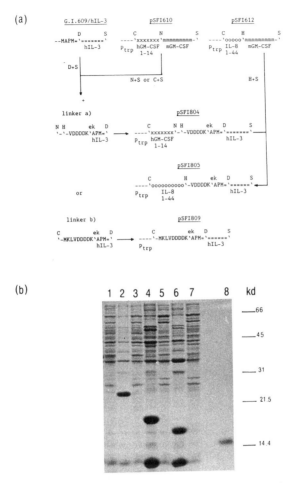

FIGURE 1. Construction of fusion plasmids: The hIL-3 gene was cut out from plasmid G.I. 609/hIL-3. The vector parts are derived from plasmids constructed for expression of a gene encoding murine granulocyte-macrophage colony-stimulating factor (mGM-CSF) by a similar gene fusion approach (unpublished data). **(a)** Schematic description of the cloning strategy. Restriction sites relevant for cloning are indicated: D = *Dra*I; S = *Sal*I; C = *Cla*I; N = *Nco*I; H = *Hind*III. Amino acid sequences of the NH$_2$-terminus of the mature hIL-3 gene as well as of the peptides encoded by the synthetic linkers a and b designed for specific cleavage by enterokinase (ek) are shown in single letter code. p$_{trp}$ = *E. coli* tryptophan promoter. Distances are not drawn to scale. **(b)** Expression of hybrid genes. Cells were disrupted and separated into fractions containing inclusion bodies (*lanes 2, 4, and 6*) or cytoplasmic proteins (*lanes 3, 5, and 7*). Proteins contained in these fractions were separated on a 12% polyacrylamide gel and stained with Coomassie brilliant blue. *Lanes 2 and 3,* pSFI 805 expression; *lanes 4 and 5,* pSFI 804 expression; *lanes 6 and 7,* pSFI 809 expression; *lane 1,* total cell extract from an isogenic strain without cloned genes (negative control); *lane 8,* purified unfused rhIL-3. The positions of standard molecular marker proteins are indicated. Sizes are given in kilodaltons (kD).

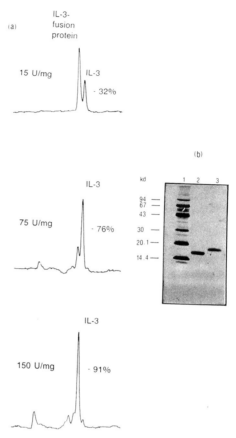

FIGURE 2. Production of homogeneous ala(1) rhIL-3 by enzymatic cleavage of a fusion protein isolated from pSFI 809 containing *E. coli* cells. (a) Reverse phase HPLC of cleavage products obtained with increasing amounts of enterokinase; (b) SDS-polyacrylamide gel electrophoresis (silver staining) of the fusion protein derived from pSFI 809 (*lane 3*) and homogeneous rhIL-3 obtained after enterokinase cleavage and further purification (*lane 2*). *Lane 3,* standard molecular weight markers. Sizes are given in kilodaltons (kD).

Expression of Hybrid Genes

The recombinant fusion plasmids were transferred into *E. coli* cells that were grown in shake flasks or a Giovanola fermentor. After induction of the tryptophan promoter by indole acrylic acid and growth under expression conditions, cells were lysed and analyzed on SDS-polyacrylamide gel. As can be seen in FIGURE 1b, proteins of the expected size could be identified. The material was found almost exclusively in inclusion bodies together with a truncated hIL-3 starting at methionine-19 of the mature hIL-3 sequence, possibly arising from an internal start of translation.

Production of Homogeneous rhIL-3 by Enterokinase Cleavage

Bacteria containing plasmid pSFI 809 were used for the production of homogeneous rhIL-3. The purification process included the following steps: Disruption of *E. coli* cells in a French pressure cell, extraction of the pellet with 8 M urea, and refolding of the fusion protein by dialysis against a redox buffer followed by ion exchange and reversed phase high performance liquid chromatography (HPLC). With the purified fusion protein the optimal enzyme/substrate ratio for enterokinase cleavage was determined (FIG. 2). Further purification by ion exchange and reversed phase HPLC yielded NH$_2$-terminally homogeneous, biologically active IL-3.

REFERENCES

1. LINDLEY, I., H. ASCHAUER, J. M. SEIFERT, C. LAM, W. BRUNOWSKY, E. KOWNATZKI, M. THELEN, P. PEVERI, B. DEWALD, V. TSCHARNER, A. WALZ & M. BAGGIOLINI. 1988. Proc. Natl. Acad. Sci. USA **85**: 9199–9203.
2. HOPP, T. P., K. S. PRICKETT, V. L. PRICE, R. T. LIBBY, C. J. MARCH, D. P. CERRETTI, D. L. URDAL & P. J. CONLON. 1988. Biotechnology **6**: 1204–1209.

A Cloned Gene for Human Transferrin

C. L. HERSHBERGER, J. L. LARSON, B. ARNOLD,
P. R. ROSTECK, JR., P. WILLIAMS, B. DeHOFF, P. DUNN,
K. L. O'NEAL, M. W. RIEMEN, P. A. TICE, R. CROFTS,
AND J. IVANCIC

Lilly Research Laboratories
A Division of Eli Lilly and Company
Lilly Corporate Center
Indianapolis, Indiana 46285

Transferrin is a circulating serum protein responsible for delivering iron to cells. Serum transferrin is a member of a protein family that includes lactoferrin, melanotransferrin,[1] and ovotransferrin.[2] It is an 80-kD glycoprotein that contains 38 cysteines that form 19 disulfide bonds.[3] The structures of lactoferrin and rabbit transferrin have been determined by X-ray crystallography.[4,5] Human transferrin evolved by gene duplication and subsequent divergence, so that the amino acid sequences in the NH_2-terminal and COOH-terminal domains are highly similar.[6] Sugars are attached to the COOH-terminal domain; however, the NH_2-terminal domain is not glycosylated.[5] The nucleotide sequence for the coding DNA compiled from partial sequences in different publications[6–8] and the amino acid sequence of the protein[9,10] are available. The nucleotide sequence for human transferrin mRNA has been published independently.[11]

Transferrin is recognized by a specific surface receptor when it contains bound iron. The iron-transferrin-receptor complex is internalized and then iron is released into the cytoplasm.[12,13] The receptor-transferrin complex recycles to the cell surface where the iron-free transferrin is released. In addition, transferrin acts as an autocrine growth factor[14] and exhibits sequence similarity to that of a transforming gene.[15,16] The species-specific[17,18] intracellular iron delivery and growth-stimulating properties of transferrin suggest several potential applications for the protein. Rapidly growing tumor cells contain unusually high concentrations of transferrin receptors on their surface because large amounts of iron are needed to sustain rapid growth.[19] Cytotoxic agents have been coupled to transferrin without affecting receptor recognition.[20] The complexes may target the agents to kill the rapidly growing tumor cells with small doses of the cytotoxins. Specific tumor targeting has been demonstrated *in vitro* and clinically.[21–23] Some mammalian pathogens use species-specific receptors for mammalian transferrin to capture iron for their own growth.[24–26] Availability of human transferrin may open the door for rational drug design and screens to find antibacterial agents that interfere with iron uptake by the pathogens. Animal cells in culture require serum for growth; however, human insulin plus human transferrin can substitute for serum to prepare serum-free growth medium.[27,28] Finally, transferrin crosses the blood-brain barrier, indicating that it might be developed into a carrier to deliver neuroactive compounds to the brain and central nervous system.[29] All of the potential uses require substantial quantities of transferrin to test and develop the applications. We chose to clone and express the human transferrin gene in *Escherchia coli* as a method to obtain large quantities of the protein.

CLONING AND CHARACTERIZATION OF THE GENE

A 44-nucleotide synthetic probe was synthesized to complement bases 41–85 in the published sequence of a partial cDNA clone of human transferrin.[7] A human liver cDNA library[30] was screened to isolate clones that were identified by colony hybridization with the synthetic probe. Twelve recombinant clones were isolated. The recombinant plasmid containing the largest DNA insert was designated pHDM99 (FIG. 1). Sequence determination revealed that pHDM99 contained the coding sequence for the entire transferrin protein including the signal peptide and mature

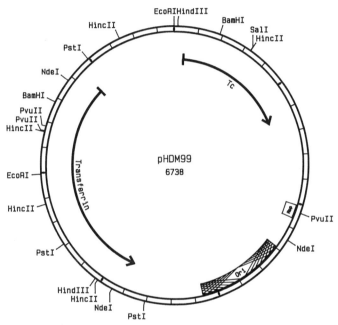

FIGURE 1. Plasmid pHDM99 contains the largest DNA insert in pBR322 that hybridized to the synthetic probe for the human transferrin gene. The *cross-hatched box* indicates the location of the *insert* on the restriction map. Oligo dG:dC tails flank the cloned insert. The *arrow* labeled Tc-r marks the position and direction of transcription for the tetracycline resistance gene on the vector.

protein, a 173-base 3′-untranslated sequence ending with polyadenosine (polyA), and a 78-base 5′-untranslated sequence (FIG. 2). The 5′-untranslated sequence contained an unusual inverted complementary sequence in the terminal 92 nucleotides. The sequence has the potential to form a stem-loop structure with 41 bp in the stem and 10 bases in the loop. Such a long inverted complementary sequence probably arose during the synthesis of cDNA from the polyA containing mRNA. A shorter inverted complementary sequence with 11 bp in the stem and 10 bp in the loop occurs in the sequence of the mRNA.[11] The first strand of the cDNA probably folded back on itself to form the 11-bp stem that served as a primer for the additional

```
        10                    30                    50                    70                    90
PstI
                                              AC  A                 C                     A
                                                                                          A
CTGCAGGGGGGGG GGGGGGGACGGCGGA GCCTCATCCTCCGGG TGCGGCGGCTGAGCAG CGAGTCxCGACTGTG CTCGCTGCTCAGCGC CGCACCCGGAGATG
                      Inverted Complement                    Loop        Inverterd Complement    Met
PolyG Cloning Linker ----------------------------------------->                  <------------------------------<----------
|------------------------|                    |-->----|            |------->
        110                  130                  150      | Exon 1  170              190


AGGCTCGCCGTGGGA GCCCTGCTGGTCTGC GCCGTCCTGGGGCTG TGTCTGGCTGTCCCT GATAAAACTGTGAGA TGGTGTGCAGTGTCG GAGCATGAGGCCACT
ArgLeuAlaValGly AlaLeuLeuValCys AlaValLeuGlyLeu CysLeuAlaValPro AspLysThrValArg TrpCysAlaValSer GluHisGluAlaThr
                                          Signal Peptide | Mature Transferrin
                                          Exon 1 | Exon 2              ATGGGCCCT-Changed Sequence to Create ApaI Site
        210                  230                  250                  270                  290                  310
        NdeI


AAGTGCCAGAGTTTC CGCGACCATATGAAA AGCGTCATTCCATCC GATGGTCCCAGTGTT GCTTGTGTGGAAGAAA GCCTCCTACCTTGAT TGCATCAGGGCCATT
LysCysGlnSerPhe ArgAspHisMetLys SerValIleProSer AspGlyProSerVal AlaCysValLysLys AlaSerTyrLeuAsp CysIleArgAlaIle
                                                                                                            Exon
        330                  350                  370                  390                  410
                                              T


GCGGCAAACGAAGCG GATGCTGTGACACTG GATGCAGAGTTTGGTG TATGATGCTTACCTG GCTCCCAATAACCTG AAGCCTGTGGTGGCA GAGTTCTATGGTCA
AlaAlaAsnGluAla AspAlaValThrLeu AspAlaGlyLeuVal TyrAspAlaTyrLeu AlaProAsnAsnLeu LysProValValAla GluPheTyrGlySer
2 | Exon 3
        430                  450                  470                  490                  510
                                              T


AAAGAGGATCCACAG ACTTTCTATTATGCT GTTGCTGTGGTGAAG AAGGATAGTGGCTTC CAGATGAACCAGCTT CGAGGCAAGAAGTCC TGCCACACGGGTCTA
LysGluAspProGln ThrPheTyrTyrAla ValAlaValValLys LysAspSerGlyPhe GlnMetAsnGlnLeu ArgGlyLysLysSer CysHisThrGlyLeu
Exon 3 | Exon 4
```

FIGURE 2. Legend on page 146.

530   550   570   590   610

PCR Primer TrfE5-1   PvuII
------------->

GGCAGGTCCGCTGGG TGGAACATCCCATA GGCTTACTTACTGT GACTTACCTGAGCCA CGTAAACCTCTTGAG AAAGCAGTGGCCAAT TTCTTCTCCGGCAGC
GlyArgSerAlaGly TrpAsnIleProIle GlyLeuLeuTyrCys AspLeuProGluPro ArgLysProLeuGlu LysAlaValAlaAsn PhePheSerGlySer
                                                                                         Exon 4 | Exon 5

630   650   670   690   710   730

PCR Primer TrfE5-2   G
<-------------------- G

TGTGCCCCTTGTGCG GATGGGACGGACTTC CCCCAGCTGTGTCAA CTGTGTCCAGGGTGT GGCTGCTCCCACCCTT AACCAATACTTCGGC TACTCAGGAGCCTTC
CysAlaProCysAla AspGlyThrAspPhe ProGlnLeuCysGln LeuCysProGlyCys GlyCysSerThrLeu AsnGlnTyrPheGly TyrSerGlyAlaPhe
                                                                                                          Exon

750   770   790   810   830

PCR Primer TrfE7-1   SacI Polymorph
------------->   C

AAGTGTCTGAAGGAT GGTGCTGGGGATGTG GCCTTTGTCAAGCAC TCGACTATATTGAG AACTTGGCAAACAAG GCTGACAGGAGGACCAG TATGAGCTGCTTTGC
LysCysLeuLysAsp GlyAlaGlyAspVal AlaPheValLysHis SerThrIlePheGlu AsnLeuAlaAsnLys AlaAspArgAspGln TyrGluLeuLeuCys
5 | Exon 6   Exon 6 | Exon 7

850   870   890   910   930

A   A

CTGGACAACACCCGG AAGCCGGTAGATGAA TACAAGGACTGCCAC TTGGCCCAGGTCCCT TCTCATACCGTCGTG GCCCAAGTATGGGC GGCAAGGAGGACTTG
LeuAspAsnThrArg LysProValAspGlu TyrLysAspCysHis LeuAlaGlnValPro SerHisThrValVal AlaArgSerMetGly GlyLysGluAspLeu

950   970   990   1010   1030

PCR Primer TrfE7-2   PCR Primer TrfE8-1
<----------------,   -------------->

ATCTGGGAGCTTCTC AACCAGGCCCAGGAA CATTTTGGCCAAAGAC AAATCAAAAGAATTC CAACTATTCAGCTCT CCTCATGGGAAGGAC CTGCTGTTTAAGGAC
IleTrpGluLeuLeu AsnGlnAlaGlnGlu HisPheGlyLysAsp LysSerLysGluPhe GlnLeuPheSerSer ProHisGlyLysAsp LeuLeuPheLysAsp
Exon 7 | Exon 8   TfD1-Gly

FIGURE 2 (*continued*). Legend on page 146.

```
1050        1070        1090        1110        1130        1150
                          A                                  PCR Primer TrfE8-2
                          A         A                        <---------------------- T---
                                    A

TCTGCCACGGGTTT TTAAAAGTCCCCCCC AGGATGATGCCAAG ATGTACCTGGCTAT GAGTATGTCACTGCC ATCCGGAATCTACGG GAAGGCACATGCCCA
SerAlaHisGlyPhe LeuLysValProPro ArgMetAspAlaLys MetTyrLeuGlyTyr GluTyrValThrAla IleArgAsnLeuArg GluGlyThrCysPro
TfDChi-Arg                      New-Asn                                                        Exon 8 | Exon

     1170           1190        1210        1230        1250
PCR Primer TrfE9-2
----------->

GAAGCCCCAACAGAT GAATGCAAGCCTGTG AAGTGGTGTGCGCTG AGCCACCACGAGAGG CTCAAGTGTGATGAG TGGAGTGTTAACAGT GTAGGGAAAATAGAG
GluAlaProThrAsp GluCysLysProVal LysTrpCysAlaLeu SerHisHisGluArg LeuLysCysAspGlu TrpSerValAsnSer ValGlyLysIleGlu
9

     1270        1290        1310        1330        1350
PvuII Polymorph  PCR Primer TrfE9-1
     T           <------- T-------|

TGTGTATCAGCAGAG ACCACCGAAGACTGC ATCGCCAAGATCATG AATGGAGAAGCTGAT GCCATGAGCTTGGAT GGAGGGTTTGTCTAC ATAGCGGCCAAGTGT
CysValSerAlaGlu ThrThrGluAspCys IleAlaLysIleMet AsnGlyGluAlaAsp AlaMetSerLeuAsp GlyGlyPheValTyr IleAlaGlyLysCys
              TfBShaw-Arg  or ArgExon9| Exon 10                                TfDEvans-Arg
         C

     1370           1390        1410        1430        1450
                                  Synthetic Probe for Colony Hybridization
                                  |----------------------------------|

GGTCTGGTGCCTGTC TTGGCAGAAAACTAC AATAAGAGCGATAAT TGTGAGGATACACCA GAGGCAGGGTATTT GCTGTAGCAGTGGTG AAGAAATCAGCTCT
GlyLeuValProVal LeuAlaGluAsnTyr AsnLysSerAspAsn CysGluAspThrPro GluAlaGlyTyrPhe AlaValAlaValVal LysLysSerAlaSer
              Exon 10 | Exon 11                                 Exon 11 | Exon 12
     1470        1490        1510        1530        1550        1570

GACCTCACCTGGGAC AATCTGAAAGGCAAG AAGTCCTCGCCATACG GCAGTTGGCAGAACC GCTGCTGGAACATC CCCATGGGCCTGCTC TACAATAAGATCAAC
AspLeuThrTrpAsp AsnLeuLysGlyLys LysSerCysHisThr AlaValGlyArgThr AlaGlyTrpAsnIle ProMetGlyLeuLeu TyrAsnLysIleAsn
```

FIGURE 2 (*continued*). Legend on page 146.

```
         1590            1610             1630             1650             1670
CACTGCAGATTGAT  GAATTTTCAGTGAA  GGTTGTGCCCCTGGG  TCTAAGAAAGACTCC  AGTCTCTGTAAGCTG  TGTATGGCTCAGGC  CTAAACCTGTTGTGAA
HisCysArgPheAsp GluPhePheSerGlu GlyCysAlaProGly  SerLysLysAspSer  SerLeuCysLysLeu  CysMetGlySerGly LeuAsnLeuCysGlu

Exon 12 | Exon 13
         1690             1710             1730             1750             1770
CCCAACAACAAAGAG  GGATACTACGGCTAC  ACAGGCGCTTCAGG  TGTCTGGTTGAGAAG  GGAGATGTGGCCTTT  GTGAAACACCAGACT  GTCCCACAGAACACT
ProAsnAsnLysGlu  GlyTyrTyrGlyTyr  ThrGlyAlaPheArg CysLeuValGluLys  GlyAspValAlaPhe  ValLysHisGlnThr  ValProGlnAsnThr
                                                                                                                  Exon
                                           Exon 13 | Exon 14
         1790             1810             1830             1850             1870

         1890             1910             1930             1950             1970             1990
GGGGGAAAAAACCCT  GATCCATGGGCTAAG  AATCTGAATGAAAAA  GACTATGAGTTGCTG  TGCCTTGATGGTACC  AGGAAACCTGTGGAG  GAGTATGCGGACTGC
GlyGlyLysAsnPro  AspProTrpAlaLys  AsnLeuAsnGluLys  AspTyrGluLeuLeu  CysLeuAspGlyThr  ArgLysProValGlu  GluTyrAlaAsnCys

14 | Exon 15
                                           HindIII
         2010             2030             2050             2070             2090
CACCTGGCCAGAGCC  CCGAATCACGCTGTG  GTCACACGGAAAGAT  AAGGAAGCTTGCGTC  CACAAGATATTACGT  CAACAGCAGCACCTA  TTTGGAAGCAACGTA
HisLeuAlaArgAla  ProAsnHisAlaVal  ValThrArgLysAsp  LysGluAlaCysVal  HisLysIleLeuArg  GlnGlnGlnHisLeu  PheGlySerAsnVal

                                                                     Exon 15 | Exon 16

ACTGACTGCTCGGGC  AACTTTTGTTTGTTC  CGGTCGGAAACCAAG  GACCTTCGTTCAGA  GATGACACAGTATGT  TTGGCCAAACTTCAT  GACAGAAACACATAT
ThrAspCysSerGly  AsnPheCysLeuPhe  ArgSerGluThrLys  AspLeuLeuPheArg  AspAspThrValCys  LeuAlaLysLeuHis  AspArgAsnThrTyr
```

FIGURE 2 (*continued*). Legend on page 146.

```
              2110          2130          2150          2170          2190
                                                                      AccI

GAAAATACTTAGGA GAAGAATATGTCAAG GCTGTTGGTAACCTG AGAAAATGCTCCACC TCATCACTCCTGGAA GCCTGCACTTCCGT AGACCTTAAAATCTC
GluLysTyrLeuGly GluGluTyrValLys AlaValGlyAsnLeu ArgLysCysSerThr SerSerLeuLeuGlu AlaCysThrPheArg ArgPro
TfB2-Gln                                                                                          | 3'-
                                                                                                  TGAGAT
   2210          2230          2250    Exon 16 | Exon 17   2270          2290

AGAGGTAGGGCTGCC ACCAAGGTGAAGATG GGAACGCAGATGATC CATGAGTTTGCCCTG GTTTCACTGCCCAA GTGGTTGTGCTAAC CACGTCTGTCTTCAC
Untranslated Sequence
CT—Changed Sequence to Create a BglII Site
   2310          2330          2350          2370          2390
                                                           PstI

AGCTCTGTGTTGCCA TGTGTGCTGAACAAA AAATAAAAATTATTA TTGATTTAAAAAAA CCCCCCCCCCCCCCC CTGCAG  2390
                                                            |--------------|
                                                            PolyC Cloning Linker
```

FIGURE 2. Sequence of cloned DNA fragment containing the human transferrin gene. The central line contains the nucleotide sequence determined in this study, whereas the next line down contains the translated amino acid sequence corresponding to human transferrin. Bases on the first line above our nucleotide sequence indicate bases that differ in the compiled nucleotide sequence. Bases on the second line above our nucleotide sequence indicate bases that differ in the sequence derived from the mRNA. *Vertical lines* show the positions of junctions between exons as described in the compiled sequence. *Labeled dashed lines* show the extent of features that are indicated by the label on the dashed line. The indicated PCR primers identify the position of primers that were used for the polymorph studies. *Solid lines* above the sequence mark the location of restriction sites that were used in the polymorph studies or in plasmid constructions described in the text. Amino acids indicated below the amino acid sequence show the identified amino acid substitutions in different transferrin alleles which are designated next to the substituted amino acid.

synthesis that terminated after 30 bases of additional sequence had been copied. Our determined nucleotide sequence in FIGURE 2 differed at 14 positions from the compiled nucleotide sequence[6-8] and at 4 positions from the mRNA sequence.[11] Five of the differences from the compiled sequence occurred in the 5'-untranslated leader and nine of the differences occurred in the coding sequence for the mature protein. However, eight of the nucleotide differences in the coding sequence were degenerate, so that they did not change the amino acid sequence predicted from the nucleotide sequence. The exception at base 1086 changed the asparagine codon, AAT, in the compiled sequence to an aspartic acid codon, GAT, in our sequence.

The published mRNA sequence for human transferrin cDNA[11] agrees with our results at position 1086 rather than the compiled sequence, but it shows sequence discrepancies at four other positions (FIG. 2). One difference occurs in the 5'-untranslated leader sequence. The three variations in the coding sequence occur in degenerate positions. They do not affect the predicted amino acid sequence.

Two restriction site polymorphisms are predicted in the compiled sequence for restriction sites that are not present in our sequence. The restriction site polymorphisms could be useful in identifying and mapping transferrin alleles. Therefore, we investigated the frequency of the polymorphs in a small sample of the human population. The sites are *Sac*I at nucleotide 833 (FIG. 2) located in exon 7 and *Pvu*II at nucleotide 1271 (FIG. 2) located in exon 9. Both the compiled sequence and Lilly's sequence contain a *Pvu*II site in exon 5. DNA was isolated from the buffy white cells of 14 human volunteers and analyzed to distinguish if Lilly's sequence or the compiled sequence predominated in a small sample of the human population. Portions of exons 5, 7, and 9 were separately amplified by polymerase chain reactions and separately analyzed to identify *Pvu*II, *Sac*I, and *Pvu*II sites, respectively. All of the examined samples contained the *Pvu*II site in exon 5, but none of the samples contained the *Sac*I site in exon 7 or the *Pvu*II site in exon 9. Therefore, 15 examples out of 15 samples including our cDNA clone and DNA from 14 volunteers did not contain the polymorphic restriction sites.

The G in Lilly's sequence and the A in the compiled sequence at base 1086, located in exon 8, predict an aspartic acid coded by our sequence and an asparagine coded by the compiled sequence. The published mRNA sequence[11] agrees with Lilly's sequence. The sequence at base 1086 was analyzed in human DNA samples to distinguish which of the sequences predominated in the natural population. A portion of exon 8 was amplified by asymmetric polymerase chain reaction[31,32] and sequenced.[33] All 14 samples contain a G at base 1086, which corresponds to Lilly's sequence and the published mRNA sequence rather than the compiled sequence at base 1086. Therefore, both of the restriction site polymorphs just analyzed and the codon discrepancy agree with our sequence and the published mRNA sequence rather than the compiled sequence.

EXPRESSION OF HUMAN TRANSFERRIN IN *ESCHERICHIA COLI*

A vector, pHKY292 (FIG. 3), was constructed so that the cloned human *trf* gene could be tested for expression with the P_L promoter of bacteriophage λ. The parental plasmid for construction of pHKY292 was equivalent to pL110/K2[34] except that it contained a different *Nde*I-*Bam*HI insert. The *Nde*I-*Bam*HI fragment in pHKY292 is a synthetic linker that was specifically designed for substitution of the human transferrin gene. pHKY292 is designed to easily exchange promoters, ribosome binding sites, first cistrons,[35,36] transcription terminators, and structural genes to examine how these elements affect gene expression. Promoters are changed by

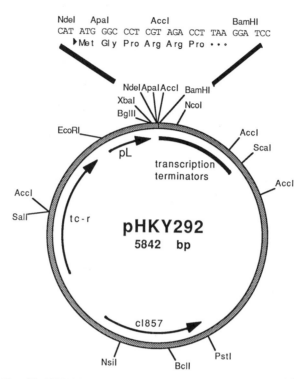

FIGURE 3. Plasmid pHKY292 is the expression vector for human transferrin. The *circle* summarizes the restriction and function map. The *arrows* labeled tc-r and cI857 identify the location and direction of transcription for the tetracycline resistance gene and λ repressor gene, respectively. The *arrow labeled pL* identifies the location and orientation of the fragment containing a transcription promoter. The *line labeled terminators* identifies the location of a DNA fragment containing transcription terminators. The *inset nucleotides* shows the sequence of the *Nde*I-*Bam*HI linker between the *Nde*I and *Bam*HI sites on the plasmid. The transferrin gene was inserted between the *Apa*I and *Acc*I sites in the synthetic linker using a multistep construction as described in the text.

substituting *Eco*RI-*Bgl*II fragments, ribosome binding sites are changed by substituting *Bgl*II-*Xba*I fragments, first cistrons are exchanged by substituting *Xba*I-*Nde*I fragments, structural genes are inserted by substituting *Nde*I-*Bam*HI fragments, and transcription terminators are exchanged by substituting *Bam*HI-*Sca*I fragments. Inclusion of the internal *Apa*I site in the synthetic *Nde*I-*Bam*HI fragment specifies the amino acids methionine, glycine, and proline as the first three amino acids of an open reading frame. Proline is the second amino acid of mature transferrin; however the natural sequence of the gene does not contain an *Apa*I site. A synthetic *Apa*I-*Nde*I linker encoding 24 amino acids to the internal *Nde*I site of transferrin was prepared to introduce an *Apa*I cleavage site into the proline codon. The reconstructed transferrin gene was cloned in place of the *Apa*I-*Bam*HI sequence in pHKY292 to generate pHKY296. The resulting sequence codes for a derivative of transferrin with methionyl-glycine replacing the valine normally present at the NH$_2$-terminus of serum transferrin. The encoded protein was designated d-Trf for derivative of transferrin. We did not perform experiments to assess whether the methionine is present or removed from the protein, which accumulates in *E. coli*. The COOH-terminal end of the gene contains the 3′-untranslated segment of the natural cDNA for transferrin. *E. coli* 294 containing pHKY296 accumulated low levels of d-Trf that could be detected in Western blots but did not form a distinct band in stained electrophoretic gels. Several *E. coli* strains (TABLE 1) were tested for expression of d-Trf from pHKY296. Production differed dramatically with the highest levels of product accumulated in the *lon* and *htpR* strain, L201. These results suggest that proteolysis exerts a major influence on the amount of d-Trf that accumulates.

An *Apa*I-*Bgl*II fragment was constructed to contain the transferrin coding sequence from the proline at codon two of serum transferrin or codon three of d-Trf through the COOH-terminal proline. A synthetic *Acc*I-*Sma*I linker was inserted at the *Acc*I site, which cleaves one codon before the C-terminal proline codon. The sequence on one strand of the linker is 5′-AGACCTTAATGAGATCTGTCGAC-CCGGG-3′. Inclusion of the *Sma*I site allowed cloning the COOH-terminal *Hind*III-*Sma*I fragment and synthetic linker into pUC19 to generate pHKY293. The COOH-terminal coding sequence could be excised on a *Hind*III-*Bgl*II fragment to construct the transferrin coding sequence on an *Apa*I-*Bgl*II fragment. The construction deleted the 3′-untranslated sequence of the cDNA. Substitution of the *Apa*I-*Bgl*II portable coding sequence for the *Apa*I-*Bam*HI fragment of pHKY292 generated pHKY320 which accumulated substantially higher levels of d-Trf than did pHKY296. The increased level of d-Trf suggests that the 3′-untranslated sequence of the natural sequence cDNA inhibits expression in *E. coli*.

Several different synthetic derivatives of the λ P_L promoter were substituted in pHKY320. The highest level of product accumulated with pHKY341 which produced approximately 15 μg/ml or 2% of the dry cell mass as d-Trf. Strains containing pHKY341 produced cytoplasmic inclusion bodies containing d-Trf. Similar levels of product accumulated in a variety of the host strains in TABLE 1, suggesting that proteolysis ceased to affect yield when the cells accumulate high levels of protein in cytoplasmic granules.

TABLE 1. *Escherichia coli* Host Strains

Strain	Genotype	Reference
JM109	*recA*21 *supE*44 *endA*1 *hsdR*17 *gyrA*96 *relA*1 *thi* δ(*lac-proAB*) F′(*traD*36 *proAB*⁺ *lacI*q *lacZ* δM15)	44
L201	*lonR*9 *htp*165 *supC*ᵗˢ (deletion of Tn*10* from LC137)	45 and Jaskunas[a]
L640	*lacY*1 *supE*44 *rfbD*1 (P1 transduction of WA704)	Jaskunas[a]
L641	*tonA*21 *lacY*1 *supE*44 *rfbD*1 *thi*-1 (P1 transduction of C600)	Jaskunas[a]
L687	*lacY*1 *supE*44 *rfbD*1 *lon cps* (from L640)	Jaskunas[a]
L693	*tonA*21 *lacY*1 *supE*44 *rfbD*1 *thi*-1 *lon cps* (from L641)	Jaskunas[a]
RV308	RV*lac*δₓ 74 *gal* ISII::OP308 *strA*	46
MM294	*supE*44 *rfbD*1 *spoT*1 *thi*-1 *endA*1 *hsdR*17	47
L695	RV*lac*δₓ 74 *gal* ISII::OP308 *fusA* (P1 transduction of RV308)	Jaskunas[a]

[a]Strain was constructed by R. Jaskunas and A. Majeski, personal communication.

BINDING TRANSFERRIN RECEPTOR

d-Trf produced in recombinant *E. coli* cells is not glycosylated because *E. coli* does not glycosylate proteins. Therefore, control experiments compared the ability of native transferrin from human serum and deglycosylated transferrin from human serum to bind human transferrin receptor. Serum transferrin was deglycosylated by treatment with peptide: *N*-glycosidase F (E. C. 3.5.152 and 3.2.2.18; Genzyme, Boston, Massachusetts). Samples were prepared in which approximately 50% of the polysaccharide was removed in one case and more than 90% was removed in the other. Mock digested native transferrin, partially deglycosylated transferrin, and fully deglycosylated transferrin showed equivalent activities in competition with serum transferrin in receptor binding assays; therefore, glycosylation is not required for receptor binding.

Recombinant transferrin was isolated and refolded using standard procedures to obtain the protein in the "native" conformation to examine its ability to bind the transferrin receptor. Granules containing recombinant transferrin were isolated by centrifugation of *E. coli* RV308/pHKY341 lysates. The granules were solubilized, clarified by centrifugation, and concentrated by diafiltration. The retentate was fractionated by size-exclusion chromatography and the appropriate fraction diluted in a pH 10.5 glycine buffer containing cysteine. Air oxidation was carried out at ambient temperature. A parallel control renaturation was performed with serum transferrin substituted for the solubilized granule preparation. Recoveries of refolded transferrin were based on the amount of activity found in the competitive receptor binding assay divided by the total transferrin protein in the samples. Receptor binding assays showed 2.0% recovery of activity with refolded serum transferrin and 1.2% recovery of activity with recombinant transferrin after folding.

DISCUSSION

Human transferrin cDNA was cloned, sequenced, and expressed in *E. coli* to produce cytoplasmic granules containing d-Trf. Recombinant transferrin was recovered in a soluble form that bound to the transferrin receptor protein. In a previous report,[37] investigators expressed a gene fragment coding for the NH_2-terminal half-molecule in baby hamster kidney cells because they were unable to express recombinant transferrin in *E. coli*. The first constructions in our experiments also failed to express significant quantities of transferrin in *E. coli*. The expression cassette encoding methionyl-glycine in place of valine at the NH_2-terminus of the coding sequence was the first construct that produced high levels of product, d-Trf, in *E. coli*. All subsequent constructions to express the product contained the encoded methionyl-glycine at the NH_2-terminus. The coding sequence at the NH_2-terminus may be critical for expression of human transferrin.

Positive results in the receptor binding assays are very encouraging because of the complexity of human transferrin. The serum protein is a glycoprotein containing 678 amino acids and 19 disulfide bonds which suggest that folding of the denatured protein to a functional conformation might be problematic. Receptor binding studies with serum transferrin and deglycosylated serum transferrin demonstrated that glycosylation was not needed for binding to the receptor. Recovery of receptor binding activity in the recombinant protein indicates that glycosylation, the NH_2-terminal valine of the mature protein, and the signal peptide are not needed for conformational folding to generate a binding domain. However, the experiments do not distinguish if receptor binding occurs in the natural configuration or if it simply reflects a fortuitous conformation recognized by the receptor. Although receptor binding is necessary for normal biological function, it may not be adequate to insure complete biological function of the protein. The recombinant protein has a recognizable receptor binding domain, but assays for iron binding, membrane translocation, iron release in the cytoplasm, or other critical functions have not been applied.

Sequence differences between Lilly's clone and the compiled sequence[6-8] came as a surprise. Only one of the nucleotide discrepancies results in an amino acid substitution; however, a limited screen of human volunteers indicates that aspartic acid predominates at amino acid position 310. It is noteworthy that the published amino acid sequence for human transferrin[9,10] contains an asparagine at position 310. The published amino acid sequence differs from our predicted amino acid sequence at eight other positions. All eight differences reflect normal ambiguities that occur during determination of amino acid sequences. Unlike the predicted sequence at amino acid 310 from the compiled nucleotide sequence, the published mRNA nucleotide sequence and our sequence predict the same amino acids at all eight positions. Two other sequence differences were readily distinguishable because they predicted restriction site polymorphisms between Lilly's clone and the compiled sequence. All 15 samples contain the sequence of our clone at both sites. These results indicate that the sequence represented in Lilly's clone predominates, at least in the small population sample. Possibly the two restriction site polymorphisms can be used in rapid screening procedures to identify the allele with asparagine at amino acid 310.

At least 30 different alleles for human transferrin are known.[38] The variant transferrins are designated Tf (electrophoretic class as a capital letter, e.g., B, C, and D; subclass designation as a subscript name or number). TfC, TfD_1, TfD_{Chi}, TfD_{Evans}, TfB_2, and TfB_{Shaw} are examples of different transferrin derivatives. The derivatives of transferrin are distinguishable by electrophoretic analysis. Many variants are believed to differ by mutations that cause substitution of different amino acids.[39] The amino acid substitutions have been identified in five mutants. TfD_1 contains an aspartic acid to glycine change at amino acid 277.[40] TfD_{Chi} contains a histidine to arginine change at amino acid 300.[41] TfB_2 contains a glycine to glutamic acid change at amino acid 652.[42] TfD_{Evans} contains a glycine to arginine change at amino acid 394.[43] TfB_{Shaw} contains an isoleucine to arginine change at amino acid 378 or 381.[38] All of these variants and apparently the wild-type TfC contain aspartic acid at amino acid 310. TfC normally predominates in the population; however, our samples were not analyzed to identify the electrophoretic variant. The predicted amino acid asparagine at amino acid 310 might represent a previously unrecognized allele with asparagine at position 310[6] instead of the aspartic acid that seems to predominate in the population. We propose the designation TfX_{Park} for the new transferrin derivative until its electrophoretic class is determined.

The published mRNA sequence[11] contains four differences from Lilly's sequence. One change occurs in the 5'-untranslated leader (FIG. 2). The three differences in the coding sequence occur in degenerate positions and do not alter the predicted amino acid sequence. Therefore, our nucleotide sequence and the published mRNA nucleotide sequence predict identical amino acid sequences for human transferrin. The compiled nucleotide sequence,[6-8] the published mRNA sequence,[11] and Lilly's sequence (this paper) contain a total of 16 nucleotide differences, which seems to represent a very high frequency of nucleotide divergence for a single gene; however, two of the three sequences agree with each other at each position that contains a nucleotide difference between the other two sequences. Interestingly, only one of the sequence variations causes an amino acid substitution.

ACKNOWLEDGMENTS

Appreciation is expressed to C. Holmes and W. Muth for growing E. coli cultures containing recombinant plasmids that code for human transferrin, B. Roberts for preparing granules to isolate human transferrin, and Eli Lilly's management for supporting the work.

REFERENCES

1. ROSE, T. M., G. D. PLOWMAN, D. B. TEPLOW, W. J. DREYER, K. E. HELLSTROM & J. P. BROWN. 1986. Primary structure of the human melanoma-associated antigen p97 (melanotransferrin) deduced from the mRNA sequence. Proc. Natl. Acad. Sci. USA **83:** 1261–1265.
2. AISEN, P. & I. LISTOWSKY. 1980. Iron transport and storage proteins. Ann. Rev. Biochem. **49:** 357–393.
3. WELCH, S. & A. SKINNER. 1989. A comparison of the structure and properties of human, rat and rabbit serum transferrin. Comp. Biochem. Physiol. **93B:** 417–424.
4. ANDERSON, B. F., H. M. BAKER, E. J. DODSON, G. E. NORRIS, S. V. RUMBALL, J. M. WATERS & E. N. BAKER. 1987. Structure of human lactoferrin at 3.2-A resolution. Proc. Natl. Acad. Sci. USA **84:** 1769–1773.
5. BAILEY, S., R. W. EVANS, R. C. GARRATT, B. GORINSKY, S. HASNAIN, C. HORSBURGH, H. JHOTI, P. F. LINDLEY, A. MYDIN, R. SARRA & J. L. WATSON. 1988. Molecular structure of serum transferrin at 3.3-A resolution. Biochemistry **27:** 5804–5812.
6. PARK, I., E. SCHAEFFER, A. SIDOLI, F. E. BARALLE, G. N. COHEN & M. M. ZAKIN. 1985. Organization of the human transferrin gene: Direct evidence that it originated by gene duplication. Proc. Natl. Acad. Sci. USA **82:** 3149–3153.
7. UZAN, G., M. FRAIN, I. PARK, C. BESMOND, G. MAESSEN, J. S. TREPAT, M. M. ZAKIN & A. KAHN. 1984. Molecular cloning and sequence analysis of cDNA for human transferrin. Biochem. Biophys. Res. Commun. **119:** 273–281.
8. SCHAEFFER, E., M. A. LUCERO, J.-M. JELTSCH, M.-C. PY, M. J. LEVIN, P. CHAMBON, G. N. COHEN & M. M. ZAKIN. 1987. Complete structure of the human transferrin gene. Comparison with analogous chicken gene and human pseudogene. Gene **56:** 109–116.
9. MACGILLIVRAY, R. T. A., E. MENDES, J. G. SCHEWALE, S. K. SINHA, J. LINEBACK-ZINS & K. BREW. 1983. The primary structure of human serum transferrin. J. Biol. Chem. **258:** 3543–3546.
10. MACGILLIVRAY, R. T. A., E. MENDEZ, S. K. SINHA, M. R. SUTTON, J. LINEBACK-ZINS & K. BREW. 1982. The complete amino acid sequence of human serum transferrin. Proc. Natl. Acad. Sci. USA **79:** 2504–2508.
11. YANG, F., J. B. LUM, J. R. MCGILL, C. M. MOORE, S. L. NAYLOR, P. H. VAN BRAGT & B. H. BOWMAN. 1984. Human transferrin: cDNA characterization and chromosomal localization. Proc. Natl. Acad. Sci. USA **81:** 2752–2756.
12. JING, S., T. SPENCER, K. MILLER, C. HOPKINS & I. S. TROWBRIDGE. 1990. Role of human transferrin receptor cytoplasmic domain in endocytosis: Localization of a specific signal sequence for internalization. J. Cell Biol. **110:** 283–294.
13. SOROKIN, L. M., E. H. MORGAN & G. C. T. YEOH. 1988. Transferrin endocytosis and iron uptake in developing myogenic cells in culture: Effects of microtubular and metabolic-inhibitors, sulfhydryl-reagents and lysosomotrophic agents. J. Cell Physiol. **137:** 483–489.
14. SHAPIRO, L. E. & N. WAGNER. 1989. Transferrin is an autocrine growth factor secreted by Reuber H-35 cells in serum-free culture. In Vitro Cell. Dev. Biol. **25:** 650–654.
15. GOUBIN, G., D. S. GOLDMAN, J. LUCE, P. E. NEIMAN & G. M. COOPER. 1983. Molecular cloning and nucleotide sequence of a transforming gene detected by transfection of chicken B-cell lymphoma DNA. Nature **302:** 114–119.
16. DIAMOND, A., J. M. DEVINE & G. M. COOPER. 1984. Nucleotide sequence of a human *blym* transforming gene activated in a Burkitt's lymphoma. Science **225:** 516–519.

17. SOROKIN, L. M. & E. H. MORGAN. 1988. Species specificity of transferrin binding, endocytosis and iron internalization by cultured chick myogenic cells. J. Comp. Physiol. B Biochem. Syst. Environ. Physiol. **158:** 559–566.

18. YOUNG, S. P. & C. GARNER. 1990. Delivery of iron to human-cells by bovine transferrin: Implications for the growth of human cells *in vitro.* Biochem. J. **265:** 587–591.

19. LASKEY, J., I. WEBB, H. M. SCHULMAN & P. PONKA. 1988. Evidence that transferrin supports cell proliferation by supplying iron for DNA synthesis. Exp. Cell Res. **176:** 87–95.

20. ADES, E. W. & G. J. CULLINAN. 1985. Cytotoxic compositions of transferrin coupled to vinca alkaloids. U.S. Patent. 4,522,750: 1–8.

21. CHEN, Y., N. WILLMOTT, J. ANDERSON & A. T. FLORENCE. 1988. Haemoglobin, transferrin and albumin/polyaspartic acid microspheres as carriers for the cytotoxic drug adriamycin. I. Ultrastructural appearance and drug content. J. Controlled Release **8:** 93–101.

22. WILLMOTT, N., Y. CHEN & A. T. FLORENCE. 1988. Haemoglobin, transferrin and albumin/polyaspartic acid microspheres as carriers for the cytotoxic drug adriamycin II. *In vitro* drug release rate. J. Controlled Release **8:** 103–109.

23. ELLIOTT, R. L., R. STJERNHOLM & M. C. ELLIOTT. 1988. Preliminary evaluation of platinum transferrin (MPTC-63) as a potential nontoxic treatment for breast cancer. Cancer Detect. Prev. **12:** 469–480.

24. WEST, S. E. H. & P. F. SPARLING. 1985. Response of *Neisseria gonorrhoeae* to iron limitation: Alterations in expression of membrane proteins without apparent siderophore production. Infect. Immun. **47:** 388–394.

25. MORTON, D. J. & P. WILLIAMS. 1989. Utilization of transferrin-bound iron by *Haemophilus* species of human and porcine origins. FEMS Microbiol. Lett. **65:** 123–128.

26. MORTON, D. J. & P. WILLIAMS. 1990. Siderophore-independent acquisition of transferrin-bound iron by *Haemophilus influenzae* type-B. J. Gen. Microbiol. **136:** 927–933.

27. WATKINS, L. F., L. R. LEWIS & A. E. LEVINE. 1990. Characterization of the synergistic effect of insulin and transferrin and regulation of their receptors on a human colon carcinoma cell line. Int. J. Cancer. **45:** 372–375.

28. NAKANISHI, Y., F. CUTTITTA, P. G. KASPRZYK, I. AVIS, S. M. STEINBERG, A. F. GAZDAR & J. L. MULSHINE. 1988. Growth factor effects on small cell lung cancer cells using a colorimetric assay: Can a transferrin-like factor mediate autocrine growth. Exp. Cell Biol. **56:** 74–85.

29. MONTEROS, A. E., L. A. PENA & J. DE VELLIS. 1989. Does transferrin have a special role in the nervous system. J. Neurosci. Res. **24:** 125–136.

30. BECKMAN, R. J., R. J. SCHMIDT, R. F. SANTERRE, J. PLUTZKY, G. R. CRABTREE & G. L. LONG. 1985. The structure and evolution of a 461 amino acid human protein C precursor and its messenger RNA, based upon the DNA sequence of cloned human liver cDNAs. Nucl. Acids Res. **13:** 5233–5247.

31. KREITMAN, M. & L. F. LANDWEBER. 1989. A strategy for producing single-stranded DNA in the polymerase chain reaction: A direct method for genome sequencing. Gene Anal. Tech. **6:** 84–88.

32. GYLLENSTEN, V. B. & H. A. ERLICH. 1989. Generation of single-stranded DNA by the polymerase chain reaction and its application to direct sequencing of the HLA-DQ(alpha) locus. Proc. Natl. Acad. Sci. USA **85:** 7652–7656.

33. HERSHBERGER, C. L. & P. R. ROSTECK, JR. 1991. Genetic and molecular studies to certify seed pools and production fermentations of industrial *Escherichia coli* strains. *In* Drug Biotechnology Regulation Scientific Basis and Practices. Y.-Y. Chiu & J. L. Guerigian, eds.: 86–101. Marcel Dekker, Inc. New York, NY.

34. WILHELM, O. G., S. R. JASKUNAS, C. J. VLAHOS & N. U. BANG. 1990. Functional properties of the recombinant kringle-2 domain of tissue plasminogen activator produced in *Escherichia coli.* J. Biol. Chem. **265:** 14606–14611.

35. SCHONER, B. E., H. M. HSIUNG, R. M. BELAGAJE, N. G. MAYNE & R. G. SCHONER. 1984. Role of mRNA translational efficiency in bovine growth hormone expression in *Escherichia coli.* Proc. Natl. Acad. Sci. USA **81:** 5403–5407.

36. SCHONER, B., R. M. BELAGAJE & R. G. SCHONER. 1990. Enhanced translational efficiency

with two-cistron expression system. *In* Methods in Enzymology: Gene Expression Technology. D. V. Goeddel, ed. **185:** 94–103. Academic Press, Inc. New York, NY.

37. FUNK, W. D., R. T. A. MACGILLIVRAY, A. B. MASON, S. A. BROWN & R. C. WOODWORTH. 1990. Expression of the amino-terminal half-molecule of human serum transferrin in cultured cells and characterization of the recombinant protein. Biochemistry **29:** 1654–1660.

38. WELCH, S. & L. LANGMEAD. 1990. A comparison of the structure and properties of normal human transferrin and a genetic variant of human transferrin. Int. J. Biochem. **22:** 275–282.

39. PETREN, S. & O. VESTERBERG. 1989. Separation of different forms of transferrin by isoelectric-focusing to detect effects on the liver caused by xenobiotics. Electrophoresis **10:** 600–604.

40. WANG, A. C. & H. E. SUTTON. 1965. Human transferrin C and D_1. Science **149:** 435–437.

41. WANG, A. C., H. E. SUTTON & P. N. HOWARD. 1967. Human transferrin C and D_{Chi}. Biochem. Genet. **1:** 55–59.

42. WANG, A. C., H. E. SUTTON & A. RIGGS. 1966. A chemical change between transferrin B_2 and C. Am. J. Hum. Gen. **18:** 454–458.

43. EVANS, R. W., A. MEILAK, A. AITKEN, K. J. PATEL, C. WONG, R. C. GARRATT & B. CHITNAVIS. 1988. Characterization of the amino acid change in a transferrin variant. Biochem. Soc. Trans. **16:** 834–835.

44. SAMBROOK, J., E. F. FRITSCH & T. MANIATIS. 1989. Molecular Cloning: A Laboratory Manual. 2nd Edition. Cold Spring Harbor Laboratory Press. Cold Spring Harbor, New York.

45. GOFF, S. A. & A. L. GOLDBERG. 1985. Production of abnormal proteins in *E. coli* stimulates transcription of *lon* and other heat shock proteins. Cell **41:** 587–595.

46. MEYER, B. J., R. MAURER & M. PTASHNE. 1980. Gene regulation at the right operator (O_R) of bacteriophage (lambda) II. O_R1, O_R2, and O_R3: Their roles in mediating the effects of repressor and cro. J. Mol. Biol. **139:** 163–194.

47. BACHMANN, B. J. 1987. Derivations and genotypes of some mutant derivatives of *Escherichia coli* K-12. *Escherichia coli* and *Salmonella typhimurium* Cellular and Molecular Biology. F. C. Neidhardt, J. L. Ingraham, B. Megasanik, K. B. Low, M. Schaechter & H. E. Umbarger, eds. **2:** 1190–1230. American Society for Microbiology. Washington, DC.

Alteration of Industrial Food and Beverage Yeasts by Recombinant DNA Technology

G. H. RANK AND W. XIAO

Department of Biology
University of Saskatchewan
Saskatoon, Saskatchewan S7N 0W0

INFORMATION CONTENT OF HAPLOID LABORATORY YEAST

Genetic information for biological traits is partitioned among three major sources of DNA: nuclear, mitochondrial, and 2 μm plasmid. The hierarchy of information content is evident from the number of base pairs (bp) of DNA found in haploid nuclear (1.5×10^7 bp), mitochondrial (85,000 bp), and 2-μm (6,318 bp) genomes. The complete 2-μm DNA sequence is known,[1] and 92% of the mitochondrial DNA sequence has been published.[2] Sequence decipherment of nuclear DNA is occurring haphazardly on genes of interest in numerous laboratories. An international effort is underway to sequence the entire nuclear genome with chromosomes I, VI, and III assigned to Canada, Japan, and Europe.[3] The immense intellectual and practical benefits that will result from nuclear genome sequencing claim the vigorous, collective support from industry, university, and government.

The nuclear genome is carried on 16 chromosomes and 1 fragment. Chromosome sizes vary from 242 to approximately 2,200 kilobases (kb). Seven hundred sixty-nine genetic loci have been mapped to yeast chromosomes, 15% of the total estimated 5,000 genes. This information is conveniently summarized in Edition 10 of the genetic map.[3]

An apt paradigm for the current and future status of chromosomal DNA analysis is found in the genetic and physical map of chromosome I published by Kaback *et al.*[4] Twenty-one genes are mapped to a 242-kb structure (FIG. 1). Assuming 2,500 bp per locus, the expectation is for approximately 100 genes, so that the current map represents 20% of the genetic information. Early gene disruption[5] and temperature-sensitive mutation[6] analyses indicated that only a small fraction of the genome (< 30%) was essential for cell growth and division. However, DNA sequence analysis of chromosome III and transcriptional mapping of chromosome I[7] indicates that the functional organization of yeast chromosomal DNA approximates contiguous functional genes with minimal intergenic regions (D. Kaback, personal communication). Of particular interest to industry is the expectation that most of the sequences identified as "function unknown" (FUN) genes have a role in cryptic physiological functions related to fermentation. Complete gene banks of individual chromosomes can readily be constructed,[8] and over 95% of the nuclear genome has been resolved by a global physical mapping strategy.[9,10] It is clear that technology is available that allows the identification and isolation of any haploid yeast gene. The industrial use of this technology will depend on the identification of useful traits with isolated sequences and the facile transfer of laboratory strain-based recombinant DNA technology to industrial *Saccharomyces cerevisiae* strains.

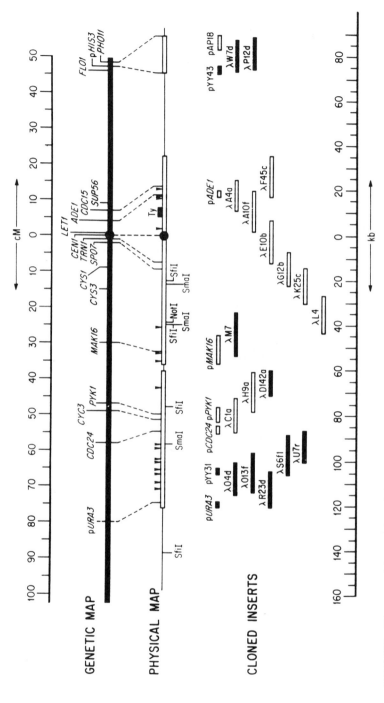

FIGURE 1. Alignment of the genetic and physical maps of chromosome *I* in *S. cerevisiae* (reproduced from Kaback *et al.*[4]). pURA3 denotes a YIp vector integrated into the left arm of chromosome *I*. Edition 10 of the genetic map (Mortimer *et al.*[3]) includes an additional seven markers.

COMPARISON OF LABORATORY AND INDUSTRIAL NUCLEAR GENOMES

The principal laboratory progenitor strain, EM93, originated as an isolate from rotting figs near Merced, California. EM93 contributes to 88% of the gene pool of strain S288C, the latter strain serving as the gene pool for virtually all current laboratory strains.[11] Laboratory strains were selected for convenient traits such as a well-defined heterothallic life cycle and good ascospore survival. In contrast, industrial *Saccharomyces* strains were selected for high performance under diverse fermentation conditions. Industrial selection resulted in isolates with traits that interfere with normal genetic analyses, such as homothallism, polyploidy, aneuploidy, poor sporulation, and low spore viability. Nevertheless, standard breeding procedures emanating from induced sporulation[12] or protoplast fusion are viable methodologies for strain improvement (reviewed by Spencer *et al.*[13] and Spencer and Spencer[14]); for example, formal genetic analysis of principal components of bakers' yeast was recently reported.[12,15]

Alternating field gel techniques are capable of resolving the chromosomes of *S. cerevisiae*.[16] The limited chromosomal length polymorphism seen in laboratory strains has been attributed to variation in copy number of the telomeric consensus sequence C_{1-3}-A.[17] Extensive polymorphism of industrial chromosomes is evident within and between strains used in different fermentation industries[18] and provides a convenient method of fingerprinting individual strains (G. Casey, Anheuser-Busch Inc., personal communication).

A detailed genetic/physical map of any industrial *Saccharomyces* chromosome has not been produced. Furthermore, the resources required for this information make such mapping unlikely in the near future. Thus, recombinant DNA manipulation of industrial strains currently relies on the homology of laboratory and industrial chromosomes. Transfer of single Carlsberg lager strain M244 chromosomes to a laboratory background identified both homologous and homeologous regions (FIG. 2). Some homeologous areas, such as *ILV3* to *HOM6* of Type I (FIG. 2), do not undergo general recombination, whereas other areas of other homeologous chromosomes (*ILV3* to *HOM6* of Type II, FIG. 2) recombine with the laboratory reference chromosome. Similar observations have been recorded for chromosomes III, V, XII, and XIII.[19] Homeologous genes function in laboratory strains because they complement haploid auxotrophic markers.[20,21] Partial sequence conservation between industrial and laboratory genes was evident from low stringency hybridization with laboratory gene probes and from partial alignment with laboratory gene restriction maps.[21] Thus, selection pressure has resulted in sufficient sequence conservation to maintain biological function, but this is often accompanied by enough genomic divergence to interfere with genetic recombination.

GENE TRANSFER VECTORS

Gene transfer and cloning vectors used in *S. cerevisiae* have been reviewed extensively.[22-25] Reviews that focus on industrial gene transfer include those of von Wettstein *et al.*,[26] Knowles and Tubb,[27] Sturley and Young,[28] and Rank *et al.*[29] A generic cartoon of a circular vector is given in FIGURE 3. Vectors contain origins of DNA replication (*ori*) and selectable markers that allow manipulation in the bacterium *Escherichia coli* and *S. cerevisiae*. A variety of unique gene-cloning sites are employed; see Parent *et al.*[24] for the detailed structure of commonly used vectors. Use of the yeast 2-μm plasmid *ori* results in a chimeric yeast episomal plasmid (YEp),

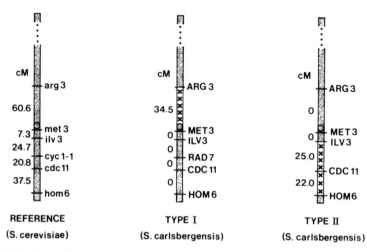

FIGURE 2. Comparison of chromosomes *X* of *S. cerevisiae* (reproduced from Casey[20]). Regions in chromosome *X* of *S. carlsbergensis* (*S. cerevisiae* M244, TABLE 1) that combine with reference chromosome *X* of the haploid laboratory strain are indicated by *crosses*.

whereas use of a chromosomal *ori* (autonomously replicating sequence, *ARS*) creates a yeast-replicating plasmid (YRp) (reviewed by Williamson[30]). In the absence of selection pressure for plasmid maintenance, YEp and YRp are rapidly lost. Vector stability is dramatically increased by the addition of chromosomal cen-

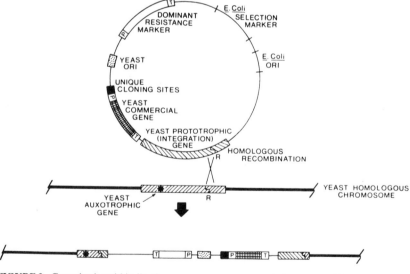

FIGURE 3. Generic plasmid indicating common sequences and chromosomal integration in an area of sequence homology (from Rank *et al.*[29]). Note that the integrant structure contains two copies of the homologous integration sequence flanking an area containing bacterial DNA (—).

tromere sequences (*CEN*). Thus, yeast centromere vectors (YCp) behave somewhat as minichromosomes; nevertheless, in the absence of selection pressure, YCp vectors are also rapidly lost during mitosis and meiosis. Yeast linear vectors YLp contain telomeres (*TEL*) in addition to *ori* and *CEN* and have increased stability when total vector size is in excess of 50 kb. Up to 200 kb of DNA have been inserted into linear vectors[31] to result in yeast artificial chromosomes (pYAC). pYACs have the capacity to clone all or major portions of industrial chromosomes, thereby providing a convenient methodology for isolation and subsequent manipulation of homeologous sequences. Although pYAC technology is extensively used in genomic cloning of diverse species, the yeast fermentation industry has not yet taken advantage of the approximately 10-fold increased cloning capacity of pYAC technology.

Specialized vectors include variable promoter/terminator regulatory combinations as well as coding sequences for protein secretion and posttranslational processing.[24,25,28]

VECTOR REQUIREMENTS OF INDUSTRIAL *SACCHAROMYCES*

As indicated in FIGURE 2 the general model of industrial genomic structure is that every gene is present in several functional copies. For any given function two or more gene sequences may exist, such as homology or homeology to the haploid laboratory sequence. Homologous and homeologous sequences are wild-type and therefore generally do not express any recognizable auxotropic or slow growth phenotypes. This prototrophic amphiploid structure creates the following major difficulties.

1. The commonly employed prototrophic selectable vector markers (*URA3, LEU2, TRP1,* and so on) cannot be used in industrial strains.

2. Recombinant DNA modifications of one gene copy (e.g., site-specific mutation and conditional promoter fusions) may be masked by other homologous or homeologous genes.

3. Homeologous gene structure may prevent general recombination with vector sequences, thereby preventing *in vivo* haploid-based gene manipulation technology (e.g., gene disruption and gene transplacement).

Practical vector requirements include a high transformation efficiency with diverse industrial strains. Transferred genes must be stably inherited without the requirement of an artificial selection protocol for plasmid maintenance.

An important additional requirement of the food and beverage industry is that recombinant yeasts be free of contaminating bacterial DNA vector sequences. This requirement results from the unsubstantiated concern that bacterial sequences *may* result in a real hazard to public health. In addition, industry is wary of adverse emotional public reaction to *any* use of recombinant DNA technology in the food industry.

Despite the aforementioned problems, technology for beneficial recombinant DNA manipulation of industrial yeast has been developed. This technology generally has modified, and built upon, gene cloning and gene transfer technology developed with haploid laboratory yeast.

DOMINANT SELECTABLE MARKERS

Dominant selectable markers fall into two categories (reviewed by Rank *et al.*[29]): (1) non-yeast genes such as *E. coli* transposon genes for resistance to chloramphenicol and the aminoglycoside G418; and (2) yeast genes conferring resistance to copper (*CUP1*) or sulfometuron methyl (*SMR1*). Utilization of non-yeast genes is facilitated by the use of yeast promoter (P) and terminator (T) sequences (FIG. 3) and by identification of new transposon genes with enhanced expression in yeast.[32] Casey *et al.*[33] found that YEp vectors such as pCP2-4-30 (FIG. 4) carrying *SMR1* genes were universally efficient in transforming a wide range of industrial yeasts (TABLE 1). *CUP1* resistance to copper[34,35] and amino-glycoside-3-phosphotransferase resistance to G418[36] have been widely used for transformation of industrial yeast.[29] In the absence of selection pressure, all plasmids are rapidly lost from the population.

STABLE INHERITANCE BY HOMOLOGOUS INTEGRATION

FIGURE 3 depicts the method of integration of circular plasmids into the yeast chromosome. The key feature of the integration process is crossing over within an area of genetic homology. Recombination frequency can be increased up to 1,000-fold by restriction endonuclease cleavage (e.g., site R of FIG. 3) of the plasmid homologous sequence prior to transformation.[37] Such cleavage also has the advantage of precisely targeting integration.

The integrant structure has the disadvantage of containing bacterial sequences flanked by a duplicate copy of the homologous sequence (FIG. 3). The bacterial sequence is undesirable *per se,* and the duplicated sequence provides for a low reversal of the integration procedure.

Nevertheless, Casey *et al.*[38] used this procedure to obtain stable high level expression of melibiase in industrial baker's yeast strains ATCC6037 and ATCC7754. An integrating plasmid (YIp) was constructed containing *SMR1* as a selectable marker and *MEL1* as the commercial gene of interest (plasmid pWX521 of FIG. 4).

TABLE 1. Natural Sulfometuron Methyl Resistance Levels and Transformation Frequencies with YEp-*SMR1* (Modified from Casey *et al.*[33])

S. cerevisiae Strain	Industrial Application	Natural Level of Resistance to Sulfometuron Methyl (SM, in µg/ml)[a]	Number of Transformants per 10 µg Plasmid DNA[b]
ATCC 18824	Ale yeast	6	800 (11)
ATCC 6037	Baker's yeast	6	3,340 (6)
ATCC 7754	Baker's yeast	6	810 (14)
ATCC 32120	Baker's yeast	8	540 (9)
ATCC 560	Distiller's yeast	3	1,640 (48)
ATCC 4110	Distiller's yeast	6	8,070 (43)
ATCC 287	Lager yeast	1	3,650 (18)
ATCC 2700	Lager yeast	6	1,440 (14)
Carlsberg Breweries-M244	Lager yeast	3	920 (5)
ATCC 26421	Sake yeast	6	6,800 (64)
ATCC 4098	Wine yeast	6	200 (28)
ATCC 4108	Wine yeast	15	580 (8)

[a]Levels of SM above this concentration on plates of SD agar completely prevent colony development on streak plates after 48 hours.
[b]The number in brackets indicates the number of resistant colonies that arose on control plates in the absence of chimeric plasmid.

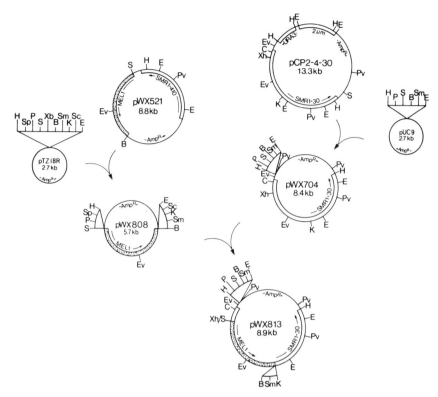

FIGURE 4. Construction of the *MEL1-SMR1*-30 cassette for gene replacement (from Xiao and Rank[40]). pCP2-4-30 is a YEp vector similar to that used in experiments summarized in TABLE 1. pWX521 is a YIp vector used to produce stable *MEL1* integrants of industrial strains (Casey *et al.*[38]). One-step gene replacement used the bacterial DNA-free *Cla*I (C)-*Hin*dIII (H) fragment of pWX813 (Xiao and Rank[40]).

pWX521 was targeted to the homologous *ILV2* locus by *Pvu*II cleavage within the *SMR1-410* selectable gene. This system has the advantage of using the selectable marker as an integration sequence and by having a convenient color assay for *MEL1* (FIG. 6). The latter doubles as a strain identification phenotype. Similar results were observed by Liljestrom-Suominen *et al.*[39]

STABLE INTEGRANTS FREE OF BACTERIAL DNA

The limitations of the standard YIp integrant structure (FIG. 3) were overcome by a one-step gene replacement strategy.[40] *MEL1* was inserted upstream (5′) of the *SMR1-30* gene. The 5′ insertion site had no apparent phenotypic function. In the final construct (FIG. 4, pWX813) *MEL1* is flanked by upstream (*Xho*I to *Cla*I) and downstream (*Kpn*I to *Hin*dIII) sequences of the *SMR1-30* locus. Standard 3′ and 5′ integration of pWX813 resulted in integrant types I and II of FIGURE 5. These structures contain bacterial DNA flanked by target homologous sequences. The

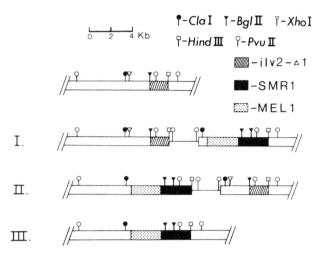

FIGURE 5. Three types of plasmid pWX813 (FIG. 4) integrant structure (from Xiao and Rank[40]). Types I and II result from the standard plasmid integration diagrammed in FIGURE 3. Type III integrant structure devoid of bacterial DNA results from a one-step replacement of host chromosome area with the C-H *MEL1-SMR1-30* fragment of pWX813 (FIG. 4).

novel technology employed a *Cla*I-*Hin*dIII digestion of pWX813, thereby releasing an *MEL1-SMR1-30* replacement fragment targeted to the host *ILV2* homologous sequence. The expected integrant structure (type III, FIG. 5) was confirmed and resulted in stable melibiase gene expression in a variety of industrial yeasts (FIG. 6). Following transformation of Carlsberg lager strain M244 it was possible to quantita-

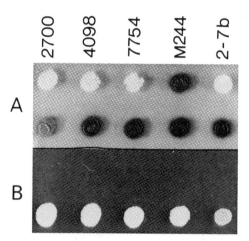

FIGURE 6. Phenotypes of industrial *Saccharomyces* strains (TABLE 1) and their pWX813 (FIGURE 4) transformants. *Upper row* in each panel contains untransformed yeasts, whereas the *lower row* contains transformants via gene-replacement technology. *Panel A* shows *MEL1* phenotype (*dark*) and *panel B* shows SM-resistant phenotypes (from Xiao and Rank[40]).

tively demonstrate the expression of both the resident and the introduced melibiase genes.[40]

An alternative method for production of bacterial DNA-free integrants uses postintegration eviction (jettison) of the flanked intervening sequence.[36] The standard jettison vector uses the mating *HO* sequence as a homologous integration locus. The commercial gene of interest is cloned within vector *HO* DNA, and integration is targeted to resident *HO* loci by site-specific restriction endonuclease cleavage. This results in the G418 selectable marker and other bacterial sequences, including *lacZ*, being flanked by *HO* loci. Jettison of bacterial sequences is identified by loss of a blue colony phenotype associated with *lacZ*. The final integrant is therefore free of all bacterial sequences, including the dominant selectable marker.

Jettison vector methodology has the disadvantage of requiring negative selection for postintegration marker loss. An attractive feature is that repeated use of the same jettison vector is possible for the introduction of additional commercial genes of interest. The one-step *SMR1* replacement method is more efficient than jettison technology. However, both methods are restricted by the use of a single sequence (*HO* or *ILV2*) for homologous integration and the restriction of all recombinant manipulations to a single vector.

COTRANSFORMATION-ASSOCIATED INTEGRATION

It has been observed that transformation with a mixed population of two vectors often results in the simultaneous cellular transformation by both vectors. Haploid yeasts were cotransformed with nonselectable linear DNA replacement fragments and a plasmid carrying a selectable prototrophic marker.[41,42] The linear fragment was integrated into 1–4% of cells transformed for the selectable marker, and the plasmid carrying the selectable marker was subsequently removed by mitotic segregation. Penttila *et al.*[35] used YEp-*CUP1* cotransformation to effect replacement of a brewer's yeast *LEU2* gene with a linearized vector carrying an *egl1* endoglucanase gene. Stable Egl⁺ integrants were obtained with enzyme levels about one tenth of that produced by multicopy vectors. We have also used cotransformation technology to integrate *MEL1* at the *ILV2* locus (unpublished observations). pWX813 (FIG. 4) was digested with *Bgl*II to remove *SMR1*-30 sequences from −190 to +630 (pWX821, FIG. 7a and b). The *Cla*I-*Hin*dIII fragment of pWX821 was then used in pCP2-4-30 (FIG. 4) cotransformation experiments by selection for sulfometuron methyl resistance. Stable MEL⁺ replacement integrants with the structure shown in FIG. 7c were observed at a frequency of approximately 0.5%. Cotransformation technology has also been used to disrupt *LEU2* in *Candida albicans*.[43] Although this methodology requires further refinement, it appears to be generally applicable. Research of industrial interest includes increasing cotransformation efficiency and reducing the frequency of unstable cotransformants.[34]

DIRECT INTRA-STRAIN GENE MANIPULATION

None of the industrial gene manipulations to date involves the *direct* isolation or modification of resident (native) genes. Haploid laboratory-based techniques allow the isolation of genes via transplacement[44] or double-strand gap repair[45]; endogenous gene inactivation (disruption) is also commonplace (ref 46, reviewed by Sturley and

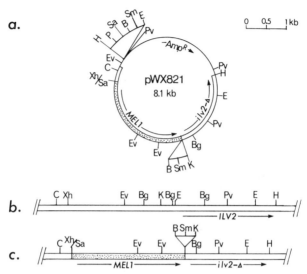

FIGURE 7. *MEL1* construct used in cotransformation gene transfer: (**a**) pWX821, vector used to transfer *MEL1;* (**b**) restriction map of *ILV2(SMR*1) wild-type sequence; (**c**) transformant structure at *ILV2* after transformation. The 5′ region of gene *SMR1-30* of pWX813 (FIG. 4), from the first upstream *Bgl*II site (−190) to a *Bgl*II site (+630) within the coding sequence (see **b**) was removed to result in an *MEL1-ilv2-Δ1* replacement cassette. A C–H fragment of pWX821 was cotransformed with pCP2-4-30 (FIG. 4), and stable *MEL1-ilv2-Δ1* integrants were selected with the structure shown in **c**.

Young[28]). When coupled with *in vitro* mutagenesis,[47] transplacement technology allows for precise nucleotide manipulation at any site within the genetic locus.

The main hindrance to industrial application of transplacement and disruption technology results from multiple copies of homologous/homeologous genes. For example, if two homologous and one homeologous sequences exist, alteration of homeologous expression requires successive gene disruption of homologous sequences followed by the isolation and/or disruption of the homeologous sequence.

Straightforward gene disruption of homologous genes should be possible by successive cotransformation integration[35,41] of 5′-3′ deleted haploid laboratory genes.[48] Manipulation of the homeologous sequence remains an unsolved problem, requiring further research. It may be possible to induce homologous/homeologous recombination by restriction endonuclease targeting to limited areas of common DNA sequence. Other approaches include efficient global cloning of homeologous DNA via pYAC vectors[31] with subsequent subcloning and *in vivo* manipulation of the gene of interest.

2-μM DNA POLYMORPHISM

Structure-function analysis of the 2 μm plasmid (FIG. 8) of haploid laboratory strains (reviewed by Futcher[49]) defined the following: (1) three *trans*-acting genes (*REP1, REP2,* and *RAF*) and one *cis*-acting locus (*STB*) involved in plasmid partition, (2) a *cis*-acting origin of DNA replication (*ori*), and (3) a site-specific recombinase gene (*FLP*) and the FLP target site (*FRT*). *FRT* is found within a 599-bp inverted repeat; *FLP-FRT* activity within the repeat results in an equilibrium between two isomeric forms of the 2-μm plasmid. The site-specific recombinase is believed to function in plasmid copy number control.[50]

Analyses of laboratory strains of common geneology indicated very little polymorphism from the published DNA sequence.[1] In sharp contrast, Xiao and Rank[51] found extensive polymorphism in 2 μm DNA from industrial *Saccharomyces* strains. Polymorphism was particularly pronounced at the *cis*-acting *STB* locus; variations included additions and deletions of the 62-bp repeat unit as well as up to 40% divergence from the *STB* consensus sequence. In addition, approximately 10% amino acid divergence was found in the associated *trans*-acting REP1 and RAF partition proteins. Of particular importance to industrial gene transfer is that polymorphic variants showed host specificity. That is, 2 μm DNA from industrial strain 1 partitioned more effectively in strain 1 than in other industrial or haploid laboratory strains. Consequently, maximal stability of 2 μm DNA recombinants in industrial strains requires that the host 2 μm plasmid sequence, rather than laboratory 2 μm plasmid DNA, be used for vector construction.

[*CIR*⁰] STRAINS DEFICIENT IN 2 μM PLASMID

The majority of *S. cerevisiae* haploid yeasts contain about 60 copies of 2-μm plasmid. Cells can spontaneously be cured of 2-μm DNA by continuous culture[52,53] or induced by use of chimeric YEp vectors maintained at high copy number by selection pressure.[54,55] However, these methods are inefficient and/or rely on prototrophic marker selection that is not applicable to industrial yeasts.

Several comparisons between isogenic [*cir*⁰] and [*cir*⁺] strains indicate approximately a 1–3% longer generation time of [*cir*⁺] strains.[52,53] Nevertheless, induced [*cir*⁰] may also show a growth disadvantage due to cryptic genomic damage during a prolonged period of YEp chimeric vector selection.[56]

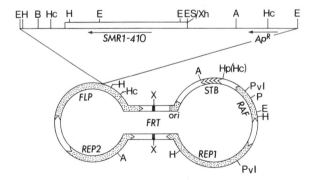

FIGURE 8. pWX823B was constructed by the insertion of *SMR1*-410 into the *FLP* gene of 2-μm DNA (lower dumbbell-shaped structure). The gene designations of 2-μm loci are indicated with the direction of transcription. *STB*, *ori*, and *FRT* are *cis*-acting sites involved in partition, DNA replication, and site-specific recombination. pWX823B was used for *in vivo* cloning of industrial 2-μm DNA (Xiao and Rank[51]) and for 2-μm DNA curing of industrial strains (Xiao and Rank[57,58]).

Xiao and Rank[57] introduced an *SMR1* dominant selectable marker into the *FLP* locus of laboratory 2 μm DNA to yield the YEp vector pWX823B (FIG. 8). This vector was generally useful in the efficient induction of [*cir°*] in both laboratory and industrial prototrophic strains; a selection protocol yielded approximately 15% [*cir°*] cells within 10 days.[57] [*cir°*] baker's yeast cells were observed to have a growth advantage[58] over isogenic [*cir⁺*] controls (FIG. 9). Although the [*cir°*] growth advantage is small, these cells may be of direct commercial benefit to the baking industry. In addition, [*cir°*] industrial yeasts are preferred hosts for specialized YEp vectors free of bacterial DNA.

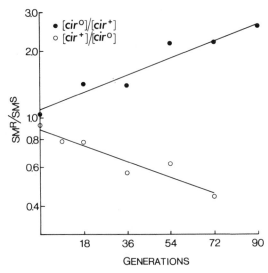

FIGURE 9. Cellular growth competition between baker's yeast ATCC7754 [*cir⁺*] and ATCC7754 [*cir°*]. Equal mixtures of cells marked by SM^R or SM^S phenotypes were allowed to replicate for the indicated number of generations. [*cir°*] cells showed a constant 0.4% growth advantage over isogenic [*cir⁺*] cells.

AUTONOMOUS VECTORS FREE OF BACTERIAL DNA

Fagan and Scott[59] constructed yeast acentric-ring plasmids (pYAR) entirely of yeast chromosomal DNA. pYARs were maintained, without selection pressure, as a relatively stable high copy number plasmid in laboratory strains. The high copy number (about 100) resulted in a vector-carried gene expression 100–300 times that of wild-type cells.[59]

Application of all yeast vectors to industrial strains employed YEp vectors with the bacterial selectable and *ori* loci flanked by direct *FRT* repeats.[60] When transformed into industrial yeast, the FLP-*FRT* recombination system efficiently excised all bacterial DNA. Insertion of foreign genes between *ori* and *STB* (FIG. 8) or within *RAF* had a minimal effect on plasmid instability. The former inserts "showed no measurable instability in any of the host strains studied." This vector system has the advantages of high copy number for high-level gene expression, absence of bacterial vector DNA, and high stability in the absence of selection pressure. Further enhancement could include the use of resident 2 μm DNA in [cir°] industrial strains,[51,57,58] thereby fine-tuning the vector to the industrial strain of interest. It should also be possible to clone bacterial and yeast selectable markers in tandem between flanking FRTs. Jettison of both selectable markers should increase plasmid stability and would also increase vector-carrying capacity.

KILLER FACTOR AND TY RETROTRANSPOSON-BASED VECTORS

Yeast killer strains harbor double-stranded RNA (dsRNA) that codes for an extracellular toxin and an immunity factor that protects infected cells. A cDNA of dsRNA cloned in a YEp vector resulted in expression of the killer phenotype in host cells.[61,62] Transformed industrial yeasts showed stable inheritance of the killer and immunity phenotype.[62] This technology has been proposed as a method of making industrial yeasts resistant to contaminant killer strains and as a system for secretion of heterologous proteins.[63]

The transposable element, Ty, is found in 30–40 copies per yeast genome. A Ty element was modified to contain a galactose-inducible transposition as well as marker gene (neo or TRP1) insertions.[64] The modified Ty was introduced into yeast on a YEp vector. Following galactose induction of transposition, Ty-neo and Ty-TRP1 were found in multiple copies throughout the nuclear genome. This method resulted in stable inheritance of up to 20 copies of the Ty cassette. An alternative approach that produced up to 200 integrated copies used insertion of a YIp vector at the rDNA locus.[65]

MITOCHONDRIAL DNA TRANSFORMATION

Natural recombination of mitochondrial DNA (mtDNA) results in the reassortment of parental markers among progeny cells (reviewed by Dujon[66]). Johnston et al.[67] used a cotransformation protocol to transform an mtDNA [oxi3] mutant. A ura3 [oxi3] cell was bombarded with tungsten projectiles coated with a mixture of chimeric plasmids containing the nuclear URA3 gene and the mtDNA [OXI3] marker. Approximately 1 in 1,000 URA+ cells were also transformed to the OXI+ phenotype. The mtDNA transformation frequency is low, and transfer of desirable mtDNA traits within a wild-type industrial background remains a future goal. Nevertheless, this biolistic process opens the way for recombinant DNA modification of the 85,000-bp mitochondrial genome. Thus, methodology has been developed for the in vitro manipulation of all three sources of genetic information, the nuclear, mitochondrial, and 2-μm genomes.

PRESENT AND FUTURE PROSPECTS

Recombinant DNA manipulations of interest generally fall into two main categories: (1) transfer of known genes of interest, and (2) direct in vivo alteration of resident loci.

Examples of cloned beneficial genes include S. cerevisiae genes such as MEL1 for the use of raffinose[38,39] and ILV5 for control of diacetyl levels in lager.[19] Isolated non-yeast genes include Aspergillus niger glucoamylase genes for production of low calorie beer[36] and *Trichoderma reesei egl1* for hydrolysis of beta-glucans that cause hazes and precipitates in beer.[35]

The transfer of beneficial *S. cerevisiae* genes free of bacterial sequences is the obvious starting point for industrial application of recombinant DNA technology to food and beverage yeasts. The main limiting factor to this type of manipulation is the dearth of identified beneficial genes. Initial identification of useful genes for a given industry is facilitated by the isolation of haploid ascospores that will mate with laboratory strains. The elegant work of Oda and Ouchi[12,15] with baker's yeasts indicates the feasibility of this approach. Once genes with a beneficial influence on an industrial phenotype have been mapped by meiotic analyses, facile cloning from genomic libraries is possible.[9,10] Subsequent beneficial *in vitro* gene modification and stable bacterial-DNA free single copy[39] or multiple copy[60] gene transfer are attractive industrial protocols.

Technology for the *in vivo* alteration of industrial genes is just beginning to emerge. Gene inactivation via deletion is certain to be generally useful. Thus, the deletion of homologous/homeologous *ILV2* genes offers the possibility of reducing diacetyl levels in lager.[19] Further developments in cotransformation regimens[38,41] in conjunction with haploid gene disruption and transplacement methodology should allow the inactivation or isolation of any industrial gene of interest.

REFERENCES

1. HARTLEY, J. L. & J. E. DONELSON. 1980. Nucleotide sequence of the yeast plasmid. Nature **286:** 860–864.
2. DE ZAMAROCZY, M. & G. BERNARDI. 1986. The primary structure of the mitochondrial genome of *Saccharomyces cerevisiae:* A review. Gene **47:** 155–177.
3. MORTIMER, R. K., D. SCHILD, C. P. CONTOPOULOU & J. KANS. 1989. Genetic map of *Saccharomyces cerevisiae,* edition 10. Yeast **5:** 321–403.
4. KABACK, D. B., H. Y. STEENSMA & P. DE JONGE. 1989. Enhanced meiotic recombination on the smallest chromosome of *Saccharomyces cerevisiae.* Proc. Natl. Acad. Sci. USA **86:** 3694–3698.
5. GOEBL, M. G. & T. D. PETES. 1986. Most of the yeast genomic sequences are not essential for cell growth and division. Cell **46:** 983–992.
6. KABACK, D. B., P. W. OELLER, H. Y. STEENSMA, J. HIRSCHMAN, D. RUEZINSKY, K. G. COLEMAN & J. R. PRINGLE. 1984. Temperature-sensitive lethal mutations on yeast chromosome I appear to define only a small number of genes. Genetics **108:** 67–90.
7. BARTON, A., H. Y. STEENSMA, W. SHAH, V. GUACCI & D. B. KABAK. 1989. The molecular organization of chromosome I. Yeast Cell Biology.: 86. Cold Spring Harbor Press. Cold Spring Harbor, NY.
8. STEENSMA, H. Y., J. C. CROWLEY & D. B. KABACK. 1987. Molecular cloning of chromosome *I* DNA from *Saccharomyces cerevisiae:* Isolation and analysis of the *CEN1-ADE1-CDC15* region. Mol. Cell. Biol. **7:** 410–419.
9. OLSON, M. V., J. E. DUTCHIK, M. Y. GRAHAM, G. M. BRODHEUR, C. HELMS, M. FRANK, M. MACCOLLIN, R. SCHEINMAN & T. FRANK. 1986. Random-clone strategy for genomic restriction mapping in yeast. Proc. Natl. Acad. Sci. USA **83:** 7826–7830.
10. LINK, A., J. E. DUTCHIK, L. RILES, G. F. CARLE & M. V. OLSON. 1988. Physical mapping of the yeast genome. Yeast **4S:** 18.
11. MORTIMER, R. K. & J. R. JOHNSTON. 1986. Genealogy of principal strains of the yeast genetic stock center. Genetics **113:** 35–43.
12. ODA, Y. & K. OUCHI. 1989. Genetic analysis of haploids from industrial strains of baker's yeast. Appl. Environ. Microbiol. **55:** 1742–1749.
13. SPENCER, J. F. T., D. M. SPENCER & N. REYNOLDS. 1988. Genetic manipulation of non-conventional yeasts by conventional and non-conventional means. J. Basic Microbiol. **28:** 321–333.

14. SPENCER, J. F. T. & D. M. SPENCER. 1983. Genetic improvement of industrial yeast. Ann. Rev. Microbiol. **37:** 121–142.
15. ODA, Y. & K. OUCHI. 1988. Principal component analysis of the characteristics desirable in baker's yeast. Appl. Environ. Microbiol. **55:** 1495–1499.
16. CARLE, G. F. & M. V. OLSON. 1985. An electrophoretic karyotype for yeast. Proc. Natl. Acad. Sci. USA **82:** 3756–3760.
17. WALMSLEY, R. M. & T. D. PETES. 1985. Genetic control of chromosome length in yeast. Proc. Natl. Acad. Sci. USA **82:** 506–510.
18. CASEY, G. P., W. XIAO & G. H. RANK. 1988. Application of pulsed field chromosome electrophoresis in the study of chromosome *XIII* and the electrophoretic karyotype of industrial strains of *Saccharomyces* yeasts. J. Inst. Brew. **94:** 239–243.
19. GJERMANSEN, C., T. NILSSON-TILLGREN, J. G. L. PETERSON, M. C. KIELLAND-BRANDT, P. SIGSGAARD & S. HOLMBERG. 1988. Towards diacetyl-less brewers' yeast. Influence of *ilv2* and *ilv5* mutations. J. Basic Microbiol. **28:** 175–183.
20. CASEY, G. P. 1986. Molecular and genetic analysis of chromosomes *X* in *Saccharomyces carlsbergensis*. Carlsberg Res. Commun. **51:** 343–362.
21. CASEY, G. P. & M. B. PEDERSON. 1988. DNA sequence polymorphisms in the genus *Saccharomyces*. V. Cloning and characterization of a *LEU2* gene from *S. carlsbergensis*. Carlsberg Res. Commun. **53:** 209–219.
22. BOTSTEIN, D. & R. W. DAVIS. 1982. Principles and practice of recombinant DNA research with yeast. *In* The Molecular Biology of the Yeast Saccharomyces: Metabolism and Gene Expression. J. N. Strathern, E. N. Jones & J. R. Broach, eds.: 607–636. Cold Spring Harbor Press. Cold Spring Harbor Laboratory, NY.
23. STRUHL, K. 1983. The new yeast genetics. Nature **305:** 391–397.
24. PARENT, S. A., C. FENIMORE & K. A. BOSTIAN. 1985. Vector systems for the expression, analysis and cloning of DNA sequences in *S. cerevisiae*. Yeast **1:** 83–138.
25. MEYERS, A. M., A. TZAGALOFF, D. M. KINNEY & C. J. LUSTY. 1986. Yeast shuttle and integrative vectors with multiple cloning sites suitable for construction of *lacZ* fusions. Gene **45:** 299–310.
26. VON WETTSTEIN, D., C. GJERMANSEN, S. HOLMBERG, M. C. KIELLAND-BRANDT, T. NILSON-TILLGREN, M. B. PEDERSEN, J. G. L. PETERSEN & P. SIGSGAARD. 1984. Genetic engineering in the improvement of brewers' yeast. M.B.A.A. Tech. Quart. **21:** 45–61.
27. KNOWLES, J. K. C. & R. S. TUBB. 1986. Recombinant DNA: Gene transfer and expression techniques with industrial yeast strains. EBC Symposium on Brewer's Yeast. Helsinki Monograph XII: 169–185.
28. STURLEY, S. L. & T. W. YOUNG. 1986. Genetic manipulation of commercial yeast strains. Biotechnol. Genet. Eng. Rev. **4:** 1–38.
29. RANK, G. H., G. CASEY & W. XIAO. 1988. Gene transfer in *Saccharomcetes* yeasts. Food Biotechnol. **2:** 1–41.
30. WILLIAMSON, D. H. 1985. The yeast ARS element, six years on: A progress report. Yeast **1:** 1–14.
31. BURKE, D. T., G. F. CARLE & M. V. OLSON. 1987. Cloning of large segments of exogenous DNA into yeast by means of artificial chromosome vectors. Science **236:** 806–812.
32. LANG-HEINRICHS, C., D. BERNDORFF, C. SEEFELDT & U. STAHL. 1989. G418 resistance in the yeast *Saccharomyces cerevisiae*: Comparison of the neomycin resistance genes from Tr5 and Tr903. Appl. Microbiol. Biotechnol. **30:** 388–394.
33. CASEY, G. P., W. XIAO & G. H. RANK. 1988. A convenient dominant selection marker for gene transfer in industrial strains of *Saccharomyces* yeast: *SMR1* encoded resistance to the herbicide sulfometuron methyl. J. Inst. Brew. **94:** 93–97.
34. HENDERSON, R. C. A., B. C. COX & R. TUBB. 1985. The transformation of brewing yeasts with a plasmid containing the gene for copper resistance. Curr. Genet. **9:** 133–138.
35. PENTTILA, M. E., M. L. SUIHKO, U. LEHTINEN, M. NIKKOLA & J. K. C. KNOWLES. 1987. Construction of brewer's yeasts secreting fungal endo-B-glucanase. Curr. Genet. **12:** 413–420.
36. YOCUM, R. R. 1986. Genetic engineering of industrial yeasts. *In* Proceedings, Bio Expo 86.: 171–180. Butterworth Publishers. Stoneham, M. M.
37. ORR-WEAVER, T. L., J. W. SZOSTAK & R. C. ROTHSTEIN. 1981. Yeast transformation: A

model system for the study of recombination. Proc. Natl. Acad. Sci. USA **78:** 6354–6358.

38. CASEY, G. P., W. XIAO & G. H. RANK. 1988. Construction of α-galactosidase-positive strains of industrial baker's (*Saccharomyces cerevisiae*) yeasts. J. Am. Soc. Brew. Chem. **46:** 67–71.

39. LILJESTROM-SUOMINEN, P. L., V. JOUTSJOKI & M. KORHOLA. 1988. Construction of a stable α-galactosidase-producing baker's yeast strain. Appl. Environ. Microbiol. **54:** 245–249.

40. XIAO, W. & G. H. RANK. 1989. The construction of recombinant industrial yeasts free of bacterial sequences by directed gene replacement into a nonessential region of the genome. Gene **76:** 99–107.

41. RUDOLPH, H., I. KOENIG-RAUSEO & A. HINNEN. 1985. One-step gene replacement in yeast by transformation. Gene **36:** 87–95.

42. SILICIANO, P. G. & K. TATCHELL. 1984. Transcription and regulatory signals at the mating type locus in yeast. Cell **37:** 969–978.

43. KELLEY, R., S. M. MILLER & M. B. KURTZ. 1988. One-step gene disruption by cotransformation to isolate double auxotrophs in *Candida albicans*. Mol. Gen. Genet. **214:** 24–31.

44. WINSTON, F., F. CHUMLEY & G. R. FINK. 1983. Eviction and transplacement of mutant genes in yeast. Methods Enzymol. **101:** 211–227.

45. ORR-WEAVER, T. L. & J. W. SZOSTAK. 1983. Yeast recombination: The association between double-strand gap repair and crossing over. Proc. Natl. Acad. Sci. USA **80:** 4417–4421.

46. ROTHSTEIN, R. J. 1983. One step gene disruption in yeast. Methods Enzymol. **101:** 202–211.

47. BOTSTEIN, D. & D. SHORTLE. 1985. Strategies and applications of *in vitro* mutagenesis. Science **229:** 1193–1201.

48. SHORTLE, D., J. E. HABER & D. BOTSTEIN. 1982. Lethal disruption of the yeast actin gene by integrative DNA transformation. Science **217:** 371–373.

49. FUTCHER, A. B. 1988. The 2 μm plasmid of *Saccharomyces cerevisiae*. Yeast **4:** 27–40.

50. FUTCHER, A. B. 1986. Copy number amplification of the 2 μm circle plasmid of *Saccharomyces cerevisiae*. J. Theor. Biol. **119:** 197–204.

51. XIAO, W. & G. H. RANK. 1990. Cloning of industrial *Saccharomyces* 2μ plasmid variants by *in vivo* site-specific recombination. Plasmid **23:** 67–70.

52. FUTCHER, A. B. & B. S. COX. 1983. Maintenance of the 2 μm circle plasmid in population of *Saccharomyces cerevisiae*. J. Bacteriol. **154:** 612–622.

53. MEAD, D. S., D. C. J. GARDNER & S. G. OLIVER. 1986. The yeast 2μ plasmid: Strategies for the survival of a selfish DNA. Mol. Gen. Genet. **205:** 417–421.

54. ERHART, E. & C. P. HOLLENBERG. 1983. The presence of a defective *LEU2* gene on 2μ DNA recombinant plasmids of *Saccharomyces cerevisiae* is responsible for curing and high copy number. J. Bacteriol. **156:** 625–635.

55. HARFORD, M. N. & M. PEETERS. 1987. Curing of endogenous 2 micron DNA in yeast by recombinant vectors. Curr. Genet. **11:** 315–319.

56. MEAD, D. J., D. C. J. GARDENER & S. G. OLIVER. 1987. Phenotypic differences between induced and spontaneous 2μ-plasmid-free segregants of *Saccharomyces cerevisiae*. Curr. Genet. **11:** 415–418.

57. XIAO, W. & G. H. RANK. 1990. An improved method for yeast 2 μm plasmid curing. Gene **88:** 241–245.

58. XIAO, W. & G. H. RANK. 1990. Curing industrial *Saccharomyces* yeasts of parasitic 2 μm DNA. J. Am. Soc. Brew. Chem. **48:** 107–110.

59. FAGAN, M. C. & J. F. SCOTT. 1985. New vectors for construction of recombinant high-copy number yeast acentric-ring plasmids. Gene **40:** 217–229.

60. CHINEY, S. A. & E. HINCHLIFFE. 1989. A novel class of vector for yeast transformation. Curr. Genet. **16:** 21–25.

61. LOLLE, S., N. SKIPPER, H. BUSSEY & D. Y. THOMAS. 1984. The expression of cDNA clones of yeast M1 double-stranded RNA in yeast confers both killer and immunity phenotypes. EMBO J. **3:** 1383–1387.

62. BUSSEY, H. & P. MEADEN. 1985. Selection and stability of yeast transformants expressing cDNA of an M1 killer toxin-immunity gene. Curr. Genet. **9:** 285–291.
63. THOMAS, D. Y., N. A. SKIPPER, P. C. K. LOU, S. LOLLE & H. BUSSEY. 1987. Production and secretion of proteins and polypeptides in yeast. Can. Patent 479062.
64. BOEKE, J. D., H. XU & G. R. FINK. 1988. A general method for the chromosomal amplification of genes in yeast. Science **239:** 280–282.
65. LOPES, T. S., J. KLOOTWIJK, A. E. VEENSTRA, P. C. VAN DER AAR, H. VAN HEERIKHUIZEN, H. A. RAUÉ & R. J. PLANTA. 1989. High-copy-number integration into the ribosomal DNA of *Saccharomyces cerevisiae:* A new vector for high-level expression. Gene **79:** 199–206.
66. DUJON, B. 1981. Mitochondrial genetics and functions. *In* The Molecular Biology of the Yeast *Saccharomyces:* Life cycle and inheritance. J. N. Strathern, E. N. Jones & J. R. Broach, eds.: 505–635. Cold Spring Harbor Press. Cold Spring Harbor Laboratory, NY.
67. JOHNSTON, S. A., P. Q. ANZIANO, K. SHARK, J. C. SANFORD & R. A. BUTOW. 1988. Mitochondrial transformation in yeast by bombardment with microprojectiles. Science **240:** 1530–1541.

Use of Yeasts in Production and Discovery of Pharmaceuticals

GURNAM S. GILL AND PHILLIP G. ZAWORSKI

Molecular Biology Research
The Upjohn Company
Kalamazoo, Michigan 49007

Rapid advances in the study of gene structure and regulation using recombinant DNA techniques have made it possible to express virtually any gene in host organisms ranging from bacteria to mammals. Numerous vector systems exist that allow heterologous genes to be introduced and expressed in host cells. Vectors either are commercially available or can be constructed to suit a particular host. The choice of a host often may be dictated by the gene product. For example, a number of glycoprotein hormones (erythropoietin, human choriogonadotropin, and follicle-stimulating hormone) require terminal sialic acid residues to accomplish their hormone action *in vivo*.[1-3] The oft-used bacterial host, *Escherichia coli*, does not glycosylate proteins. The lower eukaryotes, the yeasts, will glycosylate a protein, but do not add sialic acid residues. Therefore, if yeasts were to be used for the production of such hormones, the products would have to be modified *in vitro* for full biological activity. Alternatively, such glycoproteins have been produced in mammalian cell culture systems. Further examples on this subject can be cited, but suffice it to say that no single system may be capable of handling every gene product. For more discussion of the relative merits of various systems for heterologous protein production, the reader is referred to a recent article by Marino.[4]

Although they lack the capability to add certain specialized modifications, microbial hosts, such as *E. coli* and the yeasts, offer attractive features for heterologous protein production. Chief among these are well worked out genetics, rapid growth on inexpensive carbon sources, and the relative ease of large scale fermentations. In this article we discuss the use of yeasts for the production of polypeptides of therapeutic importance and consider the opportunities offered by yeast in drug discovery.

PRODUCTION OF PHARMACEUTICALS AND YEASTS

No matter what host is used to produce a gene product, a basic criterion to be met is that the product must have maximal biological activity. For products destined for use as therapeutics in humans, further constraints imposed by the regulatory agencies have to be met. Certain posttranslational modifications, such as acetylation, phosphorylation or glycosylation, may be necessary for optimal activity, increased *in vivo* half-life, solubility, or minimizing antigenicity of the product. For example, human superoxide dismutase, an oxygen radical scavenger, is NH_2-terminal acetylated. Analysis of the protein produced in *E. coli* and yeast has shown that the yeast version is acetylated and is identical to the human enzyme.[4] We have already cited examples of the hormone glycoproteins in which glycosylation is known to affect their *in vivo* half-lives and biological activity.[1-3] In some proteins, specific disulfide bonds may have to form for proper folding and full biological activity. An improperly folded

172

protein could elicit an immunological response. As the cytoplasmic environment is rather reducing, proteins with numerous disulfides should be targeted to the endoplasmic reticulum (ER) and secretory pathways. Furthermore, if glycosylation plays a role in proper protein folding, this aspect should be taken care of when the protein is translocated to the ER.

How do yeasts rate as hosts in meeting some of these requirements? Yeasts, being eukaryotes, can perform a number of posttranslational modifications. Yeasts have a well-characterized secretory pathway and can glycosylate and secrete proteins.[5] However, when it comes to glycosylation, yeasts apparently run into problems. In eukaryotes there are two types of glycosylation sites, N-linked and O-linked. The N-linked recognition site consists of an asparagine-X-threonine/serine triplet, X being any amino acid except proline. The carbohydrate is added to the asparagine residue. In the O-linked version, the carbohydrate moieties are added to serine/threonine residues; however, the site is not well defined. Yeasts add both types of glycosylation, that is, the N-linked and the O-linked. The core glycosylations, which are added in the ER, are identical in both yeasts and higher eukaryotes, but the secondary extensions, put on in the Golgi, differ drastically. Yeasts do not add any of the complex sugars such as sialic acid, instead, they add only mannose residues. With *Saccharomyces cerevisiae*, long mannose chains are added that can affect the enzymatic activity of a protein and its *in vivo* half-life, and make it antigenic. For example, human tissue plasminogen activator (tPA), a complex glycoprotein with 17 disulfide bridges, can be expressed and secreted by yeast. The secreted protein is heavily glycosylated and is heterogeneous in size. We found that the heavy glycosylation diminished the activity of the protein. Removal of glycosylation with endoglycosidase H resulted in a 10-fold increase in enzymatic activity. Elimination of the glycosylation site on the catalytic domain dramatically decreased the amount of secreted material, suggesting at this site a possible role for glycosylation in folding and secretion.[6] When the lack of glycosylation may not affect the biological activity of a protein, yeasts can successfully be used to produce large amounts of the desired product.[7] Human urokinase has one glycosylation site in the catalytic domain. Removal of this site did not seem to affect the activity or secretion level of the enzyme (A. Hinnen, personal communication). *S. cerevisiae* strains are available or can be constructed that do not add very long carbohydrate chains. The methylotrophic yeast *Pichia pastoris* does not seem to glycosylate the heterologous proteins as heavily as does *S. cerevisiae*. Reducing the carbohydrate chain length, however, will not, solve the problem of rapid removal of such a product from circulation or of eliciting an immune response. *In vitro* modification may circumvent the problem, but the economics of such a step will dictate the options. Apart from this disadvantage of altered glycosylation, the yeasts offer a number of advantages as hosts for the production of pharmaceuticals. We shall consider some of these aspects in the following sections.

YEAST GENETICS, VECTORS, AND PROMOTERS

Unlike *E. coli*, yeasts are devoid of known pyrogens and have been used in the food and beverage industry literally for centuries. Although a number of yeasts are being developed as hosts, *S. cerevisiae* and lately the methylotrophic yeast *P. pastoris* have been the work horses for large scale production of foreign proteins. The choice of *S. cerevisiae* seemed to be automatic, for, among the eukaryotes, it is the best characterized organism *vis-à-vis* its life cycle, genetics, and physiology.[8] Also, *S. cerevisiae* is on the FDA's GRAS (generally regarded as safe) list. Given that yeast cells can be maintained as haploids or diploids, a combination of classical and

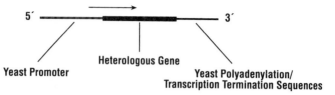

FIGURE 1. The sketch depicts an arrangement of an expression unit generally used for the production of a heterologous gene product in yeast. The heterologous gene sequences are inserted between yeast promoter and polyadenylation sequences. The *arrow* indicates the direction of transcription. If the gene codes for a secreted protein, either a yeast secretion sequence is inserted in front of the gene sequences coding for the mature protein or its endogenous leader sequences are retained if these are recognized efficiently by the yeast secretory machinery.[46]

modern genetic techniques allows the construction of strains with any desired genetic background.[8,9] A large assortment of vector types, such as integrating, multicopy episomal, centromeric, linear, or circular, are readily available or can be assembled.[9–12] Genes such as URA3, TRP1, and LEU2 have been incorporated into these vectors for selection in strains auxotrophic for these markers. Recent techniques enable manipulations of industrial (usually polyploid) *Saccharomyces* strains. Genes of interest can be introduced into these strains using dominant selectable markers such as resistance to the antibiotic G418, resistance to copper (CUPI), or resistance to sulfomoturon methyl (SMRI) either at specific loci through homologous recombination or by adding these constructs to the 2-μm circle-based vectors.[12–15] Further details about the yeast vectors can be found in the accompanying paper by Rank and Xiao and references therein.

Any heterologous gene to be expressed in yeast ought to be flanked by an efficient yeast promoter at the 5′ end and a polyadenylation/transcription terminator site at the 3′ end. Such a generic arrangement is shown in FIGURE 1. Efficient and well-characterized yeast promoters and terminators from genes that are constitutively expressed or tightly regulated have been used to express foreign proteins in yeast.

Examples of the strong constitutive promoters would be the promoters from genes involved in glycolysis, such as phosphoglycerate kinase (PGK), glyceraldehyde-3 phosphate dehydrogenase (GPD), triosephosphate isomerase (TPI), and alcohol dehydrogenase (ADHI). Some of the well-characterized regulated promoters are alcohol dehydrogenase II (ADHII), galactokinase (GALI), acid phosphatase (PHO5), and chelatin (CUPI). Studies on the anatomy of yeast promoters, apart from the presence of the canonical TATA box, have revealed the existence of *cis*-acting elements such as the upstream activating sequences (UASs) and other motifs through which *trans*-acting factors mediate their positive or negative regulatory effects on gene expression.[16,17] The GAL4 gene product, a positive regulator of genes involved in galactose metabolism, for example, recognizes specific sequences upstream of the TATA box.[18] Similarly, a sequence involved in glucose repression of the ADHII gene has been identified that, when inserted at strategic places in the ADHI promoter, makes the constitutive ADHI gene subject to glucose repression.[19] Availability of such detailed information about the yeast promoters has led to the construction of hybrid controllable promoters consisting of UASs from strong promoters and regulatory elements from regulated but relatively weaker promoters.[20–22] Such regulatable hybrid systems are very useful during large-scale fermentations when the cell growth phase can be separated from expression and synthesis of the product.

As mentioned earlier, the secretory pathway of *S. cerevisiae* has been well characterized.[5] Leader sequences from some secreted proteins, such as mating factor α1 (MFα1) gene, invertase (SUC2), and PHO5, have been engineered to target the heterologous proteins to the yeast secretory pathway.[23-25] One of the most used leaders is from the MFα1 gene. This leader sequence is 89 amino acids in length and prior to export of the desired protein from the cell is processed by the KEX2 gene product (a serine protease). The recognition/processing site for KEX2 is a pair of basic amino acids, Lys-Arg or Arg-Arg.[26] Heterologous proteins with this leader sequence get processed and secreted and usually carry no extra amino acid residues.

THE METHYLOTROPHIC YEAST *P. PASTORIS* AS HOST

So far, we have concentrated on describing *S. cerevisiae* as host. During the last few years the methylotrophic yeast *P. pastoris* has come of age as a host mainly through efforts made at Phillips Petroleum and at the Salk Institute of Biotechnology/ Industrial Associates (SIBIA). This yeast can be grown to very high cell densities (up to 130 g dry weight cells/liter) on methanol in a continuous culture system.[27] During its growth on methanol, *P. pastoris* expresses, at very high levels, a set of genes that are involved in methanol utilization. The gene products are sequestered in an organelle called the peroxisome. One of the enzymes involved in methanol utilization, alcohol oxidase (AOX1), essentially forms crystals in the peroxisomes and can constitute up to 30% of the total soluble cell protein.[28] The AOX1 gene is induced only when cells are grown on methanol. It was this strong regulatable promoter that was exploited for expressing foreign genes in *P. pastoris*. Construction of genetically marked strains and expression vectors based on the AOX1 promoter, combined with high density fermentation technology, have made *P. pastoris* an attractive system.[29,30] A number of gene products such as tissue necrosis factor (TNF) and hepatitis B surface antigen (HBSAg), have been expressed at very high levels (TABLE 1).[31-33]

TABLE 1. Some[a] of the Therapeutically Important Proteins Produced at High Levels in Yeast

Protein	Amount (g/L)	Source (Reference)
γ-interferon	2	20
Proinsulin	>1.0	21, 37
Human serum albumin	50 mg to >1.0	46
		R. Hitzeman[b]
		R. Brierley[b]
Human superoxide dismu-		
tase	>1.0	4
Tissue necrosis factor	6–10	7
GM-CSF	High levels[c]	7
Human urokinase	~0.5	A. Hinnen[b]
Hirudin	~0.5	A. Hinnen
HBSAg	>1.0	7
Streptokinase	0.1–0.3	7

[a]Many other peptides and proteins have been expressed in yeast, some probably at high levels. For further information see ref. 34.
[b]Personal communication.
[c]No amounts were quoted. It is probably in the range of hundreds of milligrams per liter.

OPTIMIZING EXPRESSION AND PRODUCTION

Although we have described the various components needed to assemble expression vectors, inserting a heterologous gene behind a yeast promoter and transforming it into a host will not necessarily lead to the production of large amounts of a protein, whether expressed cytoplasmically or targeted to the secretory pathway. A construction (strain and expression vector) that works for one gene product often does not work for the next. Additionally, each gene product can bring in its own unique properties that have to be addressed individually.[7] Despite this, certain parameters have to be considered for the expression of any gene, such as strain background, effect of expression on plasmid copy number, type of promoter, message and protein stability, and codon usage.[34–36] Various laboratories have developed unique strategies to address these potential problems. If the expression of a gene from a multicopy plasmid stresses a cell, the cell tends to compensate by lowering the copy number. Chromosomal insertions can alleviate this problem to some extent. Use of a regulated promoter could lead to further improvement by effectively separating cell growth from expression, thus allowing high cell densities to be achieved before induction of the heterologous protein. A number of proteins have been produced at high levels using such a strategy.[20,21]

Proteins that are unstable can sometimes be stabilized by fusion to another well-characterized protein that can be produced at high levels and is stable. This can make downstream processing easier. The fusion protein can be recovered the same way as the stable protein. Fusions can be engineered so that the desired moiety can be cleaved from the fusion protein enzymatically or chemically and further purified. Such schemes have successfully been employed by various groups. For example, researchers at the Chiron Corporation, using fusions between the human superoxide dismutase gene and the gene coding for proinsulin, were able to recover large amounts of proinsulin following high level expression of the chimeric protein.[21,37] A second group at the University of Oxford has exploited the biology of the yeast Ty elements. The Ty elements (~ 6 kb long) in yeast are the equivalent of retroviral genomes in higher eukaryotes. A yeast cell may have 20–30 copies of a Ty element scattered around its genome. The Ty gene products have an almost one-to-one equivalence with the retroviral proteins such as gag and pol. It has been shown that the Ty protein P1 can self-assemble to form virus-like particles (VLPs) that can be overproduced and purified.[38] Proteins of interest (e.g., human immunodeficiency virus [HIV] proteins) are produced as fusions with P1 which is able to form a VLP structure with the fused protein exposed on the particle surface. VLPs are easily purified, and the recombinant protein could be cleaved off. If the added protein is an antigen, the VLPs with the fused moiety on the outside can be used in the preparation of subunit vaccines.[39]

As mentioned earlier, the aim of producing a foreign protein in a heterologous system is to maximize quantity while maintaining quality. Proteins produced for use as pharmaceuticals have to be as authentic as possible, for if the protein is not properly processed or folded, there is always the danger of adverse side effects, such as eliciting an immune response, that could endanger the recipient. For example, cytoplasmic overexpression may result in the incomplete processing of NH_2-terminal methionine which may not normally be present in the mature protein. Yeast is known to possess a gene that codes for a methionine amino peptidase.[40] This protease could be overproduced, perhaps, when the production of the heterologous protein is turned on. Recently, another ingenious scheme has been developed whereby the gene for the protein of interest is fused at the carboxy-terminus of the yeast ubiquitin gene. The fusion protein is efficiently cleaved by the protease that processes polyubiquitin.[41,42] For the secreted protein, the hydrophobic leader sequence that targets it to the secretory pathway can be fused in such a way that when it is cleaved off in the ER or the Golgi, the heterologous protein has no extra amino acid at its NH$_2$-terminus. Also, for secreted proteins having disulfide bonds, targeting to the secretory pathway helps in proper bond formation and folding in the ER.

Whether a protein is produced in the cytoplasm or targeted for secretion, strain development and physiological manipulations in a fermenter are essential for realizing the full potential of the yeast system. Initially, a protein may be produced at low levels or secreted very poorly. A systematic study of growth parameters can make a tremendous difference to the level of production. For example, it has been shown that by using an inducible promoter and controlling nutrient addition rate and dissolved oxygen, yields of Hepatitis B surface antigen could be increased 15 to 30-fold.[43] Similar results were seen for lysozyme yields in *P. pastoris* when various media formulations were tried in high cell density fermentations.[44] Human serum albumin was secreted very poorly by some strains of *S. cerevisiae.*[45] Through genetic and physiologic manipulations of strains, a number of groups have obtained levels of secreted human serum albumin ranging from 10s to 100s of mg/L in *S. cerevisiae* and g/L in *P. pastoris* (ref. 46; R. Hitzman and R. Brierley, personal communications). Some of the proteins that have been produced at high levels in yeasts are listed in TABLE 1.

YEASTS AND DRUG DISCOVERY

Applications of recombinant DNA techniques have greatly contributed to our understanding of the basic biologic processes at the cellular and the molecular levels. The yeasts *S. cerevisiae* and *S. pombe,* in this regard, are perhaps the best studied eukaryotes. Extensive genetic and biochemical studies have produced a plethora of information regarding their cell cycle, gene structure, and regulation of the metabolic pathways. Homologues of important mammalian genes have been identified in yeast. Information of this nature can be exploited for devising targeted high volume screens to search for novel therapeutic compounds.[47,48] The initial step is usually the construction of appropriate yeast mutants whose genetic deficiencies can be complemented by their mammalian homologue.

This strategy can be extended to pathogenic organisms. For example, mammalian cells lack the thick cell walls that are essential for the survival of the fungal pathogens. The cell wall composition of *Candida albicans* is very similar to that of *S. cerevisiae.* It should therefore be possible to set-up screens that specifically target the yeast cell wall. Recently, we devised a screen that could be used to look for agents that affect cell wall biosynthesis as opposed to the cytoplasmic membrane.[49] The rationale for the screen is based on the osmotic stability imparted by the cell wall to the cell membrane. Drugs affecting the cell wall should also affect the resistance of a cell to lysis following an osmotic shock. Cell lysis could be detected by expressing an easily assayed protein such as β-galactosidase cytoplasmically. Drugs such as aculeacin A which has been shown to compromise cell wall biosynthesis by inhibiting β-glucan synthase invariably resulted in a fivefold higher release of β-galactosidase following osmotic shock than cell membrane active compounds such as nystatin.[49]

In some situations, if a heterologous protein has an easily monitored enzymatic activity, it may be possible to isolate analogs of the enzyme with more desirable properties than those of the parent molecule. As an example, we studied expression and secretion of tPA in *S. cerevisiae.* tPA is a complex multidomain glycoprotein that plays important roles in thrombolysis and tissue remodeling.[50,51] During the course of identification and isolation of yeast strains capable of secreting tPA, we developed a screening procedure that could be used to look for tPA analogs more resistant to the action of the inhibitor PAI-1. Such analogs could have longer *in vivo* half-lives. The same procedure could be adapted to screen for tPA derivatives having a higher specific activity.[6]

Numerous other possibilities exist that can be exploited. For example, it may be possible to obtain functional expression of certain mammalian receptors and develop screens for ligands that act as agonists or antagonists. It has been shown that all four subunits of the nicotinic acetylcholine receptor can be expressed in yeast.[52] The structure of the yeast α-factor receptor (STE2) is similar to that of a number of mammalian receptors, such as serotonin and muscarinic receptors.[53,54] It consists of seven membrane-spanning domains, and signal transduction is mediated by a G-protein.[55,56] It may be possible to couple the mammalian receptors to the yeast system and develop screens using the easily scorable phenotype, growth/no growth.

The list can be extended to screens that look for antiviral and anticancer agents.[48] Essential gene products in pathogens, when identified, can be expressed in yeast, and this can eventually lead to the development of a screen.

In this article we have covered, albeit briefly, features of yeasts that make these attractive as host organisms for the production of polypeptides of therapeutic importance and for devising methodologies to look for novel drugs. We believe that the latter aspect, that is, the use of yeasts for targeted drug screening, will become more and more important. Lastly, we have made no attempt to describe the use of yeasts in bioconversions.

ACKNOWLEDGMENTS

We would like to thank Sharon Mahoney, Juli Hammond, and Kathy Hiestand for careful preparation of the manuscript.

REFERENCES

1. FUKUDA, M. N., H. SASAKI, L. LOPEZ & M. FUKUDA. 1989. Blood **73:** 84–89.
2. LOWRY, P. H., G. KEIGHLEY & H. BORSOOK. 1960. Nature **185:** 102–
3. SAIRAM, M. R. & G. N. BHARGAVI. 1985. Science **229:** 65–67.
4. MARINO, M. H. 1989. BioPharm. **2:** 18–33.
5. SCHEKMAN, R. & P. NOVICK. 1982. The secretory process and yeast cell-surface assembly. *In* The Molecular Biology of the Yeast *Saccharomyces.* J. Strathern et al., eds.: 361–398. Cold Spring Harbor Laboratory. Cold Spring Harbor, NY.
6. GILL, G. S., P. G. ZAWORSKI, K. R. MAROTTI & E. F. REHBERG. 1990. Biotechnology **8:** 956–958.
7. RATNER, M. 1989. Bio/Technology **7:** 1129–1131.
8. STRATHERN, J., E. JONES & J. BROACH (eds). 1981. The Molecular Biology of the Yeast *Saccharomyces cerevisiae.* Cold Spring Harbor Laboratory. Cold Spring Harbor, NY.
9. STRUHL, K. 1983. Nature **305:** 391–397.
10. PARENT, S. A., C. FENIMORE & K. A. BOSTIAN. 1985. Yeast **1:** 83–138.
11. MYERS, A. M., A. TZAGALOFF, D. M. KINNEY & C. J. LUSTY. 1986. Gene **45:** 299–310.
12. RANK, G. H., G. CASEY & W. XIAO. 1988. Food Biotechnol. **2:** 1–41.
13. YOCUM, R. R. 1986. Genetic engineering of industrial yeasts. *In* Proceedings, Bio Expo 86.: 171–180. Butterworth Publishers. Stoneham, MA.
14. HENDERSON, R. C. A., B. C. COX & R. TUBB. 1985. Curr. Genet. **9:** 133–138.

15. CASEY, G. P., W. XIAO & G. H. RANK. 1988. J. Inst. Brew. **94:** 93–97.
16. GUARENTE, L. 1984. Cell **36:** 799–800.
17. STRUHL, K. 1987. Cell **49:** 295–297.
18. GININGER, E., S. M. VARNUM & M. PTASHNE. 1985. Cell **40:** 767–774.
19. BEIER, D. R. & E. T. YOUNG. 1983. Nature **300:** 724–728.
20. FIESCHKO, J. C., K. M. EGEN, T. RITCH, R. A. KOSKI, M. JONES & G. A. BITTER. 1987. Biotech. & Bioeng. **29:** 1113–1121.
21. COUSENS, L. S., J. R. SHUSTER, C. GALLEGOS, L. KU, M. M. STEMPIEN, M. S. URDEA, R. SANCHES-PESRADOS, A. TAYLOR & P. TEKAMP-OLSON. 1987. Gene **61:** 265–275.
22. SLEDZIEWSKI, A. Z., A. BELL, K. KELSAY & V. L. MACKAY. 1988. Bio/Technology **6:** 411–416.
23. BRAKE, A. J., J. P. MERRYWEATHER, D. G. COIT, U. A. HEBERLEIN, F. R. MASIARZ, G. T. MULLENBACH, M. S. URDEA, P. VALENZUELA & P. J. BARR. 1984. Proc. Nat'l. Acad. Sci. USA **81:** 4642–4646.
24. ZSEBO, K. S., H. S. LU, J. C. FIESCHKO, L. GOLDSTEIN, J. DAVIS, K. DUKER, S. V. SUGGS, P. S. LAI & G. A. BITTER. 1986. J. Biol. Chem. **261:** 5858–5865.
25. SMITH, R. A., M. J. DUNCAN & D. T. MOIR. 1985. Science **229:** 1219–1224.
26. JULIUS, D., A. BRAKE, L. BLAIR, R. KUNISAWA & J. THORNER. 1984. Cell **37:** 1075–1089.
27. WEGNER, E. H. 1983. U.S. Patent 4414329. U.S. Patent Office, Washington, DC.
28. COUDERC, R. & J. BARATTI. 1980. Agric. Biol. Chem. **44:** 2279–2289.
29. ELLIS, S. B., P. F. BRUST, P. J. KOUTZ, A. F. WATERS, M. M. HARPOLD & T. R. GINGERAS. Mol. Cell. Biol. **5:** 1111–1121.
30. CREGG, J. M., K. J. BARRINGER, A. Y. HESSLER & K. R. MADDEN. 1985. Mol. Cell. Biol. **5:** 3376–3385.
31. SREEKRISHNA, K., L. NELLES, R. POTENZ, J. CRUZE, P. MAZZAFERRO, W. FISH, M. FUKE, K. HOLDEN, D. PHELPS, P. WOOD & K. PARKER. 1989. Biochemistry **28:** 4117–4125.
32. CREGG, J. M., J. F. TSCHOPP, C. STILLMAN, R. SIEGEL, M. AKONG, W. S. CRAIG, R. G. BUCKHOLZ, K. R. MADDEN, P. A. KELLARIS, G. R. DAVIS, B. L. SMILEY, J. CRUZE, R. TORREGROSSA, G. VELECELEBI & G. P. THILL. Bio/Technology **5:** 479–485.
33. CREGG, J. M., M. E. DIGAN, J. F. TSCHOPP, R. A. BRIERLEY, W. S. CRAIG, G. VELECELEBI, R. S. SIEGEL & G. P. THILL. 1989. Expression of foreign genes in *Pichia pastoris. In* Genetics and Molecular Biology of Industrial Microorganisms. C. Hershberger, G. Hegeman & S. Queener, eds.: 343–352. American Society of Microbiology Publishers. Washington, D.C.
34. KINGSMAN, S. M., A. J. KINGSMAN & J. MELLOR. 1987. Trends Biotechnol. **5:** 53–57.
35. BITTER, G. A. 1987. Methods Enzymol. **152:** 673–684.
36. BITTER, G. A., K. M. EGAN, R. A. KOSKI, M. O. JONES, S. G. ELLIOT & J. C. GIFFIN. 1987. Methods Enzymol. **153:** 516–544.
37. TOTTRUP, H. V. & S. CARLSON. 1990. Biotechnol. Bioeng. **35:** 339–348.
38. ADAMS, S. E., J. MELLOR, K. GULL, R. B. SIM, M. F. TUITE, S. M. KINGSMAN & A. J. KINGSMAN. 1987. Cell **49:** 111–119.
39. ADAMS, S. E., K. M. DAWSON, K. GULL, S. M. KINGSMAN & A. J. KINGSMAN. 1987. Nature **329:** 68–70.
40. SHERMAN, F. & J. W. STEWART. 1982. Mutations altering initiation of translation of yeast iso-1-cytochrome *c*; Contrasts between the eukaryotic and prokaryotic initiation process. *In* The Molecular Biology of the Yeast *Saccharomyces:* Metabolism and Gene Expression. J. N. Strathern, E. W. Jones & J. R. Broach, eds.: 301–333. Cold Spring Harbor Laboratory. Cold Spring Harbor, NY.
41. SABIN, E. A., C. T. LEE-NG, J. R. SHUSTER & P. J. BARR. 1989. Bio/Technology **7:** 705–709.
42. OZKAYNAK, E., D. FINLEY, M-J. SOLOMON & A. VARSHAVSKY. 1987. EMBO J. **6:** 1429–1439.
43. HSIEH, J-H., K.-Y. SHIH, H-F. KUNG, M. SHIONG, L.-Y. LEE, M.-Ч. MENG, C.-C. CHANG, H.-M. LIU, S.-C. SHIH, S.-Y. LEE, T.-Y. CHOW, T.-Y. FENG, T.-T. KUO & K.-B. CHOO. 1988. Biotechnol. Bioeng. **32:** 334–340.
44. SIEGEL, R. S. & R. A. BRIERLEY. 1988. Biotechnol. Bioeng. **34:** 403–404.
45. ETCHEVERRY, T., W. FORRESTER & R. HITZEMAN. 1986. Bio/Technology **4:** 726–730.

46. SLEEP, D., G. P. BELFIELD & A. R. GODDEY. 1990. Bio/Technology **8:** 42–46.
47. BASSON, M. E., M. THORSNESS & J. RINE. 1986. Proc. Nat'l. Acad. Sci. USA **83:** 5563–5567.
48. SCHAFER, W. R., R. KIM, R. STERNE, J. THORNER, S.-H. KIM & J. RINE. 1989. Science **345:** 379–385.
49. ZAWORSKI, P. G. & G. S. GILL. 1990. Antimicrob. Agents Chemother. **34**(4): 660–662.
50. WIMAN, B. & D. COLLEN. 1978. Nature **272:** 549–550.
51. REICH, E. 1978. Activation of plasminogen: A general mechanism for producing localized extracellular proteolysis. *In* Molecular Basis of Biological Reproductive Processes. R. D. Berlin, M. Herman, L. H. Lepow & J. M. Tanzee, eds.: 155–169. Academic Press. New York, NY.
52. JANSEN, K. W., W. G. CONROY, T. CLANDIO, T. D. FOX, N. FUJITA, O. HAMILL, J. M. LINDSTROM, M. LUTHER, N. NELSON, K. A. RYAN, M. T. SWEET & G. P. HESS. 1989. J. Biol. Chem. **264:** 15022–15027.
53. HARTIG, P. R. 1989. Trends Pharmacol. Sci. **10:** 64–69.
54. PERALTA, E. G., J. W. WINSLOW, G. L. PETERSON, D. H. SMITH, A. ASHKENAZI, J. RAMACHANDRAN, M. I. SCHIMERLIK & D. J. CAPON. 1987. Science **236:** 600–605.
55. DIETZEL, C. & J. KURJAN. 1987. Cell **50:** 1001–1010.
56. MIYAJIMA, I., M. NAKAFUKU, N. NAKAYAMA, C. BRENNER, A. MIYAJIMA, K. KIBUCHI, K. ARAI, Y. KAZIRO & K. MATSUMOTO. 1987. Cell **50:** 1011–1019.

Genetic Modification of Food and Beverage Yeast

T.-M. ENARI

VTT
Technical Research Centre of Finland
SF-02150 Espoo, Finland

The first industry to use a pure microbial culture was the brewing industry. E. C. Hansen introduced the use of a brewer's yeast strain in the Carlsberg brewery in 1883. Today, pure yeast cultures are used in brewing and distilling and for the production of baker's yeast and fodder yeast. In wine fermentations pure cultures are used in some countries, whereas others still use spontaneous fermentation.

Most yeast strains used belong to the species *Saccharomyces cerevisiae*. Traditionally industrial yeasts have been improved mainly by selection. This is due partly to a lack of knowledge about the genetic background of the desirable industrial properties and partly to a lack of techniques to modify the genome of the yeasts. Industrial strains of *S. cerevisiae* mate rarely and sporulate poorly. This mating behavior has hampered both the genetic analysis of these yeasts and the breeding of yeast strains by crossing. One complicating fact is that industrial yeast strains usually are polyploid, mostly triploid or tetraploid. Strain improvement using conventional mutagenesis has generally not been successful mainly because of the polyploid nature of the strains.

Rare mating, single chromosome transfer, and spheroplast fusion are newer techniques that have been applied to strain improvement of industrial yeasts. In most cases these techniques have not been successful from an industrial point of view.

The development of recombinant DNA techniques has introduced new possibilities to modify industrial yeast strains. It is now possible to transfer single genes from foreign species to industrial yeast strains in a controlled way. Inasmuch as the genetic composition of the recipient yeast strain is not, in principle, otherwise altered, it is possible to introduce a desirable property into a good industrial yeast strain without destroying its good properties. Transformation of yeast was first achieved in 1978, and since then, techniques have been developed for industrial yeast strains.[1-3]

DEVELOPMENT OF FOOD AND BEVERAGE YEASTS

Most genetically modified industrial microorganisms are presently used for the production of single molecules such as hormones or enzymes. In these cases a maximal production of the product is essential, and all microbial properties that do not affect this are of no importance. When yeast is used in food or beverage fermentations, the situation is different because the product is not a single component but a balanced mixture of hundreds of compounds. When yeast is genetically modified for food or beverage use, it is therefore important not to disturb the normal functions of the yeast. With beer, for example, more than 200 flavor compounds are known, and most of them are produced by the yeast during fermentation. On the other hand, maximal production of the new gene product is usually not needed or not even wanted. A lower level of production is often enough, and higher production may

181

TABLE 1. Genetic Modification of Brewer's Yeast

Gene	Purpose	Source
Glucoamylase	High attenuation	*Schwanniomyces* sp.
		Saccharomyces diastaticus
		Aspergillus niger
α-Amylase	High attenuation	*Bacillus subtilis*
β-Glucanase	Filterability	*Bacillus subtilis*
		Trichoderma reesei
α-Acetolactate decarboxy-	Prevention of diactyl for-	*Enterobacter aerogenes*
lase	mation	*Klebsiella terragena*
Flocculation	Induction of flocculence	*Saccharomyces cerevisiae*
Zymocin	Resistance against wild	*Saccharomyces cerevisiae*
	yeasts	

cause disturbances. For gene technology this means that a multicopy plasmid may not be the right solution. A single gene integrated into a chromosome may give the desired level of expression and the construct is more stable. Another important aspect is the acceptability to authorities and consumers of the genetically modified organism. For this reason the foreign DNA transferred should be minimized and restricted to the required sequences only.

BREWER'S YEAST

Most of the published work on genetic modification of food and beverage yeast concerns brewer's yeast (TABLE 1). The first applications in this field were all centered on extracellular carbohydrate-degrading enzymes. Research work in these cases could build on existing knowledge of enzymes that have traditionally been industrially produced using microorganisms and that are used in the brewing industry. As the enzymes are secreted by brewer's yeast, it can also be assumed that they do not affect metabolism in the yeast cell; therefore, beer flavor should remain unchanged. Successful applications involving expression of intracellular enzymes, however, have also been reported.

AMYLOLYTIC ENZYMES

In addition to monosaccharides, brewer's yeast strains use maltose and maltotriose but not higher oligosaccharides. In normal brewer's wort, about 20% of the carbohydrates are present as such unfermentable oligosaccharides. In many types of beer these carbohydrates are wanted because they contribute to the body of the beer, but in the production of low-carbohydrate beers ("light" or "dry") they are hydrolyzed by the addition of microbial amyloglucosidase (glucoamylase) preparations. The addition of industrial enzymes can be avoided if a brewer's yeast secreting amylolytic enzymes is used.

The most interesting studies with brewer's yeast have been performed using *Saccharomyces diastaticus* and *Schwanniomyces occidentalis* as donor organisms. If a high degree of fermentation is wanted in beer fermentation, it is essential to hydrolyze both α-1,4- and α-1,6-glycosidic linkages in the residual carbohydrates.

The yeast *S. diastaticus* produces starch-degrading amyloglucosidases coded by three genes, *DEX1, DEX2,* and *DEX3* (also called *STA1, STA2,* and *STA3*).

One of these, *DEX1,* has been transferred to brewer's yeast.[4,5] In laboratory scale fermentations the transformants used 30% of the dextrins normally left in beer. The resulting beer was considered to be of sound quality with no off-flavors.

The *S. diastaticus* amyloglucosidase only hydrolyzes α-1,4-linkages but does not possess debranching activity. Consequently only one third of the remaining dextrins in normal beer are hydrolyzed. To obtain superattenuation, α-1,6-glycosidase activity also is needed. In this respect the glucoamylase of *S. occidentalis* is preferable. Furthermore, the *S. occidentalis* enzyme is inactivated by pasteurization in contrast to that of *S. diastaticus.* This is an additional technical advantage of the *S. occidentalis* enzyme. One glucoamylase gene from this yeast was coupled to the yeast *ADH* promoter and transferred to lager yeast strains[6] both by an autonomously replicating plasmid and by integration into a chromosome. Both recombinant strains produced an active enzyme that was secreted and was able to hydrolyze α-1,6-linkages. In laboratory scale fermentations, decreased dextrin levels were obtained but the hydrolysis was insufficient for practical purposes. This experiment, however, demonstrates the practicability of this approach.

GLUCANOLYTIC ENZYMES

One technical difficulty frequently encountered in beer production is slow filtration caused by β-glucans. Barley contains β-glucan, a polysaccharide consisting of β-1,4- and β-1,3-linked glucose units. During malting and mashing the β-glucan is hydrolyzed by barley β-glucanases. This activity is not always sufficient, and therefore difficulties are caused that can be remedied by adding microbial β-glucanases.

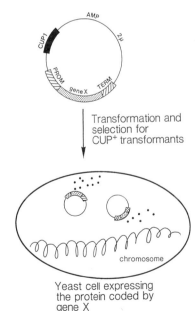

FIGURE 1. Transformation of yeast with plasmid containing copper resistance marker.

β-Glucanase genes have been cloned from *Bacillus subtilis,* from the filamentous fungus *Trichoderma reesei,* and from barley. The barley gene has been transferred to brewer's yeast,[7] but no information is available about its performance in brewing. The β-glucanase gene from *B. subtilis* has also been transferred to brewer's yeast on a multicopy plasmid.[8] Both lager and ale yeast were transformed, and the lager yeast degraded barley β-glucan more efficiently than did the ale yeast during laboratory

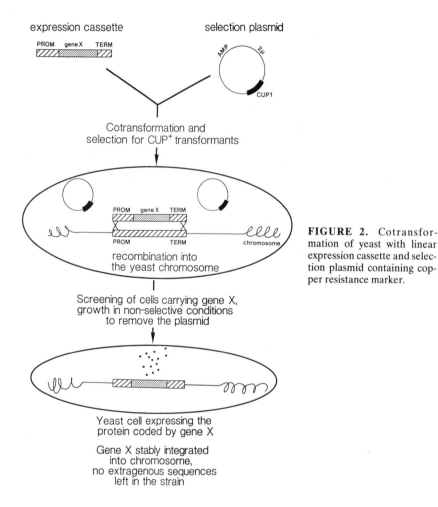

FIGURE 2. Cotransformation of yeast with linear expression cassette and selection plasmid containing copper resistance marker.

scale fermentations. The reason for this could be that the plasmid is lost more rapidly from the ale yeast than from the lager yeast.[4]

These and other experiments show that the *B. subtilis* β-glucanase can be secreted by the brewer's yeast and that the enzyme is active against barley β-glucan in fermentations. The results from these experiments have been encouraging, but nevertheless *Bacillus* enzyme may not represent the best solution to the technical

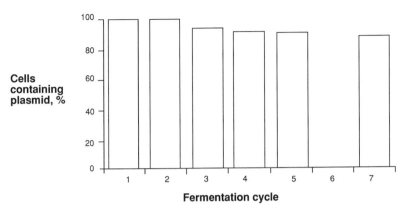

Fermentation cycle

FIGURE 3. Stability of multicopy plasmid in brewer's yeast transformed with 2 μ plasmid containing β-glucanase gene under PGK promoter.

problems in brewing. It has been reported that because of the high pH optimum of the bacterial enzyme (6.2), it loses its activity rapidly when added to wort (pH 5) or beer (pH 4.2), and as a consequence the enzyme is inactive in beer.[9]

Good results have been achieved with brewer's yeast transformed with the β-1,4-endoglucanase gene from the filamentous fungus *T. reesei*.[10,11] The cDNA copy of the gene coding for the endoglucanase 1 enzyme was transferred to a bottom-fermenting brewer's yeast strain used in commercial brewing. The transformation was made both using a multicopy plasmid (FIG. 1) with a copper resistance marker (*CUP1*) and by chromosomal integration to the *LEU2* locus of yeast. Integration was achieved using cotransformation (FIG. 2). Also in this case *CUP1* resistance marker was used in the plasmid, which was then removed by growing the cells on a nonselective medium. In both cases, yeast phosphoglycerokinase (*PGK*) and alcohol dehydrogenase (*ADH*) promoters were used.

Beer was produced on a pilot scale (50-liter fermentations) in a process that was closely similar to the one used in commercial lager beer breweries. Because the

FIGURE 4. Viscosity of beers produced using glucanolytic brewer's yeast.

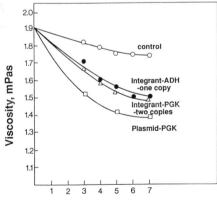

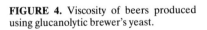

Fermentation time, days

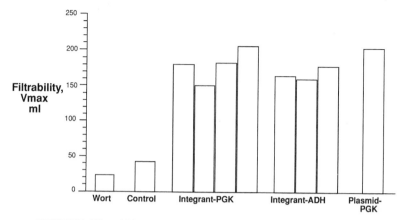

FIGURE 5. Filterability of beers produced using glucanolytic brewer's yeast.

commercial process includes reusing the yeast in several cycles, the yeast was reused seven times and the stability of the plasmid under those nonselective conditions was followed (FIG. 3). The results confirmed that bottom-fermenting brewer's yeast loses the plasmid very slowly. It would thus be possible also to use the plasmid strain in practical brewing. The plasmid strain was found to secrete about 10 times as much enzyme as the integrant strain due to the higher gene copy number. The β-glucanase secreted by the integrant strain, however, was also sufficient to hydrolyze the wort β-glucans almost completely during primary fermentation. Beer viscosity was reduced (FIG. 4), and consequently filtrability also was significantly improved (FIG. 5). The analyses and the flavor of the beers were not affected by the genetically modified yeast strains (TABLE 2).

FLOCCULENCE

At the end of fermentation, yeast cells are aggregated to flocs. This phenomenon, called flocculation, is traditionally used in beer production to remove yeast from the

TABLE 2. Analyses of Beers Produced Using Genetically Modified Brewer's Yeast Expressing β-Glucanase Activity

	Control Strain	Plasmid Strain	Control Strain	Integrant Strain
Original extract (%[w/w])	10.6	10.6	10.6	10.6
Alcohol content (%[w/w])	3.60	3.40	3.55	3.45
Apparent extract (%[w/w])	1.95	2.45	2.10	2.30
Apparent attenuation (%)	81.5	77.0	80.0	78.5
Ethyl acetate (mg/l)	10.8	16.0	13.7	18.5
Isoamyl acetate (mg/l)	0.9	1.5	1.7	2.5
n-Propanol (mg/l)	9.5	9.4	9.0	9.2
Isobutanol (mg/l)	7.1	7.6	8.0	4.3
3-Methylbutanol (mg/l)	12.4	13.1	14.8	10.0
2-Methylbutanol (mg/l)	32.6	35.1	43.6	35.3
Vicinal diketones (mg/l)	0.33	0.37	0.47	0.46

fermented beer. Bottom-fermenting yeasts used for lager beer fermentation sediment when the beer is cooled at the end of the main fermentation period, whereas top-fermenting yeasts in the production form flocs that rise to the surface and can be skimmed off. Flocculence is therefore a technically important property, and yeasts are divided into flocculent and nonflocculent powdery strains. Genetic studies on yeast flocculation were first reported by Gilliland,[12] and extensive research into flocculation has been carried out.[13] Several genes controlling flocculation have been reported. A flocculation gene was isolated from an *S. cerevisiae* strain,[14,15] and a powdery *S. cerevisiae* strain was transformed with a multicopy plasmid expressing the flocculence gene. Flocculence was conferred to the recipient and was dependent on gene dose. Furthermore, flocculence was calcium dependent and could be completely inhibited by EDTA. Brewing experiments with such transformed yeasts have not yet been reported.

ZYMOCINS

Some yeast strains called "killer yeasts" secrete zymocins, toxins that kill other yeast strains. Zymocin-producing yeasts are themselves immune to this toxin. Zymocins are encoded by a double-stranded RNA. Several different zymocins have been described.[16,17] Brewer's yeast strains lack zymocin characters, but contaminant wild yeast strains have been found in beer fermentations that kill the production strain.[18,19] Many wild yeast contaminants, however, are also sensitive to the zymocins.[20]

Conferring the zymocin character to a brewer's yeast strain would thus help prevent wild yeast contamination in breweries. A cDNA copy of the RNA encoding zymocin K1 has been transformed to brewer's yeast.[21] The cDNA was coupled to a yeast promoter and expressed on a multicopy plasmid. The transformed strains were stable and capable of killing zymocin-sensitive wild yeasts.

BREWER'S YEAST NOT PRODUCING DIACETYL

Most of the flavor compounds in beer are produced by yeast during primary fermentation. In lager beer production a secondary fermentation period, or lagering, is also necessary for flavor maturation. The main reason for this lagering period is to remove diacetyl, a very strong and, in beer, unpleasant flavor component.

During fermentation, yeast produces α-acetolactate (FIG. 6), an intermediary metabolite in the biosynthesis of valine. The surplus α-acetolactate is secreted, and at the low temperature used for fermentation (8–10°C), it is slowly spontaneously decarboxylated, yielding diacetyl. The diacetyl is then taken up by the yeast and rapidly reduced to acetoin. The lagering period is slow, from 3 weeks to several months. If the enzyme α-acetolactate decarboxylase (α-ALDC) is added to beer, the α-acetolactate is quickly converted to acetoin and the secondary fermentation period can be shortened. A gene coding for this enzyme has been cloned from bacteria by two groups, one in Japan[22] and one in Finland.[23] This gene has been transferred to brewer's yeast. We have also produced beer on a pilot scale using these new yeast strains.[24]

Several strains were constructed in which α-*ald* genes from *Klebsiella terrigena* and *Enterobacter aerogenes* were expressed. In these constructions again, as with β-glucanase, both *PGK* and *ADH* promoters were used. The genes were transferred

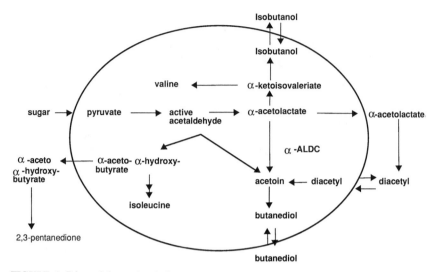

FIGURE 6. Diacetyl formation in beer production. Scheme showing α-acetolactate excreted from yeast cell, diacetyl taken up by yeast cell being converted to acetoin and transferred α-acetolactate decarboxylase (α-ALDC) preventing diacetyl formation.

on multicopy plasmids or integrated in a chromosome using a cotransformation technique.

A commercial lager brewer's yeast was used as host, and beer was produced on a pilot scale (50 liters). With the best new yeast strains, diacetyl formation was decreased well below the taste threshold, and the secondary fermentation period could be completely omitted without impairing beer analyses or flavor, as judged by a taste panel (FIG. 7).

DISTILLER'S YEAST

The production of potable and technical ethanol by distiller's yeast is easier to control than is the fermentation of beverages such as beer and wine. Because the product is distilled, it is virtually pure ethanol, and flavor problems typical for beer fermentation do not exist. The dominating technical problem is the efficiency of the process including carbohydrate utilization. Often the raw material is a starchy product. The hydrolysis of starch is a well-known technologic process, but the accessibility of the starch can be enhanced by breaking down cell wall material using cellulolytic enzymes. Good results have been obtained using preparations of microbial cellulases. The problem of transforming yeast with these cellulolytic genes is that cellulolysis is affected by a complex mixture of enzymes, and any single enzyme would perhaps not be enough to accomplish the required breakdown of cell walls.

Ethanol production has been improved using a distiller's yeast transformed with a gene coding for amyloglucosidase from *Aspergillus awamori*.[25] The transformed yeast strain was able to use over 90% of the starch present in the medium.

A waste or byproduct that can be used for ethanol production is whey. The lactose content of whey is 4–5%, but distiller's yeast cannot use it. *Kluyveromyces lactis* is a yeast species that can grow on lactose because it possesses both lactose permease and β-galactosidase needed for the uptake and hydrolysis of lactose. A large piece of DNA carrying both these genes has been transferred to *S. cerevisiae* by chromosomal integration.[26] The growth of the transformed strain, however, was slow, and the host was not an industrial yeast strain. The genes have also been transferred into a strain of distiller's yeast,[27] but results on ethanol production have not yet been reported.

Another possibility is to transform distiller's yeast with a gene coding for secreted β-galactosidase. The extracellular enzyme hydrolyzes lactose to glucose and galactose which are taken up and fermented by distiller's yeast. This approach has been taken by Kumar[28] who used the gene for extracellular β-galactosidase from *Aspergillus niger*. The yeast strain used, however, was a laboratory strain, and this approach has not yet been applied to an industrial strain.

BAKER'S YEAST

Baker's yeast is normally produced using molasses as substrate. Beet molasses contains small amounts of raffinose, 0.5–3.0% of the total sugar, which is hydrolyzed by yeast invertase to fructose and melibiose. Melibiose is a disaccharide, which is not used by baker's yeast, because the yeast does not possess α-galactosidase (melibiase). Bottom-fermenting brewer's yeast carries the *MEL1* gene encoding α-galactosidase. This gene has been cloned and transferred to baker's yeast strains.[29,30] When these strains containing both plasmid strains and integrated strains were cultivated on molasses, all melibiose was utilized.

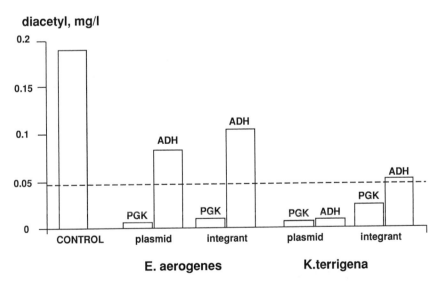

FIGURE 7. Diacetyl formation with genetically modified brewer's yeasts. The α-acetolactate decarboxylase gene from two bacteria was expressed on a multicopy plasmid or integrated into a chromosome under two different promoters. *Dotted line* represents the taste threshold of diacetyl in beer.

It was recently reported that baker's yeast has been transformed with maltose permease and maltase genes from another strain of the same species of *S. cerevisiae*. This yeast produces carbon dioxide more efficiently than does the parent strain, and its baking quality is therefore improved. The Dutch company, Gist-Brocades, has obtained approval for this baker's yeast in the UK.[31]

SAFETY CONSIDERATIONS

Genetically modified yeasts have until now not been used anywhere for the production of food or beverages. Laboratory and pilot experiments, however, have clearly demonstrated that yeast strains have been developed that would be advantageous to the industry.

Recently the first two applications of genetically modified organisms in food production have been approved by authorities. The use of chymosin produced by a genetically modified strain of *Escherichia coli* has been approved. This genetically modified organism is not used in cheese production but only to produce one of the process aids. Chymosin was approved in Australia in September 1989 and in the United States in March 1990.

The other application concerns a genetically modified baker's yeast, which was approved for use in the UK in March 1990. The food produced, bread, actually contains live, genetically modified yeast; however, no genetic material from another species has been transferred to the yeast. Although the transformation is totally intraspecies, this is the first food application of a genetically modified yeast.

A genetically modified yeast must fulfill three conditions to be used in industrial processes. The process must be: (1) technically and economically advantageous; (2) approved by authorities; and (3) acceptable to the consumer.

APPROVAL

Approval for the use of recombinant DNA technology in research and production has been intensively discussed, and attitudes have slowly changed. Regulations are presently being made, and although big differences exist between countries, some general principles are emerging.

The regulations are based on risk assessment and include rules for the acceptability of the host, the vector/insert, and the genetically modified organism. For the host the essential requirements are that it must (1) be nonpathogenic; (2) have a history of safe use; and (3) have built-in environmental limitations. These requirements are all fulfilled by the industrial yeast strains.

For the approval of genetically modified industrial yeasts, the vector and the insert used are the main targets of interest. It is recommended that the vector/insert (1) be well characterized and contain no harmful sequences; (2) be limited in size as much as possible; (3) not increase the stability of the genetically modified organism in the environment; (4) be poorly mobilizable; and (5) not transfer resistance markers to other microorganisms. Furthermore, the resulting genetically modified organism must be nonpathogenic and should be as safe as the host.

Many of the examples referred to earlier in this article are good examples of yeast strains that fulfill all these requirements. It should be possible to obtain the approval of the authorities for their industrial use.

CONSUMERS' ATTITUDES

The most difficult requirement for the use of genetically modified yeast in the production of food and beverages is the acceptability to the consumer. This is the main reason that only one such application has been reported so far even though research programs have clearly demonstrated the technical advantages of these applications. All the initial industrial applications of genetically modified microorganisms have been confined to pharmaceutical products.

REFERENCES

1. KNOWLES, J. K. C. & R. TUBB. 1987. Recombinant DNA transfer and expression techniques with industrial yeast strains. EBC Symposium on Brewer's Yeast. Helsinki, 1986. Monograph XII: 169–185. Verlag Hans Carl. Nürnberg.
2. RANK, G. H., G. CASEY & W. XIAO. 1988. Gene transfer in industrial *Saccharomyces* yeasts. Food Biotechnol. **2**: 1–41.
3. PENTTILÄ, M. & T.-M. ENARI. 1990. Genetic engineering of industrial yeasts. *In* Biotechnology, Applications and Research. Cheremisoff, P. E., ed. Technomic Publishers. Lancaster, PA. In press.
4. MEADEN, P. G. & R. S. TUBB. 1985. A plasmid vector system for the genetic manipulation of brewing strains. Proc. 20th EBC Congress. Helsinki, 219–229. IRL Press. Oxford, UK.
5. VAKERIA, D. & E. HINCHLIFFE. 1989. Amylolytic brewing yeast: Their commercial and legislative acceptability. Proc. 22nd EBC Congress. Zurich, 475–482. IRL Press. Oxford, UK.
6. LANCASHIRE, W. E., A. T. CARTER, J. J. HOWARD & R. J. WILDE. 1989. Superattenuating brewing yeast. Proc. 22nd EBC Congress. Zurich, 491–498. IRL Press. Oxford, UK.
7. VON WETTSTEIN, D. 1987. Molecular genetics in the improvements of brewer's and distiller's yeast. Antonie van Leeuwenhoek J. Microbiol. **53**: 299–305.
8. CANTWELL, B., G. BRAZIL, J. HURLEY & D. McCONNELL. 1985. Expression of the gene for the endo-β-1,3-1,4-glucanase from *Bacillus subtilis* in *Saccharomyces cerevisae*. Proc. 20th EBC Congress. Helsinki, 259–266. IRL Press. Oxford, UK.
9. HINCHLIFFE, E. 1985. β-Glucanase: The successful application of genetic engineering. J. Inst. Brew. **91**: 384–389.
10. ENARI, T.-M., J. K. C. KNOWLES, V. LEHTINEN, M. NIKKOLA, M. PENTTILÄ, M.-L. SUIHKO, S. HOME & A. VILPOLA. 1987. Glucanolytic brewer's yeast. Proc. 21st EBC Congress. Madrid, 529–536. IRL Press. Oxford, UK.
11. PENTTILÄ, M. E., M.-L. SUIHKO, V. LEHTINEN, M. NIKKOLA & J. K. C. KNOWLES. 1987. Construction of brewer's yeasts secreting fungal endo-β-glucanase. Curr. Genet. **12**: 413–420.
12. GILLILAND, R. B. 1951. The flocculation characteristics of brewing yeasts during fermentation. Proc. 2nd EBC Congress. Brighton, 35–58. IRL Press. Oxford, UK.
13. STEWART, G. G. & I. RUSSEL. 1981. *In* Brewing Science. J. R. J. Pollock, ed.: 61–92. Academic Press. London.
14. WATARI, J., Y. TAKATA, N. NISHIKAWA & K. KAMADA. 1987. Cloning of a gene controlling yeast flocculence. Proc. 21st EBC Congress. Madrid, 537–544. IRL Press. Oxford, UK.
15. WATARI, J., Y. TAKATA, M. OGAWA, N. NISHIKAWA & M. KAMIKURA. 1989. Molecular cloning of a flocculation gene in *Saccaromyces cerevisiae*. Agric. Biol. Chem. **53**: 901–903.
16. WICKNER, R. B. 1986. Double-stranded RNA replication in yeast: The killer system. Ann. Rev. Biochem. **55**: 373–395.
17. BUSSEY, H., T. VERNET & SDICU. 1988. Mutual antagonism among killer yeasts: Competition between K1 and K2 killers and a novel cDNA-based K1-K2 killer strain of *Saccharomyces cerevisiae*. Can. J. Microbiol. **34**: 38–44.

18. MAULE, A. P. & P. D. THOMAS. 1973. Strains of yeast lethal to brewery yeasts. J. Inst. Brew. **79:** 137–141.
19. TAYLOR, R. & B. KIRSOP. 1979. Occurrence of a killer strain of *Saccharomyces cerevisiae* in a batch fermentation plant. J. Inst. Brew. **85:** 325–331.
20. YOUNG, T. W. & N. P. TALBOT. 1979. The use of yeast extracellular substances to inhibit the growth of beer spoilage organisms. Proc. 17th EBC Congress. 817–830. Berlin, 1979. IRL Press. Oxford, UK.
21. BUSSEY, H. & P. MEADEN. 1985. Selection and stability of yeast transformants expressing cDNA of an M1 killer toxin-immunity gene. Curr. Genet. **9:** 285–291.
22. SONE, H., T. FUJII, K. KONDO & J. TANAKA. 1987. Molecular cloning of gene encoding α-acetolactate decarboxylase from *Enterobacter aerogenes.* J. Biotechnol. **5:** 87–91.
23. ENARI, T.-M., M. NIKKOLA, M.-L. SUIHKO, M. PENTTILÄ & J. KNOWLES. 1987. Yeast strains suitable for accelerated brewing. US Patent Appl. 07/044,244. US Patent Office. Washington, DC.
24. SUIHKO, M.-L., M. PENTTILÄ, H. SONE, S. HOME, K. BLOMQVIST, J. TANAKA, T. INOUE & J. KNOWLES. 1989. Pilot brewing with α-acetolactate decarboxylase active yeasts. Proc. 22nd EBC Congress. Zurich, 484–490. IRL Press. Oxford, UK.
25. INLOW, D., J. MCRAE & A. BEN-BASSAT. 1988. Fermentation of corn starch to ethanol with genetically engineered yeast. Biotech. Bioeng. **32:** 227–234.
26. SREEKRISHNA, K. & R. C. DICKSON. 1985. Construction of strains of *Saccharomyces cerevisiae* that grow on lactose. Proc. Natl. Acad. Sci. USA **82:** 7909–7913.
27. YOCUM, R. R. 1986. Genetic engineering of industrial yeasts. Proc. Bio. Expo 86. 171–180. Butterworth. Stoneham, MA.
28. KUMAR, V. 1988. Cloning and expression of *Aspergillus niger* β-galactosidase gene in *Saccharomyces cerevisiae.* PhD. Thesis. Univ. of London, p. 198.
29. CASEY, G. P., W. XIAO & G. H. RANK. 1988. Construction of α-galactosidase-positive strains of industrial baker's yeast (*Saccharomyces cerevisiae*). Am. Soc. Brew. Chem. **46:** 67–71.
30. LILJESTRÖM, P. L., V. JOUTSJOKI & M. KORHOLA. 1987. Stable α-galactosidase producing derivates of commercial baker's yeast. *In* Industrial Yeast Genetics. M. Korhola & H. Nevalainen, eds. **5:** 127–136. Proc. Alko-Symposium Industrial Yeast Genetics. Foundation for Biotechnical and Industrial Fermentation Research. Helsinki.
31. 1990. Genetic engineering: Modified yeast fine for food. Nature **344:** 186.

Biochemistry and Molecular Genetics of Penicillin Production in *Penicillium chrysogenum*

JUAN F. MARTÍN

University of León
Faculty of Biology
Area of Microbiology
24071 León, Spain

The biosynthetic pathways of three different β-lactam antibiotics, penicillin, cephalosporin C, and cephamycin, in *Penicillium chrysogenum, Acremonium chrysogenum, Aspergillus nidulans, Streptomyces clavuligerus,* and *Nocardia lactamdurans* (syn. *Streptomyces lactamdurans*) have largely been elucidated in the last few years. The β-lactam-thiazolidine ring nucleus of penicillin and the β-lactam-dihydrothiazine nucleus of cephalosporin are both formed by cyclization of a common precursor, δ-(L-α-aminoadipyl)-L-cysteinyl-D-valine (ACV)[1,2] (FIG. 1). ACV is cyclized by removal of four hydrogen atoms to form isopenicillin N, an intermediate having an L-α-aminoadipyl side chain attached to the nucleus, by isopenicillin N (IPN) synthase (cyclase) (IPNS). This enzyme, which was purified to homogeneity in our laboratory from extracts of *P. chrysogenum* AS-P-78, behaves as a dioxygenase. It requires O_2, Fe^{2+}, dithiothreitol (DTT), and ascorbate and has a molecular weight estimated by PAGE of 39,000.[3] High penicillin-producing strains show very high IPNS activity as compared to lower producers.[3,4]

In *P. chrysogenum,* the α-aminoadipyl side chain of IPN can be exchanged for aromatic side chains such as phenylacetyl and phenoxyacetyl groups (forming penicillin G or V) and can also be removed to yield 6-aminopenicillanic acid (6-APA). 6-APA is the major penicillin derivative that accumulates in precursor-free fermentations (FIG. 1).

The role of 6-APA in the *de novo* biosynthesis of penicillins is unclear. Demain[5] proposed that the terminal reaction of benzylpenicillin (also known as penicillin G) synthesis is most likely an exchange of the L-α-aminoadipic acid side chain of IPN for phenylacetic acid, which is activated in the form of phenylacetyl-CoA. However a two-step reaction involving 6-APA as an intermediate is also possible. Interconversions between penicillin G and 6-APA or between two hydrophobic penicillins by mycelial suspensions of *P. chrysogenum* have been described.[6] Reports on the presence of an acylase in *P. chrysogenum,* which removes the side chain of benzylpenicillin to yield 6-APA, and of an acyltransferase, which catalyzes a direct exchange of side chains between IPN and solvent-soluble penicillin (or between such penicillins and 6-APA), are contradictory. It was shown later that an enzyme(s) in crude extracts of *P. chrysogenum* catalyzes the formation of benzylpenicillin from IPN, as well as that from 6-APA, in the presence of phenylacetyl-CoA.[7]

Very little is known about the kinetics of penicillin overproduction. Large increases in penicillin titers (several hundredfold) have been achieved by empiric mutagenesis of *P. chrysogenum.* These mutations probably affected either the structural genes encoding the penicillin biosynthetic enzymes or the regulatory sequences involved in the control of expression of structural genes. However, at the molecular level, the nature of the mutations that have led to increased penicillin yields remains unknown. Cloning and analysis of penicillin biosynthetic genes from industrial strains as compared to those of low-producing strains will provide evidence of whether high penicillin-producing strains contain either higher levels of biosynthetic enzymes or mutated more active enzymes. Transcriptional initiation and polyadenylation sequences are involved in regulation of gene expression in filamentous fungi.[8,9] The study of these regulatory sequences in different β-lactam-producing fungi is therefore of utmost interest.

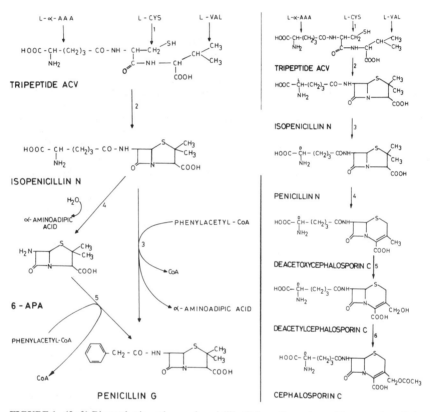

FIGURE 1. (*Left*) Biosynthetic pathway of penicillin G from the amino acids L-α-aminoadipic, L-cysteine, and L-valine. 1. ACV synthetase. 2. Isopenicillin N synthase. 3. Isopenicillin N acyltransferase. 4. Isopenicillin N amidase (6-APA forming). 5. 6-APA acyltransferase. (*Right*) Biosynthetic pathway of cephalosporin C from the same component amino acids. 1. ACV synthetase. 2. Isopenicillin N synthase. 3. Isopenicillin N epimerase. 4. Deacetoxycephalosporin C synthase. 5. Deacetoxycephalosporin C hydroxylase. 6. Deacetylcephalosporin C acetyltransferase. Note that the two initial steps are identical in both biosynthetic pathways. From Martín and Liras.[2]

RESULTS AND DISCUSSION

Purification to Homogeneity and Characterization of Acyl Coenzyme A:6-Aminopenicillanic Acid Acyltransferase

The acyl coenzyme A:6-aminopenicillanic acid acyltransferase of *P. chrysogenum* AS-P-78 was purified to homogeneity, as concluded by sodium dodecyl sulfate-polyacrylamide gel electrophoresis and isoelectric focusing. The active form of the enzyme is a monomer with a molecular weight of 30,000 ± 1,000 and a pI of about 5.5. The optimal pH and temperature were 8.0 and 25°C, respectively. This enzyme converts 6-APA into penicillin using phenylacetyl-CoA or phenoxyacetyl-CoA as acyl donors. The pure enzyme showed a high specificity and affinity for 6-APA and did not accept benzylpenicillin, 7-aminocephalosporanic acid, cephalosporin C, or isocephalosporin C as substrates. The acylCoA:6-APA acyltransferase required dithiothreitol or other thiol-containing compounds, and it was protected by thiol-containing reagents against thermal inactivation. The acyltransferase was inhibited by several divalent and trivalent cations and by *p*-chloromercuribenzoate and *N*-ethylmaleimide. The activity was absent in four different mutants that were blocked in penicillin biosynthesis.[10]

High-Frequency Transformation of P. chrysogenum

High-frequency transformation of penicillin-producing strains of *P. chrysogenum* has been obtained using plasmid vectors carrying the *pyr*4 gene of *N. crassa* and the *pyr*G gene of *P. chrysogenum* as selectable markers.[11]

Transformation to stable uracil independence is associated with acquisition of transforming DNA in high molecular forms that are not simple multimers of the vector. A transformant strain showed a fivefold higher orotidine-5'-monophosphate decarboxylase activity (coded by the *pyr*4 gene) than did the wild-type strain, indicating that the genes carried in the transforming vector are efficiently expressed in *Penicillium*.[11]

More recently a transformation system based on the use of the bleomycin-phleomycin resistance gene (*ble*) as selectable dominant marker has been constructed in our laboratory.[12]

Cloning, Sequence Analysis, and Transcriptional Study of the Isopenicillin N Synthase of P. chrysogenum AS-P-78

A gene (*pcb*C) encoding the isopenicillin N synthase of *P. chrysogenum* AS-P-78 was cloned in a 3.9-kb *Sal*I fragment using a probe corresponding to the amino terminal end of the enzyme. The gene was identified by expression in *P. chrysogenum* after transformation. The *Sal*I fragment was trimmed down to a 1.3-kb *Nco*I-*Bgl*II fragment that contained an ORF of 996 nucleotides encoding a polypeptide of 331 amino acids with an M_r of about 38,000.[13] The predicted polypeptide encoded by the *pcb*C gene of strain AS-P-78 contains a tyrosine at position 194, whereas the gene of the high penicillin-producing strains 23X-80-269-37-2[14] shows an isoleucine at the same position. The *pcb*C gene is expressed in *Escherichia coli* minicells using the P_L promoter of phage λ and does not contain introns.

The deduced amino acid sequence of the isopenicillin N synthase of *P. chrysogenum* is very similar to that of *Streptomyces griseus*,[15] *N. lactamdurans*,[16] and other actinomycetes and filamentous fungi (reviewed by Martín *et al.*[17]) (FIG. 2).

Cloning of the Acyl-CoA:6-APA Acyltransferase Gene

The acyl-CoA:6-APA acyltransferase (*pen*DE) gene from *P. chrysogenum* AS-P-78 has been cloned in a 2.4-kb *Hind*III-*Sal*I fragment (FIG. 3). The fragment has been completely sequenced. It is encoded in an ORF of 357 amino acids encoding a

```
MK****MPSAEVPTIDVSPLFGDDAQEKVRVGQEINKACRGSGFFYAANHGVDVQRLQDVVNEFHRTMSPQEKYDLAIHAYNKNNS*HVRNGYMAIEG      94
M****NRHADVPVIDISGLSGNDMDVKKDIAARIDRACRGSGFFYAANHGVDLAALQKFTDWHMAMSAEEKWELAIRAYNPANP*RNRNGYMAVEG        93
MPIP*MLPAHVPTIDISPLSGGDADDKKRVAQEINKACRESGFFYASHGIDVQILKDVVNEFHRTMTDEEKYDLAINVNKNNP*RTRNGYMAVKG        96
HPIL*MPSAEVPTIDISPLSGGDDAKAKQRVAQEINKAARGSGFFYASNHGVDVQLLQDVVNEFHRNMSDQEKHDLAINAYNKDNP*HVRNGYKAIKG     96
MPVL*MPSADVPTIDISPLFGTDPDARAHVARQINEACRGSGFFYASHHGIDVRRLQDVVNEFHRTMTDQEKHDLAIHAYNENNS*HVRNGYMARPG      96
MPVL*MPSAHVPT.IDISPLFGTDAAAKKRVAEEIHGACRGSGFFYATHHGVDVQQLQDVVNEFHGAMTDQEHDLAIHAVPDNP*HVRNGYKAVPG        96
MGSVFVPVANVPRIDVSPLFGDDKEKKLEVARAIDAASRDTGFFYAVNHGVDLPWLSRETNKFHMSITDEEKWQLAIRAVNKEHESQIRAGYLPIPG     98
MAST***PKANVPKIDVSPLFGDNMEKMKVARAIDAASRDTGFFYAVNHGVDVKRLSNKTREFHFSITDEEKWDLAIRAVNKEHQDQIRAGYLSIPE     96
MGSV**SKANVPKIDVSPLFGDDQAAKMRVAQQIDAASRDTGFFYAVNHGINVQRLSQKYTKFHMSITPEKWDLAIRAYNKEHQDQVRAGYLSIPG      96

KKAVESFCYLNFSSEDHPEIKAGTPMHEVISWPDEEKHPSFRPFCEYVWTMHRLSKVL*MRGFALALGKDERFFEPELKEADTLSSVSL*IRYPYL      190
KKANESFCYLNFSFDADHATIKAGIPSHEVIWPDEARHPGMRRFYEAVFSDVFDVAAVI*LRGFALALGREESFFERHFSMDDTLSAVSL*IRYPFI     189
KKAVESWCVLNPSSEDHPQIRSGTPMHEGNIWPDEKRHQRFRPFCEDVRDVFSLSKVL*MRGFVLALGRKPDFDASLSLADTLSAVI*IRYPVI        192
KKAVESFCVLNFSSDDHPMIKSETPMHEVILWPDEEKHPRFRPFCEDVYRQLLRLSTVI*MRGYALALGRREDFDEALAEADTLSSVSI*IRYPYI      192
RKTVESWCYLNFSFGEDHPMIAAGTPMHEVNLWPDEERHPDFRSFGEQYYREVFRLSKVLLLRGFALALGKPEFFENEVTEEDTLSASV.MIRYPYL     194
RKAVESFCYLNPDFGEDHPTIKAGTPMHEVNLWPDEERHPRFRPFCEGVYRQMLKLSTVL*MRGLALALGRPEHFDAALAEQDSLSSVS*IRYPYI      196
KKAVESFCVLLNFSFSPDHPRIKEPTPMHEVNVWPDDAKHPGFRAFAIKYTWDVFGLSSAV*LRGYALALGRDEDFTRHSRRDTTLSSVV*IRYPYL     194
KKAVESFCYLNPNFKPDHPLLQSKIFPTHEVNVWPDEKKHPGFREFABQYWDVFGLSSAL*LRGVALALGKEEDFSRHFKKEDALSSVVL*IRYPYL     198
KKAVESFCYLNFNFTPDHPRIQAKTFTHEVNVWPDEKFPGFQDEFAEQVYWDVFGLSSAL*LKGYALALGKEENFARHFKPDDTLASVVL*IRYPYL     196

EDYF*P*VKTGPDGEKLSFEDHFDVSMITVYLFQTQVQNLQVETVDGWRDLPTSDTDFEVNACYLGHIFNDYFPSFIHKVKFVNAERISLPFFPHAGQ   286
ENVF*F*IKLGPDGEKLSFEHHQDVSLLTVLYQTAIFNLQVETAEGYLDIPVSDEHFEVNCEYMAHITNGYYPAVFHRVKYINAERISLPFFANLSH     285
EDYF*P*VKTGPDGTKLSFEDHLDVSMITVLFGTEVQNLQVETADGWQDLPTSGENFEVNCGTVMGYLFNDYFPAPNHRVKFINAERISLPFFLHAGH    288
EEY*P*VKTGADGTKLSFEDHLDVSMITVLFQTEVQNLQVETVDGWQDIPRSDEDFELVCNCMGHIFHDYFPANHRVKVNAERISLPFFLHAGQ        288
DFFEAAIKTGPDGTRLSFEDHLDVSMITVLFQTEVQNLQVETVDGWQSLPTSGENFIINCGTYLGYLFNDYFPAPNHRVKVNAERISLPFFLHAGQ      292
EEYF*P*VKTGPDGQLLSFEDHLDVSMITVLFQTQVQNLQVETVDGWRDIPTSENDFIVNCGTVMAHVTNDYFPAPNHRVKFVNAERISLPFFLNGGH    288
DPYFEPAIKTADDGTKLSFEWHEDVIITVLYQSDVQNLQVKTPQGWQDIQADDTGFIINGGSYMAHIIDDYPAFIHRVKFWVNEERQSIPFFVNLGW    292
NFYFPAAIKTAEDGTKLSFEWHEDVSIITVIYQSDVANLQVEMPQGYLDIEADDNAYIVNCGSYMAHITNNIYPAFIHRVKWVNEEROSIPFFVNLGF   290
DFYFEAAIKTAADGTKLSFEWHEDVSLIITVLYQSNVQNLQVETAAGYQDIEADDTGYLINCGSYMAHLENNYYKAPIHRVKWVNAEROSLPFFVNLGY    290

HTLIEPFFPDGAEPEG***KQGN*EAVRYGDYLNHGLLHSLIVKNGQT                                                        328
ASAIDFFAPPYAPPG****GN*PTVSYGDYLQHGLLDLIRANGQT                                                           326
TTVMEPFSP****EDTRGKELN*PPVRYGDYLQQASNALIIAKNGQT                                                          329
NSVIEPFVP****EGAAGTVKN*PTTSYGEYLQHGLRALIVKNGQT                                                           329
NSVMKPFHP****EDTGDRKLN*PAVTYGEYLQEGFHALIAKRVQT                                                           331
EAVIEPFVP****EGASEEVRN*EALSYGDYLQHGLRALIVKNGQT                                                           329
EDTIQPWDPATAKDGAKDAAKDKPAISYGEVLQGGLRGLINKNGQT                                                           338
NDTVQPWDP***SKEDGKT****DQRPISYGDYLQNGLVSLINKNGQT                                                         331
DSVIDRFDF**REPNGKS***DREPLSYGDYLQNGLVSLKINKNGQT                                                          331
```

N. lactamdurans
Flavobacterium
S. griseus
S. jumonjinensis
S. lipmanii
S. clavuligerus
C. acremonium
P. chrysogenum
A. nidulans

FIGURE 2. Comparative analysis of the deduced amino acid sequences of the *P. chrysogenum* IPNS and eight other IPNSs. Gaps (*asterisks*) have been introduced to obtain maximal alignment. Shaded amino acids are conserved in all nine polypeptides. The two cysteines indicated by *arrowheads* are conserved in all IPNSs except in *N. lactamdurans*.[16] The amino acid residues are indicated by *numbers* at the right.

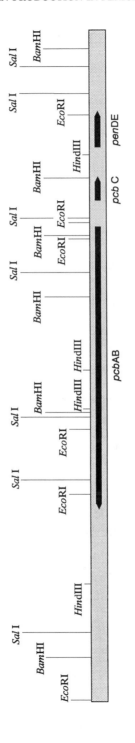

FIGURE 3. Restriction endonuclease map of a cloned *P. chrysogenum* DNA fragment showing the position and orientation of the three genes *pcb*AB, *pcb*C, and *pen*DE involved in penicillin biosynthesis. The *pcb*AB and *pcb*C genes do not contain introns, whereas the *pen*DE gene shows three introns. Note that the *pcb*AB gene is expressed in opposite orientation to *pcb*C and *pen*DE.

protein of molecular weight 39943.[18] The gene contains three introns in the 5' region of the open reading frame. A very interesting result was the finding that the gene encoded a pre-acyltransferase of 39.9 kD that is later processed into two subunits of 29 and 11 kD, of which the first seems to be the enzymatically active component.[10] The *pen*DE gene of *P. chrysogenum* is very similar to that of *A. nidulans* and contains the same three introns.[19]

Cloning and Characterization of the ACV Synthetase Gene

Penicillins, cephalosporins, and cephamycins are β-lactam antibiotics formed by condensation of L-α-aminoadipic acid (an intermediate in the lysine biosynthetic pathway in fungi), L-cysteine, and L-valine (see review by Martín and Liras[2]). The three amino acids are linked together to form the tripeptide δ(L-α-aminoadipyl)-L-cysteinyl-D-valine (ACV) which is the first common intermediate of penicillin and cephalosporin.

Little information is available on enzymes involved in ACV tripeptide biosynthesis. Cell-free systems catalyzing ACV formation have been described for *C. acremonium* and in *Streptomyes clavuligerus*. A multifunctional peptide synthetase that catalyzes the formation of ACV has recently been purified from *A. nidulans*. Inasmuch as the molecular weight of the ACV synthetase of *A. nidulans* is reported to have a relative mass of 220 kD, at least 7 kb of DNA would be required to encode this multifunctional peptide synthetase.

P. chrysogenum DNA fragments cloned in EMBL3 or cosmid vectors from the upstream region of the *pcb*C-*pen*DE cluster carry a gene (*pcb*AB) that complemented the deficiency of α-aminoadipyl-cysteinyl-valine synthetase of mutants *npe*5 and *npe*10 and restored penicillin production to mutant *npe*5. A protein of about 250 kD was observed in SDS-PAGE gels of cell-free extracts of complemented strains that was absent in the *npe*5 and *npe*10 mutants but exists in the parental strains from which the mutants were obtained. Transcriptional mapping studies showed the presence of one long transcript of about 11.5 kb that hybridized with several probes internal to the *pcb*AB gene, and two small transcripts of 1.15 kb that hybridized with the *pcb*C or the *pen*DE gene, respectively. The transcription initiation and termination regions of the *pcb*AB gene were mapped by hybridization with several small probes.[20] The region has been completely sequenced. It includes an ORF of 11,376 nucleotides that encodes a protein with a deduced M_r of 425,971 (FIG. 3). Three repeated domains were found in the α-aminoadipyl-cysteinyl-valine synthetase that has high homology with the gramicidin synthetase I and tyrocidine synthetase I.[20]

The ACV synthetase gene of *A. chrysogenum* (syn. *C. acremonium*) is very similar to that of *P. chrysogenum*.[12] It contains an open reading frame of 11,136 bp encoding a protein of 3,712 amino acids with a deduced M_r of 414,791.[12]

The Three Genes Involved in Penicillin Biosynthesis Are Clustered in a 17.6-KB DNA Fragment in the Genome of P. chrysogenum

Forty-four of 46 recombinant phages isolated from an EMBL3 library of *P. chrysogenum* DNA which gave positive hybridization with probes corresponding to the amino acid carboxyl terminal region of the isopenicillin N synthase (*pcb*C) gene also gave a positive signal when hybridized with a probe corresponding to the amino terminal end of the acyl-CoA:6-APA acyltransferase gene (*pen*DE). A 5.1-kb *Sal*I DNA fragment common to nine randomly chosen recombinant phages showed strong hybridization with both the *pcb*C and *pen*DE probes. Linkage of both genes was also observed in three plasmids isolated from an independently constructed library of *P. chrysogenum* DNA in the bifunctional (*E. coli-P. chrysogenum*) shuttle vector pDJB3. Both genes are expressed in the same orientation from different promoters as concluded from subcloning experiments. They have been sequenced. A *P. chrysogenum* mutant *npe*8 *pyr*G which is altered in the 6-APA-acyltransferase was complemented by transformation with DNA of either recombinant phages or pDJB3-derived plasmids that contained the *pcb*C-*pen*DE cluster. The *pcb*AB is linked to the *pcb*C and *pen*DE genes and is transcribed in the opposite orientation to them. Linkage of the three genes is of great interest because β-lactam biosynthetic genes appear to have derived from *Streptomyces,* where clustering of antibiotic biosynthetic genes is a well-known phenomenon.[21]

A 35-KB DNA Fragment Carrying All the Penicillin Biosynthetic Genes Is Amplified in High Penicillin-producing Strains of P. chrysogenum

A 19.5-kb DNA fragment of *P. chrysogenum* AS-P-78 DNA was cloned in EMBL3 phage vector. The ACV synthetase (*pcb*AB), isopenicillin N synthase (*pcb*C), and acyl-CoA:6-APA acyltransferase (*pen*DE) genes of *P. chrysogenum* were located in the cloned DNA. The 19.5-kb DNA fragment was mapped by double and triple digestions with several endonucleases and the *pcb*AB, *pcb*C, and *pen*DE genes were located by hybridization with probes corresponding to internal fragments of each gene. A low penicillin-producing strain (*P. chrysogenum* Wis-54-1255) and two high-producing strains (AS-P-78 and P2) showed hybridizing fragments of identical size in their chromosomes. By dot-blot hybridization of serial dilutions of total DNA of the three strains it was shown that the intensity of all hybridizing bands was much higher in strain AS-P-78 and P2 than in Wis 54-1255. Hybridization of overlapping DNA inserts in different phages with probes corresponding to fragments that mapped upstream or downstream of the *pcb*C-*pen*DE region revealed that a fragment of at least 35-kb DNA has been amplified in the high penicillin-producing strains.[22] The amplified region did not include the previously cloned *pyr*G gene[23] which encodes OMP-decarboxylase, an enzyme involved in pyrimidine biosynthesis, which is used as a selective marker in fungal cloning vectors.

FUTURE OUTLOOK

The development of the basic tools of molecular genetics in β-lactam-producing fungi has provided the groundwork for rapid progress in our understanding of the structural and functional organization of the genes involved in the biosynthesis of β-lactam antibiotics.[17] The genes from *A. chrysogenum* and *P. chrysogenum* are now available and are being studied by directed *in vitro* mutation. Following the cloning of the ACVS, IPNS, and AAT genes of *P. chrysogenum,* other genes of the cephalosporin biosynthetic pathways will soon be cloned. Gene amplification of some of these genes is already being studied in our laboratory. More efficient transformation systems using integrative vectors carrying dominant selection markers will be developed. Other "primary" and "secondary" promoters will need to be isolated and characterized for coupling to the structural genes. The first overproducing strains obtained by gene cloning will be available in a few years.[24]

(*Note added in proof:* Incorporation of material in the article was completed in December 1990. For an update, see the article by Martín, J.F. 1991. Cluster of genes for the biosynthesis of antibiotics: Regulatory genes and overexpression of pharmaceuticals. J. Indust. Microbiol. In press.)

REFERENCES

1. MARTÍN, J. F. & P. LIRAS. 1985. Biosynthesis of β-lactam antibiotics: Design and construction of overproducing strains. Trends Biotechnol. **3:** 39–44.
2. MARTÍN, J. F. & P. LIRAS. 1989. Enzymes involved in the biosynthesis of β-lactam antibiotics. *In* Advances in Biotechnology. A. Fiechter, ed. Vol. 39: 153–187. Springer Verlag. Heidelberg.
3. RAMOS, F. R., M. J. LOPEZ-NIETO & J. F. MARTIN. 1985. Isopenicillin N synthetase of *Penicillium chrysogenum,* an enzyme that converts δ-(L-α-aminoadipyl)-L-cysteinyl-D-valine to isopenicillin N. Antimicrob. Agents Chemother. **27:** 380–387.
4. MARTIN, J. F., B. DIEZ, E. ALVAREZ, J. L. BARREDO & J. M. CANTORAL. 1987. Development of a transformation system in *Penicillium chrysogenum:* Cloning of genes involved in penicillin biosynthesis. *In* Genetics of Industrial Microorganisms. M. Alacevick, D. Hranueli & Z. Toman, eds.: 297–308. Pliva. Zagreb. Yugoslavia.
5. DEMAIN, A. L. 1983. Biosynthesis of β-lactam antibiotics. *In* Antibiotics containing the β-lactam structure I. A. L. Demain & N. A. Solomon, eds.: 189–228. Springer-Verlag. Berlín.
6. MEESCHAERT, B. & H. EYSSEN. 1981. Interconversion of penicillins by mycelium of *Penicillium chrysogenum.* FEMS Microbiol. Lett. **10:** 115–118.
7. FAWCETT, P. A., J. J. USHER & E. P. ABRAHAM. 1975. Behaviour of tritium labelled isopenicillin N and 6-aminopenicillanic acid as potential penicillin precursors in an extract of *Penicillium chrysogenum.* Biochem. J. **151:** 741–746.
8. MULLANEY, E. J., J. E. HAMER, K. A. ROBERTI, M. M. YELTON & W. E. TIMBERLAKE. 1985. Primary structure of the *trp*C gene from *Aspergillus nidulans.* Mol. Gen. Genet. **199:** 37–45.
9. BALLANCE, D. J. 1986. Sequences important for gene expression in filamentous fungi. Yeast **2:** 229–236.
10. ALVAREZ, E., J. M. CANTORAL, J. L. BARREDO, B. DIEZ & J. F. MARTIN. 1987. Purification to homogeneity and characterization of acyl coenzyme A:6-aminopenicillanic acid acyltransferase of *Penicillium chrysogenum.* Antimicrob. Agents Chemother. **31:** 1675–1682.
11. CANTORAL, J. M., B. DIEZ, J. L. BARREDO, E. ALVAREZ & J. F. MARTIN. 1987. High frequency transformation of *Penicillium chrysogenum.* Biotechnology **5:** 494–497.
12. GUTIERREZ, S., B. DIEZ, E. MONTENEGRO & J. F. MARTIN. 1991. Characterization of the *Cephalosporium acremonium pcb*AB gene encoding α-aminoadipyl-cysteinyl-valine synthetase, a large multidomain peptide synthetase: Linkage to the *pcb*C gene as a cluster of early cephalosporin-biosynthetic genes and evidence of multiple functional domains. J. Bacteriol. **173:** 2354–2365.
13. BARREDO, J. L., J. M. CANTORAL, E. ALVAREZ, B. DIEZ & J. F. MARTIN. 1989. Cloning, sequencing and transcriptional study of the isopenicillin N synthase of *Penicillium chrysogenum* AS-P-78. Molec. Gen. Genet. **216:** 91–98.
14. CARR, L. G., P. L. SKATRUD, M. E. SCHEETZ, III, S. W. QUEENER & T. D. INGOLIA. 1986. Cloning and expression of the isopenicillin N synthetase gene from *Penicillium chrysogenum.* Gene **48:** 257–266.
15. GARCIA-DOMÍNGUEZ, M., P. LIRAS & J. F. MARTIN. 1991. Cloning and characterization of the isopenicillin N synthase gene of *Streptomyces griseus* NRRL 3851 and studies of expression and complementation of the cephamycin pathway in *Streptomyces clavuligerus.* Antimicrob. Agents Chemother. **35:** 44–52.
16. COQUE, J. J. R., J. F. MARTIN, J. G. CALZADA & P. LIRAS. 1991. The cephamycin biosynthetic genes *pcb*AB, encoding a large multidomain peptide synthetase, and *pcb*C of *Nocardia lactamdurans* are clustered together in an organization different from the

same genes in *Acremonium chrysogenum* and *Penicillium chrysogenum.* Mol. Microbiol. **5:** 1125–1133.

17. MARTÍN, J. F., T. INGOLIA & S. QUEENER. 1991. Molecular genetics of β-lactam antibiotic biosynthesis. *In* Molecular Industrial Mycology. S. A. Leong, ed.: 149–196. Marcel Dekker Inc. New York, NY.

18. BARREDO, J. L., P. VAN SOLINGEN, B. DIEZ, J. M. CANTORAL, A. KATTEVILDER, E. B. SMAAL, M. A. M. GROENEN, A. E. VEENSTRA & J. F. MARTIN. 1989. Cloning and characterization of the acyl-coenzyme A:6-aminopenicillanic-acid-acyltransferase gene of *Penicillium chrysogenum.* Gene **83:** 291–300.

19. MONTENEGRO, E., J. L. BARREDO, S. GUTIERREZ, B. DIEZ, E. ALVAREZ & J. F. MARTIN. 1990. Cloning, characterization of the acyl-CoA:6-aminopenicillanic acid acyltransferase gene of *Aspergillus nidulans* and linkage to the isopenicillin N synthase gene. Mol. Gen. Genet. **221:** 322–330.

20. DIEZ, B., S. GUTIERREZ, J. L. BARREDO, P. VAN SOLINGEN, L. H. M. VAN DER VOORT & J. F. MARTÍN. 1990. Identification and characterization of the *pcb*AB gene encoding the α-aminoadipyl-cysteinyl-valine synthetase and linkage to the *pcb*C and *pen*DE genes. J. Biol. Chem. **265:** 16358–16365.

21. MARTIN, J. F. & P. LIRAS. 1989. Organization and expression of genes involved in the biosynthesis of antibiotics and other secondary metabolites. Ann. Rev. Microbiol. **43:** 173–206.

22. BARREDO, J. L., B. DIEZ, E. ALVAREZ & J. F. MARTIN. 1989. Large amplification of a 35-kb DNA fragment carrying two penicillin biosynthetic genes in high penicillin producing strains of *Penicillium chrysogenum.* Curr. Genet. **16:** 453–459.

23. CANTORAL, J. M., J. L. BARREDO, E. ALVAREZ, B. DIEZ & J. F. MARTIN. 1988. Nucleotide sequence of the *Penicillium chrysogenum pyr*G (orotidine-5′-phosphate decarboxylase) gene. Nucleic Acid Res. **16:** 8177.

24. MARTIN, J. F. 1987. Cloning of genes involved in penicillin and cephalosporin biosynthesis. Trends Biotechnol. **5:** 306–308.

Improvements in Fungal Product Synthesis by Recombinant DNA Technology

D. B. FINKELSTEIN,[a] C. L. SOLIDAY, M. S. CRAWFORD,
J. RAMBOSEK, P. C. McADA, AND J. LEACH

Panlabs Incorporated
11804 North Creek Parkway South
Bothell, Washington 98011-8805

Filamentous fungi are an important source of products ranging from industrial enzymes to antibiotics and food additives. The development of commercially viable fermentation processes for the manufacture of such products often requires a protracted period of strain improvement and process development. With the advent of recombinant DNA methodology, the time required for such process development can often be foreshortened.

The application of recombinant technology to the filamentous fungi offers a powerful set of additional techniques that can be applied to a strain improvement program. It should be noted, however, that the use of these new techniques by no means rules out the application of classic strain improvement procedures. Indeed, the combination of recombinant and classic techniques has a synergistic effect.

TRANSFORMATION OF FILAMENTOUS FUNGI

In order to successfully transform any organism three requirements must be met. First a method must be devised to introduce the vector into the organism. Next, a method is required for selecting those organisms that are carrying the introduced DNA. Finally, to avoid abortive transformation, a means must be provided to allow the DNA to replicate.

In the initial transformation of a nonfilamentous fungus, the yeast *Saccharomyces cerevisiae*, these criteria were readily satisfied by mixing homologous DNA (a cloned yeast *LEU2* gene) capable of complementing a defined genetic lesion (and thus providing a nutritionally selective phenotype) with spheroplasts in the presence of the fusogen polyethylene glycol and calcium ions.[1] Examination of the recovered transformants revealed that the introduced yeast DNA could readily integrate into the genome by a mechanism involving homologous recombination. Later studies have demonstrated that cleavage of vector DNA within a sequence that shares homology with the host genome results in a significant increase in the frequency with which a vector can transform yeast.[2] With the subsequent discovery that many sequences (termed ARS) allow a vector to replicate autonomously, it became clear that integration was not the only means by which introduced DNA could be propagated in *Saccharomyces.*[2]

[a]Address for correspondence: Dr. David B. Finkelstein, Panlabs, Inc., 11804 North Creek Parkway South, Bothell, WA 98011-8805.

Following the yeast paradigm, the initial transformation of a filamentous fungus (namely, *Neurospora crassa*) also used homologous DNA (the cloned *qa2* gene) that was capable of complementing a defined genetic lesion.[3] Although the successful transformants were shown to contain DNA integrated into the host genome, it was observed in this experiment that integration had occurred via nonhomologous recombination in almost half the transformants. Indeed, as has since been shown with a variety of filamentous fungi, vectors lacking any detectable homology with the host organism can still integrate into the genome.[4] Another difference that has been noted between the filamentous fungi and their nonfilamentous yeast cousins is that with few exceptions, it has not been possible to isolate DNA sequences that are capable of functioning as ARS elements.

SELECTIVE MARKERS FOR TRANSFORMATION OF WILD-TYPE FUNGI

Although only negative results have been obtained in attempts to express yeast genes in filamentous fungi, among the filamentous fungi themselves, genes appear to be expressed across a broad host range.[4] Taken together with the lack of pressure for sequence homology between vector and host organism, the broad host range of fungal genes suggests that many vectors that were initially constructed for use with a single fungal species have in fact turned out to be broad host range integrating vectors.

A variety of vectors have been developed that can be used for the transformation of wild-type filamentous fungi. The first class of vectors includes those in which a fungal promoter has been spliced to a prockaryotic gene encoding resistance to a fungicidal compound. Genes that have been employed for this purpose include the aminoglycoside and hygromycin phosphotransferases (*aph* and *hph*, respectively) as well as the bleomycin binding protein *ble*. The result of these manipulations is a series of broad host range vectors conferring resistance to G418, hygromycin, and bleomycin (phleomycin), respectively.[5–7]

An additional series of vectors for transformation of wild-type fungi incorporate fungal benomyl-resistant beta tubulin genes that have been cloned from mutants that are resistant to this fungicide.[8] Given the conservation of beta tubulins, it is not surprising that these cloned genes readily confer benomyl resistance, even when expressed across species lines. As (semi) dominant benomyl-resistant mutants can readily be isolated in a variety of fungi, homologous vectors can be developed for use when the introduction of a heterologous gene may not be desirable.

A final example of a marker that has found widespread use in transformation of wild-type fungi is the acetamidase gene of *Aspergillus nidulans*. This gene often allows heterologous organisms that can use acetate as a sole carbon source to grow on acetamide.[9]

TRANSFORMATION BY COMPLEMENTATION OF MUTANTS

In some instances it may be desirable to use a homologous gene as a selective marker for the transformation of the filamentous fungi. As the lack of a readily manipulatable genetic system for many fungi limits the available supply of characterized mutant organisms that may serve as hosts for recombinant manipulation, a number of selection systems have been developed that allow the simple positive selection of organisms harboring mutations in readily complementable genes. For example, fungi with lesions in orotidylate decarboxylase, nitrate reductase, and ATP sulfurylase may be readily isolated as a subclass of mutants selected for resistance to 5-fluoroorotate, chlorate, and selenate, respectively.[10–12] With available cloned genes, these mutants are appropriate hosts for transformation. Although these systems require an additional step, relative to the use of genes that can be selected when introduced into wild-type organisms, they do provide a technique for transformation of organisms that may be resistant to the dominant agents just described or in which the introduction of a positive resistance gene is not desired.

CHARACTERISTICS OF FUNGAL TRANSFORMANTS

For a recombinant fungus to have a practical value in an industrial setting, it is important that the organism be stable when introduced into a large volume fermenter. Given the fact that 100,000 L fermentation vessels are quite common in the production of antibiotics and industrial enzymes, it is clear why strain stability is considered an important property. As many fermentations have already been optimized using complex media, the introduction of a new recombinant strain that requires the application of a particular selective pressure to maintain strain stability is often impractical.

As already noted herein, most transformation of the filamentous fungi occurs via illegitimate recombination into the host genome. Fortunately for the industrial microbiologist, once integrated, the vector DNA is most often found to be mitotically stable, even in the absence of any selective pressure.[5]

Interestingly, transformants are often found to contain multiple copies of vector DNA. In many instances these multiple copies of vector DNA are integrated in the form of tandem repeats. These multicopy transformants usually show the same degree of mitotic stability in the absence of selective pressure as their single copy counterparts. Depending on the vector and the organism, copy numbers greater than 100 have been reported.[9] Thus, the lack of autonomous vectors for filamentous fungi causes no serious problems for strain improvement.

STRAIN IMPROVEMENT FOR ENZYME PRODUCTION

Given the characteristics of fungal transformants, we thought to improve the production of an industrial enzyme by increasing the dosage for the gene encoding the desired enzyme. We carried out experiments to test this hypothesis using the glucoamylase gene of *Aspergillus niger,* an enzyme used in processing corn starch to high fructose corn syrup.[13] When cultivated on a simple growth medium using soluble starch as a carbon source, the type strain of *A. niger* secretes approximately 0.5 g of glucoamylase per liter.

Following transformation of the *A. niger* type strain with a vector containing a cloned copy of the *A. niger* glucoamylase gene, transformants were identified that contained multiple copies of the vector integrated in tandem into the genome. Interestingly, of more than 50 transformants examined in detail, none had plasmid integrated at the normal chromosomal locus.[13] The level of glucoamylase produced by the various transformants was generally correlated with the increased dosage of the introduced gene, suggesting that a *trans*-acting inducer (if present) is not severely limiting to gene expression. Similarly, the extra copies of the glucoamylase gene were regulated in the same fashion as the normal chromosomal gene (i.e., the basal level of glucoamylase production did not increase with increased gene dosage), suggesting the absence of any *trans*-acting repressor of the glucoamylase gene. It is worth noting that classic process development procedures could be used with the transformants. Glucoamylase yield increases (which scaled up when the organisms were grown in a stirred tank fermenter) were obtained when the transformed organisms were grown on a commercial production medium.[13]

As many of the *Aspergillus* transformants each contained a single tandem cluster of glucoamylase genes, the gene dosage as well as the level of enzyme production could be furthered increased by making use of the parasexual cycle of *Aspergillus* to cross different transformants. By such an approach it was possible to isolate an organism containing more than 60 copies of the glucoamylase gene (in two gene clusters).[13] As previously observed with transformants containing one tandem gene cluster, organisms containing two gene clusters were also completely stable when fermented in the absence of selective pressure.

YIELD IMPROVEMENT FOR MULTIGENIC PRODUCTS

In principle, the recombinant methodology described herein can also be applied to secondary metabolite production. For example, when an enzyme that is limiting to product formation can be identified, the rational approach of increasing the dosage of the biosynthetic gene that acts on an accumulated pathway intermediate can result in product yield increases.[14] Indeed, given that the systematic screening of a random cosmid gene bank may require less effort than purification and cloning of a specific gene, it may well be more rational to take an empiric approach to increasing the dosage of a postulated specific limiting gene.

Limitations to productivity are not always due to a limiting gene in the biosynthetic pathway. For example, oxygen availability can effect productivity.[15] Although oxygen availability clearly can be altered by altering the morphology of an organism,[16] the set of genes that affect morphology is, at best, poorly defined. In such a situation it is difficult to describe a time-effective rational approach to identify the "limiting" gene (if indeed it is a single gene) that is affecting cell morphology. In this case, random screening of shotgun clones provides the only rational recombinant approach to strain improvement.

Empiric screening of recombinant organisms need not be limited to those strain improvement problems that require an increase in gene dosage. Shotgun screening of gene banks constructed with DNA obtained from a progenitor organism may provide a solution to the loss of vigor that is often associated with production organisms. Inasmuch as each shotgun clone contains a relatively small portion of the genome, empiric screening of transformants is expected to be more effective than techniques such as protoplast fusion, at eliminating a harmful mutation while retaining the many desired mutations in the production organism.

One advantage of employing recombinant methodology over the use of classic mutation/selection for strain improvement is that transformants are "tagged" with vector sequences. As it is possible to recover a vector that has integrated into the host genome,[17] the application of recombinant technology affords the possibility of further manipulating the recovered gene.

REFERENCES

1. HINNEN, A., J. B. HICKS & G. R. FINK. 1978. Transformation of yeast. Proc. Natl. Acad. Sci. USA **80:** 1053–1057.

2. BOTSTEIN, D. & R. W. DAVIS. 1982. Principles and practice of recombinant DNA research with yeast. *In* The Molecular Biology of the Yeast *Saccharomyces.* Metabolism and Gene Expression. J. N. Strathern, E. W. Jones & J. R. Broach, eds.: 607–636. Cold Spring Harbor Laboratory. New York.

3. CASE, M. E., M. SCHWEIZER, S. R. KUSHNER & N. H. GILES. 1979. Efficient transformation of *Neurospora crassa* utilizing hybrid plasmid DNA. Proc. Natl. Acad. Sci. USA **76:** 5259–5263.

4. RAMBOSEK, J. & J. LEACH. 1987. Recombinant DNA in filamentous fungi: Progress and prospects. CRC Crit. Rev. Biotech. **6:** 357–393.

5. FINKELSTEIN, D. B., J. A. RAMBOSEK, J. LEACH, R. E. WILSON, A. E. LARSON, P. C. MCADA, C. L. SOLIDAY & C. BALL. 1986. Genetic transformation and protein secretion in industrial filamentous fungi. *In* Fifth International Symposium on the Genetics of Industrial Microorganisms. M. Alacevic, D. Hranueli & Z. Toman, eds.: 101–110. Pliva. Zagreb.

6. TURGEON, B. G., R. C. GARBER & O. C. YODER. 1987. Development of a fungal transformation system based on selection of sequences with promoter activity. Mol. Cell. Biol. **7:** 3297–3305.

7. KOLAR, M., P. J. PUNT, C. A. M. J. J. VAN DEN HONDEL & H. SCHWAB. 1988. Transformation of *Penicillium chrysogenum* using dominant selection markers and expression of an *Escherichia coli lacZ* fusion gene. Gene **62:** 127–134.

8. ORBACH, M. J., E. B. PORRO & C. YANOFSKY. 1986. Cloning and characterization of the gene for B-tubulin from a benomyl-resistant mutant of *Neurospora crassa* and its use as a dominant selectable marker. Mol. Cell. Biol. **6:** 2452–2461.

9. KELLEY, M. K. & M. J. HYNES. 1985. Transformation of *Aspergillus niger* by the *amdS* gene of *Aspergillus nidulans.* EMBO J. **4:** 475–479.

10. VAN HARTINGSVELDT, W., I. E. MATTERN, C. M. J. VAN ZEIJL, P. H. POUWELS & C. A. M. J. J. VAN DEN HONDEL. 1987. Development of a homologous transformation system for *Aspergillus niger* based on the *pyrG* gene. Mol. Gen. Genet. **206:** 71–75.

11. MALARDIER, L., M. J. DABOUSSI, J. JULIEN, F. ROUSSEL, C. SCAZZOCCHIO & Y. BRYGOO. 1989. Cloning of the nitrate reductase gene (*niaD*) of *Aspergillus nidulans* and its use for transformation of *Fusarium oxysporium.* Gene **78:** 147–156.

12. BUXTON, F. P., D. I. GWYNNE & R. W. DAVIES. 1989. Cloning of a new bidirectionally selectable marker for *Aspergillus* strains. Gene **84:** 329–334.

13. FINKELSTEIN, D. B., J. RAMBOSEK, M. S. CRAWFORD, C. L. SOLIDAY, P. C. MCADA & J. LEACH. 1989. Protein secretion in *Aspergillus niger. In* Genetics and Molecular Biology of Industrial Microorganisms. C. L. Hershberger, S. W. Queener & G. Hegeman, eds.: 295–300. American Society for Microbiology. Washington DC.

14. SKATRUD, P. L., A. J. TIETZ, T. D. INGOLIA, C. A. CANTWELL, D. L. FISHER, J. L. CHAPMAN & S. W. QUEENER. 1989. Use of recombinant DNA to improve production of cephalosporin C by *Cephalosporium acremonium.* Bio/Technology **7:** 477–485.

15. BALL, C. 1980. A discussion of microbial genetics in fermentation process development with particular reference to fungi producing high yields of antibiotics. Folia Microbiol. **25:** 524–531.

16. LEIN, J. 1984. The Panlabs penicillin strain improvement program. *In* Overproduction of Microbial Metabolites. Z. Vanek & Z. Hostalek, eds.: 105–139. Butterworths. Boston, MA.

17. YELTON, M. M., W. E. TIMBERLAKE & C. A. M. J. J. VAN DEN HONDEL. 1985. A cosmid for selecting genes by complementation in *Aspergillus nidulans:* Selection of the developmentally regulated *yA* locus. Proc. Natl. Acad. Sci. USA **82:** 834–838.

Systems and Approaches for Expression and Secretion of Heterologous Proteins in the Filamentous Fungus *Aspergillus niger* var. *awamori*

Current Status

RANDY M. BERKA

Genencor International, Inc.
South San Francisco, California 94080

Among the wide variety of prokaryotic and eukaryotic hosts available for the production of heterologous gene products, certain species of filamentous fungi possess features that make them exceptionally attractive. These include (1) capability to secrete large quantities of protein in culture, (2) a long history of safe use for production of industrial enzymes, antibiotics, and biochemicals, and (3) extensive fermentation experience for low-cost, large-scale production. For the last few years our laboratory has focused on the development of *Aspergillus niger* var. *awamori* (also called *A. awamori*) as a host organism for the expression and secretion of a heterologous gene product, bovine chymosin (rennin), an aspartyl protease used commercially in cheese manufacturing. For decades *A. awamori* has been used as a source of enzymes for starch processing and in the production of some oriental foods. Recently, standard techniques of molecular genetics such as DNA-mediated transformation and gene disruption and replacement have been adapted from the genetically well-characterized species *A. nidulans* and applied to *A. awamori*.

Chymosin is extracted from the fourth stomach of unweaned calves, where it has evolved for the purpose of cleaving κ-casein in milk. During cheese-making, this proteolytic cleavage results in coagulation of the milk. Chymosin is secreted as a zymogen precursor (preprochymosin) with a 16 amino acid signal peptide that is removed during secretion and a 42 amino acid propeptide that is autocatalytically cleaved at low pH.[1,2] Expression of chymosin in microorganisms has been a goal of several biotechnology companies because of its commercial application and frequent limited availability.

Several years ago we described the controlled expression and secretion of bovine chymosin in a laboratory strain of *Aspergillus nidulans*.[3] In that study, expression vectors were constructed in which the transcriptional and translational control regions of the *A. niger glaA* (glucoamylase) gene were functionally coupled to either prochymosin or preprochymosin cDNA sequences (FIG. 1). Three of the plasmid constructions involved joining of the prochymosin coding sequences to *glaA* sequences either at the glucoamylase signal peptide cleavage site (pGRG1), at the glucoamylase propeptide cleavage site (pGRG2), or after 11 codons of the mature glucoamylase (pGRG4). The fourth construction involved joining preprochymosin sequences directly to the *glaA* promoter (pGRG3). Transformants generated with each of the four expression vectors secreted polypeptides enzymatically and immunologically indistinguishable from authentic bovine chymosin. Integration of these

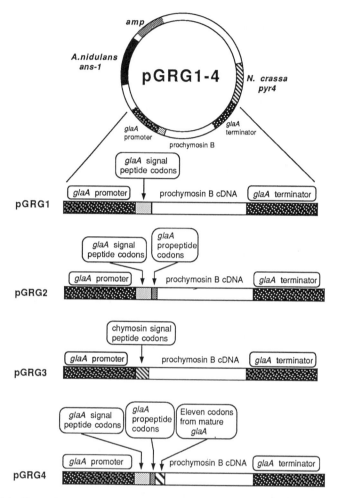

FIGURE 1. Chymosin expression cassettes contained in plasmids pGRG1 through pGRG4. Precise fusions of the DNA sequences were derived by oligonucleotide-directed mutagenesis. The *Escherichia coli* replication origin, ampicillin resistance determinant (*amp*), *Neurospora crassa pyr4* gene, and *A. nidulans ans-1* sequence were derived from plasmid pDJB3.[6] The glucoamylase (*glaA*) promoter and terminator sequences were isolated from *A. niger* genomic DNA.

vectors was predominantly by nonhomologous recombination and at varying copy number in the genome. Surprisingly, no strict correlation seemed to exist between expression level and the number of integrated copies of the expression vector. Although the extracellular yields of chymosin produced by *A. nidulans* were very modest (20–150 μg/g dry weight mycelia; approximately 1–5 mg/L), the results encouraged us to attempt a similar series of experiments in a production strain of *A.*

niger var. *awamori* that had been selected for its ability to produce large amounts of glucoamylase.

We observed that chymosin production directed from the expression vectors pGRG1 and pGRG3 in *A. awamori* was similar to that for *A. nidulans* in the following respects: (1) Transformants generated with either vector secreted active chymosin into the medium of starch-induced cultures, but at a low level (1–5 mg/L compared to 3 g/L of *A. awamori* glucoamylase); (2) Integration of the expression vectors occurred at random sites in the genome and with varying copy number; (3) There was no apparent correlation between chymosin expression levels and the number of integrated gene copies in the transformants. Interestingly, hybridization analysis of steady-state chymosin mRNA levels suggested that the major barriers to higher enzyme yields in *A. awamori* were posttranscriptional.

One major problem we encountered was proteolytic degradation of chymosin by an endogenous host aspartyl protease (aspergillopepsin A). This problem was overcome by cloning of the gene encoding the host protease (*pepA*) and elimination of its synthesis by chromosomal gene replacement.[4] Deletion of the aspartyl protease gene resulted in a significant improvement in the yield of extracellular chymosin (approximately threefold [i.e., 15 mg/L] when combined with improvements in culture medium). However, a second major problem was found to be inefficient secretion of the heterologous chymosin gene product. In many transformants, greater than 90% of the chymosin synthesized was found to be cell-associated. To alleviate this problem two molecular biology strategies were employed.

Although most of the extracellular proteins produced by *A. awamori* are glycoproteins, we noted that only a small percentage of the chymosin made by *A. awamori* is glycosylated. Thus, we elected to modify our pGRG3 expression unit by introducing a glycosylation site into chymosin with the hope that the resulting *glycochymosin* would be more efficiently secreted or more stable in *A. awamori* cultures than in native chymosin. Using techniques of site-directed mutagenesis, we introduced a consensus N-linked glycosylation site into chymosin at a position (Ser$_{74}$ → Asn$_{74}$, His$_{76}$ → Ser$_{76}$) homologous to one of two glycosylation sites found in the aspartyl protease of *Mucor miehei* (mucor rennin). Surprisingly, the yields of extracellular chymosin produced among transformants derived from this vector increased approximately tenfold over those of the wild-type chymosin vector, and in most cases more than 90% of the enzyme was extracellular. However, we also observed that the specific activity of glycochymosin was reduced to about 20% of the wild-type value. Most of this activity could be recovered by treatment with endoglycosidase H.

A second strategy to improve chymosin production involved construction of an expression vector that could direct the expression of a full-length glucoamylase-prochymosin fusion protein.[5] The rationale for this strategy assumes that if there are sequences internal to glucoamylase that promote efficient translation and secretion, they might be employed in the synthesis of a glucoamylase-prochymosin fusion protein that might subsequently be cleaved during autocatalytic activation of the prochymosin moiety (FIG. 2). Compared to previous results with nonfusion expression vectors, the levels of active extracellular chymosin were high (e.g., up to 150 mg/L of active chymosin were obtained from a strain that had not been deleted for the endogenous aspartyl protease). In addition, very little chymosin was detected intracellularly. By immunoblot analysis in a time-course experiment, the primary translation product (i.e., fusion protein) appeared in the culture medium in the first 2 days if the medium was buffered strongly at pH 6. At later times or at lower pH values, the majority of chymosin was in the mature active form. Processing of the fusion protein is inhibited by pepstatin, and therefore we suspect that chymosin may be autocatalytically released.

It seems likely that the improvements in production of chymosin obtained by introduction of a glycosylation site or by synthesis of a glucoamylase-prochymosin fusion are the result of increased secretion efficiency. It is also conceivable that future research efforts in this area may uncover critical steps in the secretory pathway

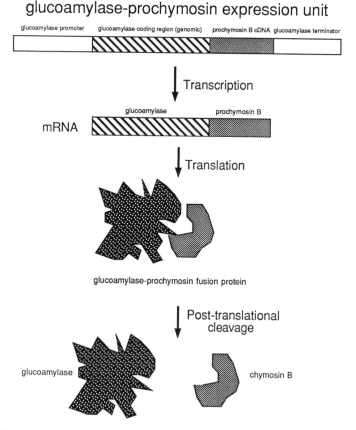

glucoamylase-prochymosin expression unit

FIGURE 2. Diagram of the expression cassette encoding a glucoamylase-prochymosin gene fusion showing expected transcription, translation, and posttranslational cleavage products.

of filamentous fungi that could be manipulated to improve these organisms as hosts for the secretion of heterologous gene products.[7]

REFERENCES

1. FOLTMANN, B. 1970. Prochymosin and chymosin (prorennin and rennin). Methods Enzymol. **19:** 421–436.

2. HARRIS, T. J. R., P. A. LOWE, P. G. THOMAS, M. A. W. EATON, T. A. MILLICAN, T. P. PATEL,
    C. C. BOSE, N. H. CAREY & M. T. DOEL. 1982. Molecular cloning and nucleotide
    sequence of cDNA coding for calf preprochymosin. Nucl. Acids Res. **10:** 2177–2187.
3. CULLEN, D., G. L. GRAY, L. J. WILSON, K. J. HAYENGA, M. H. LAMSA, M. W. REY, S.
    NORTON & R. M. BERKA. 1987. Controlled expression and secretion of bovine chymosin
    in *Aspergillus nidulans.* Bio/Technology **5:** 369–376.
4. BERKA, R. M., M. WARD, L. J. WILSON, K. J. HAYENGA, K. H. KODAMA, L. P. CARLOMAGNO
    & S. A. THOMPSON. 1990. Molecular cloning and deletion of the gene encoding as-
    pergillopepsin A from *Aspergillus awamori.* Gene **86:** 153–162.
5. WARD, M., L. J. WILSON, K. H. KODAMA, M. W. REY & R. M. BERKA. 1990. Improved
    production of chymosin in *Aspergillus* by expression as a glucoamylase-chymosin fusion.
    Bio/Technology **8:** 435–440.
6. TURNER, G. & D. J. BALLANCE. 1985. Cloning and transformation in *Aspergillus. In* Gene
    Manipulations in Fungi. J. W. Bennett & L. A. Lasure, eds.: 259–278. Academic Press.
    New York, NY.
7. LEONG, S. A. & R. M. BERKA (eds.). 1991. Molecular Industrial Mycology: Systems and
    Applications for Filamentons Fungi. M. Dekker. New York, NY.

Expression of a Human Monoclonal Anti-HIV-1 Antibody in CHO Cells[a]

FLORIAN RÜKER, VERONIKA EBERT, JOHANN KOHL,
FRANZ STEINDL, HEIDELINDE RIEGLER, AND
HERMANN KATINGER

Institut für Angewandte Mikrobiologie
Universität für Bodenkultur
Nußdorfer Lände 11
A-1190 Vienna, Austria

The main production system for monoclonal antibodies is conventional hybridoma cell culture. Although the technology for rodent hybridomas is well developed, this is not the case for human antibody-producing hybrid cell lines, where one has to deal with low antibody titers in the culture supernatants together with a high degree of clonal instability.[1]

These facts, together with the advent of antibody engineering, such as the construction of chimeric antibodies and the humanization of rodent antibodies,[2-5] has stimulated research directed towards the development of expression systems suitable for the expression of cloned antibody genes.

The logical host cells for antibody genes are myeloma cells. Several successful attempts have been undertaken to express engineered antibody genes under the control of both immunoglobulin gene-specific and non-specific promoters.[6-8] In all these cases, genomic immunoglobulin genes including introns have been used. Other papers describe the use of cDNA for the expression of antibodies in lymphoid cells[9] or in COS cells.[10]

To decrease the size of the expression plasmids and to facilitate the direct expression cloning of new antibody genes using the polymerase chain reaction, we have decided to use the cDNAs of the heavy and the light chain of a human monoclonal antibody in our attempts to establish an expression system in the universal, heterologous cell line CHO which was shown before to be capable of producing functional immunoglobulins from cloned antibody genes.[11] In comparison with hybridoma cells, CHO cells offer the specific advantage of being a well-recognized standard cell line for the production of biologicals from both a technological as well as a regulatory point of view.

It has been shown that gene amplification in CHO cells, using the dihydrofolate reductase (dhfr)-methotrexate system, can lead to high expression levels of recombinant proteins.[12-14] However, as the process of amplification is time consuming and does not always lead to cell lines that stably maintain the amplified state and the high expression levels, it is desirable to use an expression system that leads to high levels of expression even without long-term amplification and selection of clones. Examples of such expression systems have been described.[15]

We have tested three different mammalian cell expression vectors for their

[a]This work was supported in part by the Fonds zur Förderung der wissenschaftlichen Forschung, Project Nos. P6540 and P7556.

212

suitability for high level expression of antibody cDNA, using a human monoclonal anti-HIV-1 antibody as a model.

EXPERIMENTAL

The human heterohybridoma cell line 3D6, which produces a monoclonal antibody specific to HIV-1 gp41, has been described before.[16–19] Poly A^+-RNA was isolated from this cell line and used as a substrate for cDNA synthesis following routine procedures.[20] The cDNA library was ligated with *Eco*RI linkers and subsequently cloned in the plasmid pUC 19. Full-length clones of the heavy and the light chain were identified by colony hybridization using constant region-specific oligonucleotide probes, followed by restriction analysis and sequencing. The sequences of the clones have been published elsewhere.[21]

Pairs of expression plasmids were constructed by introducing the cDNA of the heavy and the light chain, respectively, into the mammalian expression vectors pRcRSV, pRcCMV (both Invitrogen Inc.) and pMAMgpt (Clontech Ltd.), respectively. See FIGURE 1 for a schematic representation of the vectors. The resulting plasmid pairs are:

pRcRSV3D6HC and pRcRSV3D6LC,
pRcCMV3D6HC and pRcCMV3D6LC, and
pMAMgpt3D6HC and pMAMgpt3D6LC.

CHO cells were grown under standard conditions in DMEM-HAM's F12 (1:1) medium containing 5% fetal calf serum.[22] The respective plasmid pairs were cotransfected into CHO cells by electroporation[23] using a BioRad Gene Pulser. 10^7 cells were washed with phosphate-buffered saline solution (PBS) and finally resuspended in 0.8 ml of the same buffer. Thirty micrograms of each, a heavy chain and a light chain plasmid (circular form), were added, and the samples were incubated on ice for 10 minutes. The samples were then transferred to the electroporation chambers (electrode distance 0.4 cm), and a single electrical pulse with an initial field strength of 1 kV was discharged from the 25 uF capacitor. After a further 10 minutes of incubation on ice, the cells were resuspended in culture medium. The next day, the selective agent (1 mg/ml G418 for pRcRSV and pRcCMV transformants, 3 µg/ml mycophenolic acid for pMAMgpt transformants) was added. Selection medium was changed every 4 days, and transformed clones were selected at a rate of approximately 10^{-4}.

Clones were screened by two independent ELISA assays specific to human Ig kappa light chains and to human Ig gamma heavy chains, respectively. Only clones expressing equimolar amounts of both chains were further subcloned and characterized.

For each vector pair, 30–40 independent clones were assayed for antibody productivity: cells were grown in 25 cm² culture flasks until a maximum cell density of approximately 2×10^5 cells/cm² was reached. Cell density was monitored by counting cells after trypsinizing them from flasks grown in parallel. From that point, medium was changed every day for 1 week, and the antibody titer in the culture supernatant was determined by anti-human IgG ELISA. In pMAMgpt transfectants, dexamethasone was added for induction of the MMTV LTR to a concentration of 0.1 mM to the

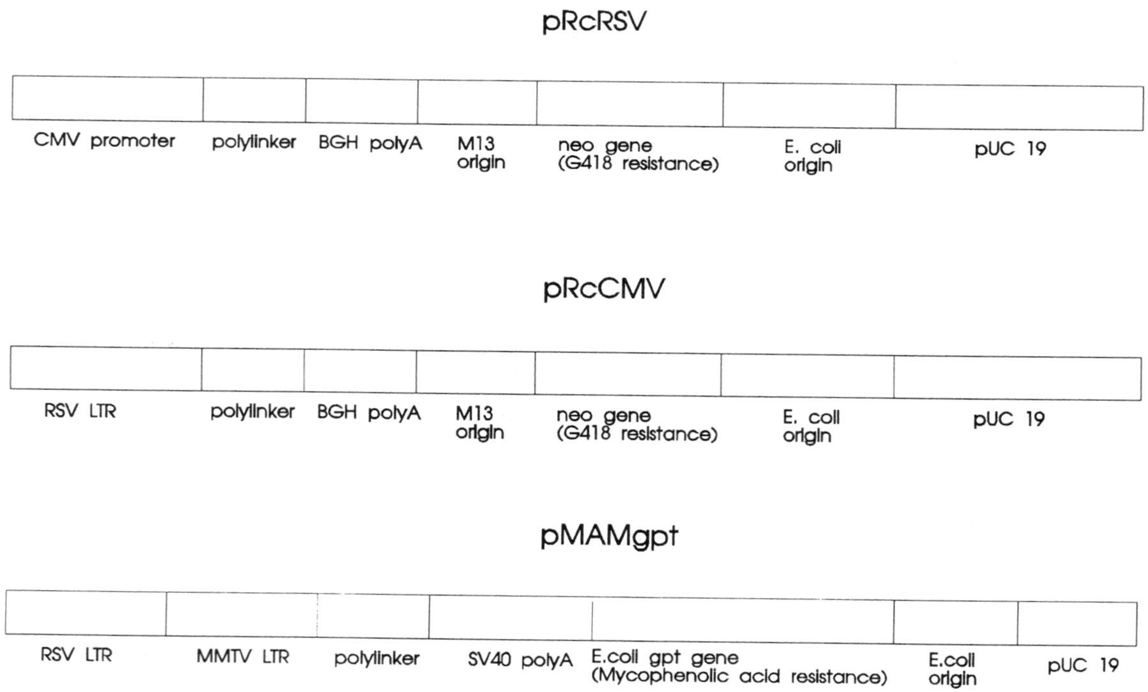

FIGURE 1. Schematic representation of the expression vectors used in this study. CMV = human cytomegalovirus; BGH polyA = bovine growth hormone polyadenylation region; RSV = Rous sarcoma virus; MMTV = mouse mammary tumor virus; SV40 = simian virus 40.

production medium. From these results, the antibody productivity of the various clones (μg IgG$/10^6$ cells × 24 hours) was calculated.

The specific reactivity of the CHO-produced antibody was compared with that of the hybridoma-produced antibody by HIV-1 gp160 specific ELISA.[17,18]

Samples of the CHO-produced antibody were further tested for antigen binding by HIV-1 specific immunoblot assay (Novopath, BioRad). To assess the clonal stability of the transfected CHO cells, cells were cultivated in the absence of selective pressure over a period of 3 months. Antibody titer in the supernatant was monitored weekly.

RESULTS

Antibody productivity mediated by the different expression vectors is shown in FIGURE 2. The values represent the amount of antibody secreted per 10^6 cells in 24 hours. For each vector, the five best producers out of 30–40 tested clones are shown.

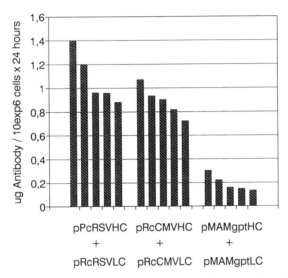

FIGURE 2. Antibody secretion by transfected CHO cells. The expression values for the five best clones out of 40 tested are given for each of the respective plasmid pairs.

The recombinant antibody produced under the control of RSV LTR was tested for its ability to bind to HIV-1 gp41. As positive control, antibody produced by the original hybridoma cell line, 3D6, was used. As shown in FIGURE 3, both the CHO-derived and the hybridoma-derived antibody bind equally well to microtiter plates coated with HIV-1 proteins.

FIGURE 4 shows the result of the HIV-1 specific immunoblot assay. It can be seen that the CHO- and the hybridoma-derived antibody show the same binding specificity. As the epitope of 3D6 is located on HIV-1 gp41, binding to both gp41 and gp160 can be observed. A light cross-reaction with gp120 is also shown.

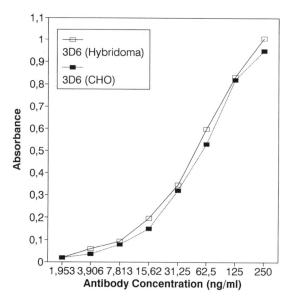

FIGURE 3. Comparison of hybridoma-derived and CHO-derived 3D6 antibody in an HIV-1 specific ELISA.

During the long-term cultivation experiment in the absence of selective pressure, no decrease in antibody productivity was observed, indicating a high degree of clonal stability of the transfected CHO cells.

DISCUSSION

We have obtained expression of a functional human monoclonal antibody in CHO cells. As structural genes, we used cDNA clones, which were transcribed under the control of three different promoters. Although the MMTV LTR showed low efficiency in driving immunoglobulin expression in CHO cells, both CMV IE promoter and RSV LTR gave reasonably high titers. When these cells are grown in a high cell density cultivation system (e.g., a fluidized bed reactor filled with porous microcarriers), antibody titer in the supernatant should reach levels ranging between 30 and 80 μg/ml, which would make this expression system compatible or even superior to conventional hybridoma culture. We are presently investigating the behavior of one of the CHO transfectants described here under such high cell density cultivation conditions. Cotransfection with dhfr and subsequent amplification of the transfected genes by increasing concentrations of methotrexate should lead to a further increase in productivity.

During the screening process, we found several clones that produced heavy and light chains in unbalanced amounts. In the expression system used here, no transcriptional or translational fine tuning of antibody expression can be expected as is observed in B-cells and also in hybridomas. Although we used exactly the same regulatory signals for both chains in each of the three plasmid pairs, it cannot be excluded that unfavorable events during the random integration process of the plasmids into the genome of the CHO cells eventually caused differences in the efficiency of transcription from the two transfected genes. It is also possible that sequences around the translational start site are recognized with different efficiency by CHO cells as compared to B-cells or hybridoma cells. Experiments to elucidate the reasons for the sometimes unbalanced expression of the two chains are presently underway.

In comparison with the culture of hybridoma cells, the antibody production system described here has several advantages. From a regulatory point of view CHO cells are considered acceptable for the production of biologicals. They do not pose potential risks of contamination with human viruses as do hybrids of primary human lymphocytes. Furthermore, large scale cultivation techniques for CHO cells, including high cell density culture in protein-poor or protein-free media, are well established, thereby eliminating the risk of contamination with serum proteins, increasing the specific titer in the cell culture supernatant, and greatly facilitating the downstream processing.[15]

Another advantage of the production system described herein is that in contrast to the high degree of clonal instability that is observed with human hybridoma cells, the transfected CHO cells are stable over long periods of growth.

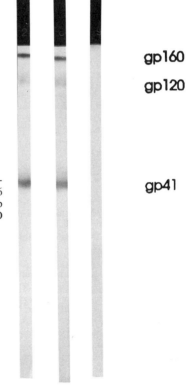

FIGURE 4. Detection of HIV-1 proteins by hybridoma-derived (*lane 1*) and CHO-derived (*lane 2*) 3D6 antibody by immunoblot assay. (*Lane 3*) Test strip incubated with supernatant of untransfected CHO cells (negative control).

SUMMARY

The cDNA coding for the light and heavy chains, respectively, of the human monoclonal antibody 3D6 (IgG1, kappa), which binds specifically to human immunodeficiency virus-1 (HIV-1) gp41, was inserted into three different mammalian expression vectors and transfected into Chinese hamster ovary (CHO) cells. Transcription was under the control of Rous sarcoma virus long terminal repeat (RSV LTR), human cytomegalovirus major immediate early (CMV IE) promoter, and mouse mammary tumor virus long terminal repeat (MMTV LTR), respectively. Antibody productivity was monitored in the supernatants of selected clones. The binding characteristics of the CHO-derived antibody to HIV-1 gp41 were found to be identical to that of the original antibody produced by hybridoma cells.

ACKNOWLEDGMENTS

We would like to thank Ali Assadian and Werner Fuchs for analytical work.

REFERENCES

1. STUDIER, F. W. & B. A. MOFFATT. 1986. Use of bacteriophage T7 RNA polymerase to direct selective high-level expression of cloned genes. J. Mol. Biol. **189**: 113.
2. SHARON, J., M. L. GEFTER, T. MANSER, S. L. MORRISON, V. T. OI & M. PTASHNE. 1984. Expression of a v-heavy.c-kappa chimaeric protein in mouse myeloma cells. Nature (Lond.) **309**: 364–367.
3. MORRISON, S. L. 1985. Transfectomas provide novel chimeric antibodies. Science **229**: 1202–1207.
4. JONES, P. T., P. H. DEAR, J. FOOTE, M. S. NEUBERGER & G. WINTER. 1986. Replacing the complementarity-determining regions in a human antibody with those from a mouse. Nature (Lond.) **321**: 522–525.
5. RIECHMANN, L., M. CLARK, H. WALDMANN & G. WINTER. 1988. Reshaping human antibodies for therapy. Nature (Lond.) **332**: 323–327.
6. DORAI, H. & G. P. MOORE. 1987. The effect of dihydrofolate reductase-mediated gene amplification on the expression of transfected immunoglobulin genes. J. Immunol. **139**: 4232–4241.
7. HENDRICKS, M. B., C. A. LUCHETTE & M. J. BANKER. 1989. Enhanced expression of an immunoglobulin-based vector in myeloma cells mediated by coamplification with a mutant dihydrofolate reductase gene. Bio/Technology **7**: 1271–1274.
8. GILLIES, S. D., K. M. LO & J. WESOLOWSKI. 1989. High-Level expression of chimeric antibodies using adapted cDNA variable region cassettes. J. Immunol. Methods **125**: 191–202.
9. LIU, A. Y., P. W. MACK, C. I. CHAMPION & R. R. ROBINSON. 1987. Expression of mouse:human immunoglobulin cDNA in lymphoid cells. Gene **54**: 33.
10. WHITTLE, N., J. ADAIR, C. LLOYD, L. JENKINS, L. DEVINE, J. SCHLOM, A. RAUBITSCHEK, D. COLCHER & M. BODMER. 1987. Expression in COS cells of a mouse-human chimeric B72.3 antibody. Prot. Eng. **1**: 499.
11. NEUMAIER, M., L. SHIVELY, F. S. CHEN, F. J. GAIDA, C. ILGEN, R. J. PAXTON, J. E. SHIVELY & A. D. RIGGS. 1990. Cloning of the genes for T84.66, an antibody that has a high specificity and affinity for carcinoembryonic antigen, and expression of chimeric human mouse T84.66 genes in myeloma and Chinese hamster ovary cells. Cancer Res. **50**: 2128–2134.
12. McCORMICK, F., M. TRAHEY, M. INNIS, B. DIECKMANN & G. RINGOLD. 1984. Inducible expression of amplified human beta interferon genes in CHO cells. Mol. Cell. Biol. **4**: 166–172.

13. KAUFMAN, R. J., L. C. WASLEY, A. J. SPILIOTES, S. D. GOSSELS, S. A. LATT, G. R. LARSEN & R. M. KAY. 1985. Coamplification and coexpression of human tissue-type plasminogen activator and murine dihydrofolate reductase sequences in Chinese hamster ovary cells. Mol. Cell. Biol. **5:** 1750–1759.

14. KAUFMAN, R. J., P. MURTHA, D. E. INGOLIA, C.-Y. YEUNG & R. E. KELLEMS. 1986. Selection and amplification of heterologous genes encoding adenosine deaminase in mammalian cells. Proc. Natl. Acad. Sci. USA **80:** 3136–3140.

15. FRIEDMAN, J. S., C. L. COFER, C. L. ANDERSON, J. A. KUSHNER, P. P. GRAY, G. E. CHAPMAN, M. C. STUART, L. LAZARUS, J. SHINE & P. J. KUSHNER. 1989. High expression in mammalian cells without amplification. Bio/Technology **7:** 359–362.

16. GRUNOW, R., S. JAHN, T. PORSTMANN, S. T. KIESSIG, H. STEINKELLNER, F. STEINDL, D. MATTANOVICH, L. GÜRTLER, F. DEINHARDT, H. KATINGER & R. VON BAEHR. 1988. The high efficiency, human B cell immortalizing heteromyeloma CB-F7. J. Immunol. Methods **106:** 257.

17. DÖPEL, S.-H., T. PORSTMANN, R. GRUNOW, A. JUNGBAUER & R. V. BAEHR. 1989. Application of a human monoclonal antibody in a rapid competitive anti-HIV ELISA. J. Immunol. Methods **116:** 229–233.

18. DÖPEL, S.-H., T. PORSTMANN, P. HENKLEIN & R. VON BAEHR. 1989. Fine mapping of an immunodominant region of the transmembrane protein of the human immunodeficiency virus (HIV-1) by different peptides and their use in anti-HIV ELISA. J. Virol. Methods **25:** 167–178.

19. JUNGBAUER, A., C. TAUER, E. WENISCH, F. STEINDL, M. PURTSCHER, M. REITER, F. UNTERLUGGAUER, A. BUCHACHER, K. UHL & H. KATINGER. 1989. Pilot scale production of a human monoclonal antibody against human immunodeficiency virus HIV-1. J. Biochem. Biophys. Meth. **19:** 223–240.

20. MANIATIS, T., E. F. FRITSCH & J. SAMBROOK. 1982. Molecular Cloning, A Laboratory Manual. Cold Spring Harbor Laboratory. Cold Spring Harbor, NY.

21. FELGENHAUER, M., J. KOHL & F. RÜKER. 1990. Nucleotide sequences of the cDNAs encoding the V-regions of H- and L-chains of a human monoclonal antibody specific to HIV-1-gp41. Nucl. Acids Res. **18:** 4927.

22. GLOVER, D. M. 1987. DNA Cloning. A Practical Approach, Vols. 1, 2 and 3. IRL Press. Oxford, Washington, DC.

23. RÜKER, F., W. LIEGL, D. MATTANOVICH, S. REITER, G. HIMMLER, A. JUNGBAUER & H. KATINGER. 1987. Electroporative gene transfer (electroporation): A method for strain improvement of animal cells. Bioelectrochem. Bioenerget. **17:** 253–257.

Swinepox Virus as a Vector for the Delivery of Immunogens

P. L. FOLEY,[a,b] P. S. PAUL,[c] R. L. LEVINGS,[a]
S. K. HANSON,[a] AND L. A. MIDDLE[a]

[a]National Veterinary Services Laboratories
Ames, Iowa 50010

[c]Veterinary Medical Research Institute
Ames, Iowa 50010

Vaccinia virus recombinants containing foreign gene inserts have frequently been used to express both RNA virus and DNA virus proteins *in vivo*.[1] However, the concerns regarding vaccinia's broad host range, residual immunity, and potential for harm to the immunosuppressed have led us to consider swinepox virus as an alternative pox viral vector. Swinepox virus is host specific and has limited pathogenic potential.[2] In addition, results from recombinant technology applied to other pox viruses, for example, fowlpox virus and raccoon poxvirus, have been promising.[3]

In anticipation of the need for recombinant virion selection, our immediate objective was to determine a standardized method for propagation and quantitation of swinepox virus on a thymidine kinase (TK)-negative cell line. We then hoped to provide a preliminary indication as to whether sufficient homology exists between the TK sequences of vaccinia and swinepox virus to permit homologous recombination between the two viruses, so that expression of foreign gene inserts could be achieved through the use of a plasmid transfer vector containing the chimeric genes flanked by vaccinia promoters and TK sequences. These efforts are part of an ongoing investigation into the feasibility of using swinepox virus as a live vector for the expression of immunogenic proteins. Our plans are to insert porcine rotavirus VP7 genes into swinepox virus, select for recombinants using bromodeoxyuridine (BUdR) and mycophenolic acid resistance, measure VP7 expression *in vitro,* then test the immunogenicity of this potential recombinant rotavirus vaccine in swine. The following text summarizes the results obtained thus far.

MATERIALS AND METHODS

Virus and Cell Lines

The Kasza strain of swinepox virus was obtained from the American Type Culture Collection (ATCC) and expanded on primary swine kidney cells as has previously been described.[2] Attempts were then made to passage the virus on four mammalian TK-negative cell lines, also obtained from ATCC, specifically, human osteosarcoma (143B), Syrian hamster kidney (ts13), BALB/c mouse embryo fibroblast (B2-1), and rat embryo fibroblast (Rat-2) cells. Also tried were a Madin-Darby bovine kidney (MDBK) cell line, for which a TK-negative variant already exists,[4] and

---

[b]To whom correspondence should be addressed.

a porcine kidney cell line (PK-15), from which we developed a TK-negative variant, using BUdR in a method previously described.[4]

Attempts at Viral Adaptation

To promote adaptation of swinepox to each of the cell lines, swinepox inoculum was added to each of two 25-cm^2 flasks of cells. One of these flasks was harvested at 3–4 days and used, along with a fresh swinepox inoculum, to inoculate a new set of flasks. The second 25-cm^2 flask was observed for 2 weeks to allow cytopathic effect (CPE) to develop. When viral growth was suggested by CPE, only harvest from infected flasks was used to inoculate the new set of flasks. Characteristic CPE and direct and indirect fluorescent antibody assays determined the success of each cell line in supporting swinepox replication.

Initial Assessment of Homology between the TK Genes of Swinepox and Vaccinia Virus

Swinepox and vaccinia viral DNA was extracted, restriction enzyme-digested, subjected to electrophoresis, and Southern blotted using standard procedures.[5] Two pUC plasmids,[b] one containing the entire vaccinia *Hind* III J fragment and the other containing vaccinia TK gene sequences, the *Escherichia coli* gpt gene, and a multiple cloning site were labeled with biotin-14-dATP and used to probe restriction enzyme-digested vaccinia and swinepox DNA.[6]

RESULTS

None of the four mammalian TK-negative cell lines or the MDBK cell line supported swinepox virus growth. However, both the PK-15 cell line and the TK-negative PK-15 variant permitted viral replication, producing characteristic CPE and intracytoplasmic immunofluorescence. This will provide a selection method for TK-negative recombinant swinepox virions.

The restriction enzyme patterns for swinepox and vaccinia virus DNA were distinct, with considerable variation in pattern using numerous enzymes. Following hybridization with a labeled probe under high stringency conditions (68°C, low salt), only vaccinia virus sequences hybridized to the probe. With lower stringency (42°C, higher salt), hybridization was detected in both viruses.

DISCUSSION AND SUMMARY

Cross-hybridization, even at low stringency, suggests that sufficient homology may exist between the TK genes of the two viruses to allow homologous recombination to occur between the vaccinia virus TK sequences of the pUC transfer vector and viral TK sequences found in swinepox virus-infected TK-negative PK-15 cells. The expression of some phenotypic marker such as β-galactosidase may provide further

[b]Generously provided by Dr. Bernard Moss, Laboratory of Viral Diseases, National Institute of Allergy and Infectious Diseases, Bethesda, Maryland 20892.

indication of this possibility. We will attempt to insert several cDNA clones of the porcine rotavirus genes encoding the major immunogenic protein VP7 of serotypes 4 and 5 into a plasmid transfer vector containing vaccinia promoters and antibiotic resistance genes.

REFERENCES

1. HRUBY, D. E. 1990. Vaccinia virus vectors: New strategies for producing recombinant vaccines. Clin. Microbiol. Rev. **3:** 153–170.
2. KASZA, L., E. H. BOHL & D. O. JONES. 1960. Isolation and cultivation of swine pox virus in primary cell cultures of swine origin. Am. J. Vet. Res. **21:** 269–273.
3. TARTAGLIA, J., S. PINCUS & E. PAOLETTI. 1990. Poxvirus-based vectors as vaccine candidates. Crit. Rev. Immunol. **10:** 13–30.
4. BELLO, L. J., J. C. WHITBECK & W. C. LAWRENCE. 1987. Map location of the thymidine kinase gene of bovine herpesvirus 1. J. Virol. **61:** 4023–4025.
5. Current Protocols in Molecular Biology. 1987. Preparation and Analysis of DNA. Greene Publishing Associates. New York, NY.
6. BluGENE: Nonradioactive Nucleic Acid Detection System. BRL Cat. No. 8279SA. Bethesda Research Laboratories. Gaithersburg, MD.

Plant Genetic Transformation
for Virus Resistance

ROGER N. BEACHY

Department of Biology
Washington University
St. Louis, Missouri 63130

During the last 5 years several different strategies have been taken to produce plants via genetic transformation that resist virus diseases. The strategies most widely undertaken include expressing genes to encode antisense RNAs, to produce satellite RNAs, and to produce virus capsid proteins. Of these three general approaches the latter has been most extensively developed, and initial field trials indicate that there will be applications to agriculture in the near future.

ANTISENSE RNAs AND SATELLITE RNAs

Expression of viral antisense RNAs has resulted in very limited success in restricting virus infection and/or disease development. A variety of antisense sequences have been expressed in plants in such studies. Cuozzo *et al.,*[1] Hemenway *et al.,*[2] and Powell-Abel *et al.*[3] expressed sequences complementary to viral coat protein in RNAs and reported very limited protection against infection. Powell-Abel *et al.*[4] demonstrated that the observed low level of resistance against tobacco mosaic virus (TMV) was due to the presence of sequences complementary to the replicase binding site on the antisense RNA. By contrast, the expression of genes encoding antisense RNAs representing non-coat protein sequences of TMV[5] or cucumber mosaic virus (CMV) had no protective effect.[6] It is unclear if significant improvements can be made to develop antisense gene constructions that confer effective levels of resistance.

Several reports have described virus disease resistance as a result of the introduction of genes that encode satellite RNAs. Satellite RNAs require a helper virus for their replication and may be replicated to high levels. In some cases this results in decreased replication of the helper virus and decreased severity of the disease.[7,8] In other cases the satellite RNA has no effect on replication of the helper, but it attenuates disease severity. In still other cases, the satellite RNA can exacerbate symptoms of disease.[9]

Transgenic plants have been developed that express genes encoding satellite RNAs of CMV[10] and tobacco ringspot virus.[11] When these plants were inoculated with the corresponding helper virus, they were less susceptible to disease than were control plants. These plants either did not develop symptoms or produced only mild symptoms. Features that may preclude the widespread use of satellite RNAs in agriculture include the variable nature of the satellite RNAs (i.e., satellites can attenuate or exacerbate symptoms, depending on the host and the helper virus), and the fact that satellite RNAs have not been identified for the majority of plant viruses. In such cases a nonnatural satellite would need to be created.

COAT PROTEIN-MEDIATED RESISTANCE

In contrast, widespread success has been achieved in producing virus-resistant plants through genetic transformation and the expression of genes that encode viral coat proteins. The first report by Powell-Abel et al.[3] documented that expression of a gene encoding the coat protein of TMV in transgenic tobacco plants confers significant levels of resistance to infection by TMV. Since that time a number of research groups have reported that this approach can be used to produce resistance against different classes of virus in plants as divergent as alfalfa, cucumber, and tomato. (See ref. 12 for a complete review of the field.) Viruses for which resistance has been achieved to date contain single-stranded, (+) sense RNAs. The viruses, however, differ markedly in structure, genome organization, and mechanisms of gene expression. The phenotypes associated with CP-mediated resistance include reduced numbers of sites of infection, reduced rate of development of disease symptoms or no symptoms on infected plants, and reduced accumulation of virus in infected plants. These features usually make it possible for the resistant plant to grow with few if any symptoms and to produce crop yields similar to those of noninfected plants.

MECHANISM(S) OF COAT PROTEIN-MEDIATED RESISTANCE

Studies have been undertaken to elucidate the cellular and molecular basis of CP-mediated resistance, but the precise nature of resistance is not yet determined. However, significant observations have been made, some of which will be described.

1. Resistance requires the accumulation of coat protein per se rather than CP mRNA.[13,14] In most but not all cases, increased levels of coat protein confer greater levels of resistance.

2. CP-mediated resistance, in the cases studied thus far, is highly effective in protoplasts isolated from transgenic plants.[15–17] These types of experimental results demonstrate that resistance is effective at the single-cell level.

3. CP-mediated resistant plants are less likely to be infected when inoculated with challenge virus than are nontransgenic plants and, if infected, either do not develop symptoms, have reduced severity of disease, or develop symptoms after a significant delay.

4. CP-mediated resistance is relatively narrow in its effect, that is, coat protein confers resistance against the virus from which the CP sequence was obtained and closely related viruses, with little or no resistance against distantly related viruses or viruses in different taxonomic groups.[18–20] The degree of relatedness between "protecting" CP and the CP of the "challenge" virus that determines the extent of resistance has not yet been determined. Limited experimental evidence suggests that in the tobamoviruses[20] and the potyviruses[21] substantial resistance occurs when the CP sequences are identical at 60% or more of the amino acid residues.

5. In most, but not all cases, CP-mediated resistance is overcome by inoculation with viral RNA (rather than the virus). This implies that resistance is not the result of encapsidation of the challenger's RNA by the CP in the transgenic plants. Furthermore, treating virus particles (in this case, TMV) at elevated pH for enough time to initiate the process of disassembly and virus replication is sufficient to overcome CP-mediated resistance.[15,22] This suggests that CP-mediated resistance acts by blocking an early stage in infection, the stage at which viral RNA is released from the virion.[16] In CP-mediated resistance against TMV in tobacco, expressing the CP gene

from a nominally constitutive transcriptional promoter provides substantially greater whole plant protection than does expression from a promoter that is regulated by light.[23] This apparently reflects the effect of the CP on infection as well as systemic spread of the infection.

Although we apparently have learned considerable details about the mechanism(s) that confers the phenotype of CP-mediated resistance, a considerable amount remains unknown. By better understanding CP-mediated resistance we hope to expand and improve its applications.

FIELD TESTING VIRUS-RESISTANT PLANTS

To date only a limited number of published reports describe the results of field trials that tested CP-mediated resistance, although others have been carried out. (See ref. 12 for a more complete discussion of the nonpublished data from other field trials.) The first report described the testing of tomato plants resistant to TMV.[24] In this test tomato lines that expressed the TMV CP gene were highly resistant to purposeful as well as casual infection by TMV, and the crop was protected from loss of yield due to infection. Plants that did not express the CP gene had 25–35% yield loss due to virus infection. Since that time other experiments with these and other lines of virus-resistant tomatoes have been carried out in Florida and Illinois by scientists at Monsanto Company and, in some cases, with seed company partners. These early field trials have demonstrated that CP-mediated resistance can reduce the incidence and severity of disease and have pointed the way to designing methods to improve field performance of tomato plants with CP-mediated resistance.

Tests with transgenic potato plants that are resistant to potato viruses X and/or Y have been equally successful.[25] The important features of these studies include the fact that the potato lines tested were derived from a cultivar that is used commercially, Russet Burbank. One plant line carried two genes that give resistance to two different viruses. Equally as important is the fact that these plants were resistant to viruses introduced either mechanically or by aphids, the natural vector.

Although these initial field trials were instructive and demonstrated the efficacy of CP-mediated resistance, more extensive trials in a variety of locales and environmental conditions are required before final judgments can be made about the role of CP-mediated resistance to maintain high levels of productivity in potatoes and tomatoes. As more types of plants are protected by the use of this technology, the better will be our ability to fully assess the implications of genetic transformation in crop improvement and its compatibility with more classical types of crop breeding.

REFERENCES

1. CUOZZO, M., K. M. O'CONNELL, W. KANIEWSKI, R. FANG, N. CHUA & N. E. TUMER. 1988. Viral protection in transgenic plants expressing the cucumber mosaic virus coat protein or its antisense RNA. Bio/technology 6: 383–389.
2. HEMENWAY, C., N. E. TUMER, P. A. POWELL & R. N. BEACHY. 1989. Genetically engineered cross protection against TMV interferes with initial infection and long distance spread of the virus. I. K. Vasil & J. Schell, eds.: 406–423. Academic Press. San Diego, California.
3. POWELL-ABEL, P. A., R. S. NELSON, BARUN DE, N. HOFFMANN, S. G. ROGERS, R. T.

FRALEY & R. N. BEACHY. 1986. Delay of disease development in transgenic plants that express the tobacco mosaic virus coat protein gene. Science **232:** 738–743.

4. POWELL, P. A., D. M. STARK, P. SANDERS & R. N. BEACHY. 1989. Protection against tobacco mosaic virus in transgenic plants that express TMV antisense RNA. Proc. Natl. Acad. Sci. USA **86:** 6949–6952.

5. BEACHY, R. N., D. M. STARK, C. M. DEOM, M. J. OLIVER & R. T. FRALEY. 1987. Expression of sequences of tobacco mosaic virus in transgenic plants and their role in disease resistance. *In* Tailoring Genes for Crop Improvement. G. Bruening, J. Harada, T. Kosuge, & J. Hallaender, eds.: 169–180. Plenum Publishing Corp. New York, NY.

6. REZAIAN, M. A., K. G. M. SKENE & J. G. ELLIS. 1988. Antisense RNAs of cucumber mosaic virus in transgenic plants assessed for the control of the virus. Plant Mol. Biol. **11:** 463–471.

7. COLLMER, C. W., M. E. TOUSIGNANT & J. M. KAPER. 1983. Cucumber mosaic virus-associated RNA 5: X. The complete nucleotide sequence of a CARNA 5 incapable of inducing tomato necrosis. Virology **127:** 230–234.

8. GERLACH, W. L., J. M. BUZAYAN, I. R. SCHEIDER & G. BREUNING. 1986. Satellite tobacco ringspot virus RNA: Biological activity of DNA clones and their *in vitro* transcripts. Virology **151:** 172–185.

9. KAPER, J. M. 1984. Plant disease regulation by virus-dependent satellitelike replicating RNAs. *In* Control of Virus Diseases. E. Kivistak, ed.: 317–343. Marcel Dekka, Inc. New York, NY.

10. HARRISON, B. D., M. A. MAYO & D. C. BAULCOMBE. 1987. Virus resistance in transgenic plants that express cucumber mosaic virus satellite RNA. Nature **334:** 799–802.

11. GERLACH, W. L., D. LLEWELLYN & J. HASELOFF. 1987. Construction of a plant disease resistance gene from the satellite RNA of tobacco ringspot virus. Nature **328:** 802–805.

12. BEACHY, R. N., S. LOESCH-FRIES & N. E. TUMER. 1990. Coat protein-mediated resistance against virus infection. Ann. Rev. Phytopathol. **28:** 451–474.

13. VAN DUN, C. M., B. OVERBUIN, L. VAN VLOTEN-DOTING & J. F. BOL. 1988. Transgenic tobacco expressing tobacco streak virus or mutated alfalfa mosaic virus coat protein does not cross-protect against alfalfa mosaic virus infection. Virology **164:** 383–389.

14. POWELL, P. A., P. R. SANDERS, N. TUMER & R. N. BEACHY. 1990. Protection against tobacco mosaic virus infection in transgenic plants requires accumulation of capsid protein rather than coat protein RNA sequences. Virology **175:** 124–130.

15. REGISTER, J. C., III, & R. N. BEACHY. 1988. Resistance to TMV in transgenic plants results from interference with an early event in infection. Virology **166:** 524–532.

16. REGISTER, J. C., III, P. A. POWELL & R. N. BEACHY. 1989. Genetically engineered cross protection against TMV interferes with initial infection and long distance spread of the virus. *In* Molecular Biology of Plant-Pathogen Interactions. B. Staskowicz, P. Ahlquist & O. C. Yoder, eds.: 269–281. Alan R. Liss. New York, NY.

17. LOESCH-FRIES, L. S., E. HALK, D. MERLO, N. JARVIS, S. NELSON, K. KRAHN & L. BURHOP. 1987. Expression of alfalfa mosaic virus coat protein gene and anti-sense cDNA in transformed tobacco tissue. *In* Molecular Strategies for Crop Protection. C. J. Arntzen & C. Ryan, eds.: 221–234. Alan R. Liss. Los Angeles, CA.

18. LOESCH-FRIES, L. S., D. MERLO, T. ZIRNEN, L. BURHOP, K. HILL, K. KRAHN, N. JARVIS, S. NELSON & E. HALK. 1987. Expression of alfalfa mosaic virus RNA 4 in transgenic plants confers virus resistance. EMBO. **6:** 1845–1851.

19. ANDERSON E. J., D. M. STARK, R. S. NELSON, N. E. TUMER & R. N. BEACHY. 1989. Transgenic plants that express coat protein gene of TMV or A1MV interfere with disease development of nono-related viruses. Phytopathology **12:** 1284–1290.

20. NEJIDAT, A. & R. N. BEACHY. 1990. Transgenic tobacco plants expressing a coat protein gene of tobacco mosaic virus are resistant to some other tobamoviruses. Mol. Plant-Microbe Interactions, in press.

21. STARK, D. M. & R. N. BEACHY. 1989. Protection against potyvirus infection in transgenic plants: Evidence for broad spectrum resistance. Bio/Technology **7:** 1257–1262.

22. REGISTER, J. C., III & R. N. BEACHY. 1989. A transient protoplast assay for capsid protein-mediated protection: Effect of capsid protein aggregation state on protection against tobacco mosaic virus. Virology **173:** 656–663.

23. CLARK, W. G., J. C. REGISTER, III, A. NEJIDAT, D. A. EICHHOLTZ, P. R. SANDERS, R. T. FRALEY & R. N. BEACHY. 1990. Tissue specific expression of the TMV coat protein in transgenic tobacco plants affects the level of coat protein mediated virus protection. Virology **179:** 640–647.

24. NELSON, R. S., S. M. MCCORMICK, X. DELANNY, P. DUBÉ, J. LAYTON, E. J. ANDERSON, M. KANIEWSKA, R. K. PROKSCH, R. B. HORSCH, S. G. ROGERS, R. T. FRALEY & R. N. BEACHY. 1988. Virus tolerance, plant growth, and field performance of transgenic tomato plants expressing coat protein from tobacco mosaic virus. Bio/Technology **6:** 403–409.

25. LAWSON, C., W. KANIEWSKI, L. HALEY, R. ROZMAN, C. NEWELL, P. SANDERS & N. E. TUMER. 1990. Engineering resistance to mixed virus infection in a commercial potato cultivar: Resistance to potato virus X and potato virus Y in transgenic Russet Burbank. Bio/Technology **8:** 127–134.

Arabidopsis as a Model System for Studying Plant Disease Resistance Mechanisms

ROGER INNES,[a,d] ANDREW BENT,[a]
MAUREEN WHALEN,[a,b,c] AND BRIAN STASKAWICZ[a]

[a]Department of Plant Pathology
University of California
Berkeley, California 94720

[b]Department of Biology
Colby College
Waterville, Maine 04901

We are using the small mustard *Arabidopsis thaliana* as a model host to identify plant genes required for disease resistance. The basic mechanisms that plants use to defend themselves against pathogens are poorly understood. A clearer understanding of these mechanisms will enable development of crop plants that are resistant to pathogens that now must be controlled by chemical pesticides. We chose *A. thaliana* for these studies because genes in this plant can be mapped, cloned, and characterized faster than those in any other plant.[1]

Resistance of a given crop cultivar to a specific pathogen is a heritable trait. In many cases, inheritance of resistance can be attributed to a single plant "resistance gene" (R).[2,3] It has been proposed that plant R genes are required for recognition of specific components of pathogens (reviewed in refs. 4 and 5). Production of these pathogen components is controlled by "avirulence" genes of the pathogen. A plant is resistant when it has an R gene that enables recognition of a pathogen component produced by a corresponding avirulence gene. Resistance of a plant to a specific pathogen breaks down when the pathogen modifies or mutates its avirulence gene so that the pathogen evades detection by the plant. We wish to determine how R genes function to confer resistance. To date, no R gene has been molecularly cloned from a plant. We are attempting to clone *A. thaliana* R genes that confer resistance to the bacterial plant pathogen *Pseudomonas syringae* pv. *tomato*.

To identify and map *A. thaliana* R genes, it was first necessary to identify corresponding avirulence genes. We have identified several avirulence genes from *P. syringae* pv. *tomato* that induce a resistant reaction on *A. thaliana*.[6] These avirulence genes were identified by taking a genomic library from a *P. syringae* pv. *tomato* strain that is avirulent (recognized) on *A. thaliana* ecotype Col-0 and introducing the library into a *P. syringae* pv. *tomato* strain that is virulent on *A. thaliana* Col-0. (An ecotype is a wild isolate of a plant species collected from a specific geographic location.) Conversion of the virulent strain to avirulence on *A. thaliana* Col-0 by the cloned DNA indicated the presence of an avirulence gene. We have characterized one of these clones in detail. The avirulence activity was localized to a 1.5 kb pair piece of DNA[6]; this locus has been designated *avrRpt2*. Insertion of the transcriptional

---

[c]PRESENT ADDRESS: Department of Biology, Colby College, Waterville, ME 04901.
[d]PRESENT ADDRESS: Department of Biology, Indiana University, Bloomington, IN 47405.

terminator Ω in the middle of this 1.5-kb clone abolished avirulence activity,[6] indicating that transcription through this region is necessary for expression of avirulence.[7] We used the intact *avrRpt2* clone to construct isogenic strains of *P. syringae* pv. *tomato* that differed only by the presence of this 1.5-kb region. We are now using these isogenic strains to identify and map the corresponding *A. thaliana* R gene.

Genetic mapping of the *A. thaliana* R gene corresponding to *avrRpt2* requires identification of *A. thaliana* ecotypes that are susceptible to *P. syringae* pv. *tomato* strains expressing *avrRpt2*. We screened more than 30 *A. thaliana* ecotypes for susceptibility to *P. syringae* pv. *tomato* strain DC3000 (*avrRpt2*).[6] The majority of these ecotypes were resistant, but we identified two ecotypes, Po-1 and Hs-0, that were phenotypically susceptible. Measurements of bacterial growth inside infected plant leaves confirmed that ecotype Col-0 was resistant to strain DC3000 (*avrRpt2*) and that ecotype Po-1 was susceptible to this strain.[6] We crossed the resistant and susceptible ecotypes to determine if resistance corresponding to each avirulence gene is inherited as a single gene. Tests of the F1 generation indicate that resistance corresponding to our cloned avirulence gene *avrRpt2* is inherited as a dominant trait.

One of the long-range goals of our research is to test whether R genes can be functionally transferred across species barriers. For example, will *A. thaliana* R genes function in crop plants such as soybean? Recent work demonstrating that single avirulence genes can induce resistant reactions on diverse species of plants suggests that this may indeed be possible.[8,9] If *avrRpt2* could convert other *P. syringae* pathovars to avirulence on their respective host plants, this would suggest that these plants contain resistance genes equivalent to R genes of *A. thaliana*. We transferred *avrRpt2* into the virulent soybean pathogen *P. syringae* pv. *glycinea* strain A29-2 and tested for conversion to avirulence on specific soybean cultivars. Inoculations were done at high bacterial cell densities (10^8 cfu/ml) to assay for induction of a hypersensitive reaction (HR). An HR is typically produced by soybean plants during a resistant response.[10] On soybean cultivars Centennial, Flambeau, and Harosoy a light but distinct HR became visible 30 hours after inoculation with *P. syringae* pv. *glycinea* strain A29-2 carrying *avrRpt2*.[6] On cultivars Acme and Norchief no HR was detected, and water-soaking phenotypes typical of a susceptible response appeared approximately 72 hours after inoculation. Inoculation with wild-type strain A29-2 produced water-soaking symptoms on all cultivars. Bacterial growth curves performed on cultivars Acme and Centennial confirmed that visible phenotypes correlated with reduction in bacterial cell growth. On cultivar Centennial, growth of a strain expressing *avrRpt2* was reduced approximately 50-fold after 5 days relative to growth of a strain carrying this locus interrupted with the Ω insertion.

Because the interaction between the end products of avirulence genes and resistance genes is specific, the results described herein suggest that resistant soybean cultivars such as Centennial may have a resistance gene functionally equivalent to the gene in *A. thaliana* ecotype Col-0. The implied similarity of resistance mechanisms between *A. thaliana* and soybean suggests that *A. thaliana* resistance genes can be used to isolate homologous loci from soybean. More generally, it may be possible to expand the resistance of crop plants by transformation using *A. thaliana* resistance genes.

REFERENCES

1. MEYEROWITZ, E. M. 1989. Arabidopsis, a useful weed. Cell **56:** 263–269.
2. ELLINGBOE, A. H. 1981. Changing concepts in host-pathogen genetics. Ann. Rev. Phytopathol. **19:** 125–143.

3. FLOR, H. 1971. Current status of the gene-for-gene concept. Ann. Rev. Phytopathol. **9:** 275–296.
4. ELLINGBOE, A. 1982. Genetical aspects of active defense. *In* Active Defense Mechanisms in Plants. R. K. S. Wood, ed.: 179–192. Plenum Press. New York.
5. KEEN, N. T. 1982. Specific recognition in gene-for-gene host-parasite systems. Adv. Plant Pathol. **1:** 35–81.
6. WHALEN, M., R. INNES, A. BENT & B. STASKAWICZ. 1991. Identification of *Pseudomonas syringae* pathogens of *Arabidopsis thaliana* and a bacterial locus determining avirulence on both arabidopsis and soybean. Plant Cell **3:** 49–59.
7. PRENTKI, P. & H. M. KRISCH. 1984. *In vitro* insertional mutagenesis with a selectable DNA fragment. Gene **29:** 303–313.
8. WHALEN, M. C., R. E. STALL & B. J. STASKAWICZ. 1988. Characterization of a gene from a tomato pathogen determining hypersensitive resistance in a non-host species and genetic analysis of this resistance in bean. Proc. Natl. Acad. Sci. USA **85:** 6743–6747.
9. KOBAYASHI, D. Y., S. J. TAMAKI & N. T. KEEN. 1990. Molecular characterization of avirulence gene D from *Pseudomonas syringae* pv. *tomato*. Mol. Plant-Microbe Interact. **3:** 94–102.
10. STASKAWICZ, B. J., D. DAHLBECK & N. T. KEEN. 1984. Cloned avirulence gene of *Pseudomonas syringae* pv. *glycinea* determines race-specific incompatibility on *Glycine max* (L.) Merr. Proc. Natl. Acad. Sci. USA **81:** 6024–6028.

Gene Expression in Insects

Biotechnical Applications

LOIS K. MILLER[a]

Departments of Entomology and Genetics
University of Georgia
Athens, Georgia 30602

Vectors that are capable of driving high-level expression of heterologous genes in insect cells have been developed.[1-4] Insect cells provide a higher eukaryotic cell environment that can supply a variety of posttranslational modification pathways and subcellular environments similar to but not necessarily identical to those of mammalian cells. Because of their utility and power, insect-based expression systems have been adopted by a vast number of research laboratories with the primary goal of high-level production of biologically active eukaryotic proteins.

The most widely used insect-based expression systems are baculovirus expression systems[2-4] that offer numerous advantages from a research perspective: (1) extremely high levels of foreign gene expression so that biologically active gene products are easily identified and purified; (2) helper-independent recombinant virus vectors that can stably carry large inserts of foreign DNA; (3) relatively rapid vector construction and determination of expression levels; and (4) temporally regulated expression that allows production of proteins affecting cell viability. Current baculovirus vector development includes improving the nature and increasing the number of promoters used to drive heterologous gene expression, facilitating the identification of recombinant viruses, manipulating the type and level of posttranslational modification of the heterologous gene product, and expanding the ability of the vectors to be used in insect larvae as well as insect cell culture. Other approaches to insect cell-based expression are also being developed and include the use of stably transformed insect cell lines and regulatable promoters.[1]

Industrial scale application of insect-based expression systems was initially limited by lack of knowledge concerning insect cell culture requirements and serum-free media. The recent development of protein-free insect cell culture media, the reduction of sheer stress by the addition of polyols such as Pluronic F-68, and the modeling of fermenter systems capable of mass insect cell culture set the stage for full industrial implementation of insect expression systems.[5,6] Although some industrial engineers prefer stably transformed cells as a basis for long-term continuous protein production rather than the batch production methods commonly used for baculovirus-based expression system, batch protein production methods may be essential for production of some short-lived or cytolytic proteins.

Although cell culture is a preferred source of biomedical protein products that require FDA approval, the large scale production of proteins for research and other industrial applications may use sericulture-like protein production techniques (i.e., use of insect larvae) in the case of baculovirus-based expression systems.[3] American industry has been skeptical of an insect-based approach, and chemical engineers/

[a] Address for correspondence: Dr. Lois K. Miller, Department of Entomology, University of Georgia, Athens, GA 30602.

protein production specialists are not familiar with such production technology. However, it is likely that Asian and third-world companies will capitalize on insectary-based production systems. Such systems have been, and are likely to continue to be, the most inexpensive source of baculovirus-expressed proteins during the current decade. However, expense and uniformity or quality of gene products may be independent issues that must be considered on an individual product basis.

Baculoviruses also have application as biological forms of insect pest control.[7–9] Their use as pesticides will likely expand considerably as genetically improved baculoviruses are developed[10–16] and as increased attention is centered on the problems of chemical pesticide residues in the environment. The development of insect cell culture facilities and/or insectaries for both baculovirus pesticide production and mass scale (pounds to tons scale) protein production could be an attractive industrial prospect.

ACKNOWLEDGMENT

I thank the National Institutes of Health, Institute of Allergy and Infectious Diseases, for continuous support of basic research on molecular baculovirology in my laboratory over the past 15 years.

REFERENCES

1. JOHANSEN, H., A. VAN DER STRATEN, R. SWEET, E. OTTO, G. MARONI & M. ROSENBERG. 1989. Regulated expression at high copy number allows production of a growth inhibitory oncogene product in *Drosophila* Schneider cells. Genes & Dev. 3: 882–889.
2. LUCKOW, V. A. & M. D. SUMMERS. 1988. Trends in the development of baculovirus expression vectors. Bio/technology 6: 47–55.
3. MAEDA, S. 1989. Expression of foreign genes in insects using baculovirus vectors. Annu. Rev. Entomol. 34: 351–370.
4. MILLER, L. K. 1988. Baculoviruses as gene expression vectors. Annu. Rev. Microbiol. 42: 177–199.
5. MAIORELLA, B., D. INLOW, A. SHAUGAR & D. HARANO. 1988. Large-scale insect cell-culture for recombinant protein production. Bio/Technology 6: 1406–1410.
6. WU, J., G. KING, A. J. DAUGULIS, P. FAULKNER, D. H. BONE & M. F. A. GOOSEN. 1989. Engineering aspects of insect cell suspension culture: A review. Appl. Microbiol. Biotechnol. 32: 249–255.
7. CARTER, J. B. 1984. Viruses as pest-control agents. Biotechnol. Genet. Eng. Rev. 1: 375–419.
8. GRANADOS, R. R. & B. A. FEDERICI. (Eds.) 1986. The Biology of Baculoviruses, Vol. II. CRC Press, Inc. Boca Raton, FL.
9. MILLER, L. K., A. J. LINGG & L. A. BULLA, JR. 1983. Bacterial, viral and fungal insecticides. Science 219: 715–721.
10. CARBONELL, L. & L. K. MILLER. 1985. Baculovirus-mediated expression of bacterial genes in dipteran and mammalian cells. J. Virol. 56: 153–160.
11. KEELEY, L. L., T. K. HAYES & J. Y. BRADFIELD. 1989. Insect neuroendocrinology: Its past; its present; future opportunities. *In* Insect Neuroendocrinology: Its Past; Its Present; Future Opportunities. A. B. Borkovec & E. P. Masler, eds. Humana Press Inc.
12. TOMALSKI, M. D., & L. K. MILLER. 1991. Insect paralysis by baculovirus-mediated expression of a mite neurotoxin gene. Nature 352: 82–85.
13. STEWART, L. M. D., M. HIRST, M. L. FERBER, A. T. MERRYWEATHER, P. J. CAYLEY &

R. D. POSSEE. 1991. Construction of an improved baculovirus insecticide containing an insect-selective toxin gene. Nature **352:** 85–88.

14. O'REILLY, D. R., & L. K. MILLER. 1991. Improvement of a baculovirus pesticide by deletion of the *egt* gene. Bio/Technology **9:** 1086–1089.

15. MAEDA, S., S. L. VOLRATH, T. N. HANZLIK, S. A. HARPER, K. MAJIMA, D. W. MADDOX, B. D. HAMMOCK & E. FOWLER. 1991. Insecticidal effects of an insect-specific neurotoxin expressed by a recombinant baculovirus. Virology **184:** 777–780.

16. MCCUTCHEN, B. F., P. V. CHONDARY, R. CRENSHAW, D. MADDOX, S. G. KAMITA, N. PALEKAR, S. VOLRATH, E. FOWLER, B. D. HAMMOCK & S. MAEDA. 1991. Development of a recombinant baculovirus expressing an insect-selective neurotoxin: Potential for pest control. Bio/Technology **9:** 848–852.

Expression of Proteins with Insecticidal Activities Using Baculovirus Vectors

R. D. POSSEE,[a] B. C. BONNING, AND
A. T. MERRYWEATHER

NERC Institute of Virology and Environmental Microbiology
Mansfield Road
Oxford, OX1 3SR UK

Baculoviruses are a safe alternative to the use of chemical pesticides for the control of insect pest species. These large DNA (about 130 kbp) viruses are specific for arthropods; safety studies have shown that they do not represent a hazard to species such as mammals or other vertebrates, invertebrates, or plants. Details of the infection process and the control of virus gene expression have been reviewed,[1,2] and some of the baculoviruses that have been used in field control studies are reported elsewhere.[3,4] The baculovirus genome is packaged in a rod-shaped nucleocapsid structure that is enclosed by a lipoprotein envelope. Nuclear polyhedrosis and granulosis viruses (NPV and GV, respectively) are further protected by an occlusion body consisting largely of a single protein of about 30 kD (polyhedrin or granulin). The NPVs package several virus particles in each occlusion body (polyhedron), but each GV occlusion body (granule) only contains a single virus particle. The occlusion bodies serve to protect the virus particles within the environment between susceptible insect larvae. For the remainder of this paper we will only consider the NPVs.

Polyhedra are ingested by the insect as it feeds on contaminated diet (foliage, etc.). These structures are dissolved in the alkaline pH of the midgut, releasing infectious virus particles to the lumen. These virions fuse with the gut epithelial cells and initiate infection in the insect. The virus procedes through a biphasic replication cycle in each cell. At first, virus particles bud from the plasma membrane and spread infection to other cells. Later, the virus particles are occluded by polyhedrin protein within the nucleus to form occlusion bodies. This infection process continues until the insect becomes packed with polyhedra and eventually disintegrates, releasing virus that may infect other larvae.

The problem with baculoviruses as control agents is the time required to kill the host insect. Chemicals frequently work in a matter of hours, producing a satisfying result for the crop grower. Viruses, however, may take several days (7–10) to produce the same effect. In this time the larvae continue to feed and cause damage. If the crop depends on cosmetic appearance for economic value (such as fruit), the producer is unlikely to tolerate this circumstance. Hence, the widespread use of baculoviruses has not become common.

The answer to the slow rate of action of baculoviruses may be to modify the virus using genetic engineering techniques. These viruses already serve as high level expression vectors of foreign genes,[5,6] and so it is feasible to insert coding sequences for insecticidal proteins into the virus genome, infect the insect, and test for an enhancement of the effect of the virus on the host. Candidate genes for inserting into baculoviruses are insect-specific toxins, such as the *Bacillus thuringiensis* δ-endotoxin

[a] Corresponding author.

234

and insect-encoded hormones or enzymes involved in the control of the larva developmental processes, such as juvenile hormone esterase.

The gram-positive bacterium, *B. thuringiensis,* produces crystalline inclusions during sporulation that contain an insect-specific endotoxin.[7] These crystals, when consumed by the larvae, dissolve to release the 130 kD protoxin, which is cleaved by gut proteases to an active form of 68 kD. This toxin interacts with the cells lining the gut wall, causing lysis and leakage of the gut contents into the hemocoel. The insect quickly ceases feeding and dies.

In the last larval instar the levels of juvenile hormone (JH) (which maintain the insect in the caterpillar stage) are dramatically reduced. Associated with this is an equally significant increase in the levels of juvenile hormone esterase (JHE). The JHE hydrolyses the chemically stable, conjugated methyl ester to the biologically inactive JH acid.[8] It was proposed that expression of JHE at an earlier stage in the larval life cycle would induce premature morphogenesis and probably death. In this report we review the production of recombinant *Autographa californica* nuclear polyhedrosis virus (AcNPV) containing either the bacterial endotoxin or the JHE coding sequences and present data describing the effect of the modified viruses on the target host.

MATERIALS AND METHODS

Production of Recombinant Baculoviruses Containing Foreign Genes

The derivation of Ac(PH$^+$)Bt(AcNPV with the *B. thuringiensis* endotoxin) has been described previously.[9] The modified virus insecticide, AcUW2(B).JHE, was constructed by inserting the JHE coding sequences,[10] under the control of a copy of the p10 promoter, upstream of the intact polyhedrin gene. The genomic organization of this virus, therefore, closely ressembled that of Ac(PH$^+$)Bt.

Assessing Biological Activity of Recombinant Viruses

Ac(PH$^+$)Bt. Spodoptera frugiperda cells infected with Ac(PH$^+$)Bt or AcNPV (10 plaque-forming units per cell) were harvested at 30 hours postinfection and then applied to a diet surface. Third instar larvae of the cabbage looper (*Trichoplusia ni*) were placed on the contaminated diet and observed over a period of days.

AcUW2(B).JHE. Mid-second instar *T. ni* were fed dilutions of purified polyhedra of AcUW2(B).JHE, using small plugs of synthetic diet. After 48 hours those insects that had consumed all of the diet were transferred to individual containers with fresh diet and weighed 5 and 7 days later.

RESULTS

Biological Activity of Baculovirus-Expressed B. thuringiensis δ-*Endotoxin*

Earlier studies had shown that the lethal dose (LD)$_{50}$ value for Ac(PH$^+$)Bt was very similar to that for wild-type AcNPV.[9] This indicated that the *de novo* expression of endotoxin in insects had no effect on the host. The biological activity of the endotoxin produced by Ac(PH$^+$)Bt in insect cells was assessed by feeding diet contaminated with endotoxin to *T. ni* larvae (FIG 1). Insects fed untreated diet

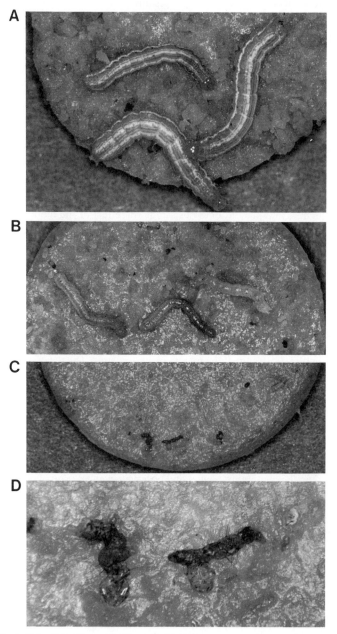

FIGURE 1. Anti-feeding effect of *B. thuringiensis* endotoxin on *T. ni* larvae. The four panels show insects fed diet contaminated with: (**A**) uninfected *S. frugiperda* cells; (**B**) AcNPV-infected cells; and (**C** and **D**) Ac(PH⁺)Bt-infected cells (**D** is an enlargement of **C**).

developed normally (FIG 1A). Those insects fed diet contaminated with AcNPV-infected *S. frugiperda* cells grew normally for 5 days and then showed symptoms of virus infection (FIG 1B). The larvae fed cells infected with Ac(PH⁺)Bt refused the contaminated diet and died within 4 days without feeding (FIG. 1C and D). This experiment clearly demonstrated that the endotoxin produced by the virus had biological activity when fed to susceptible insects. Similar results have been reported by other workers.[11]

Biological Activity of AcUW2(B) JHE

Second instar *T. ni* fed 110 or 330 polyhedra from AcNPV or AcUW2(B).JHE were weighed 5 days after infection (FIG. 2). The results show that the larvae fed the

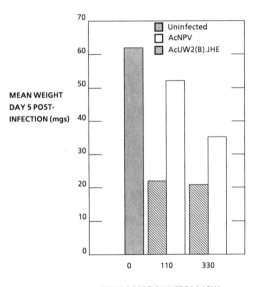

FIGURE 2. Mean weights 5 days after infection of second instar *T. ni* larvae infected with unmodified AcNPV or AcUW2(B).JHE (110 or 330 polyhedra per insect). For comparison, the average weight of uninfected larvae is shown.

recombinant virus did not gain weight at the same rate as did the larvae fed control virus. This effect is confined to first or early second instar larvae; later stages in the larval life cycle do not appear to be affected in this way.

DISCUSSION

Baculoviruses may be modified using genetic engineering techniques to incorporate sequences encoding insecticidal proteins. In this report we have described the effects on the insect larvae of expressing *B. thuringiensis* δ-endotoxin or JHE. Both proteins had a clear biological effect on the host; the bacterial toxin prevented feeding, whereas the JHE reduced the weight gain of the infected host.

The effect of the endotoxin was only evident when the protein contaminated virus preparations; purified polyhedra had LD_{50} values very similar to those of unmodified viruses. This suggests that the endotoxin produced in the insect does not have an effect. This is not surprising because the protein was expressed as a protoxin and remained intracellular; in this form it would be inaccessible to the gut lumen and thus unable to have any effect. The problem might be overcome by expressing the truncated (active) form of the toxin as a secretory product.

The effects of JHE are encouraging because here it is evident that the protein expressed within the host has modified the effect of the virus on the host. The reduction in weight was not dramatic, but it is encouraging for future work to increase the levels of JHE in the host and so improve this effect.

Genetically modified baculoviruses must eventually be tested in field trials. This goal must be preceded by extensive safety tests to monitor the effect of these agents on non-target species. For instance, is the host range of the virus altered by the expression of foreign insecticidal proteins? Can the virus recombine with other baculoviruses and exchange genetic information? Only when such questions have been thoroughly investigated to the satisfaction of the appropriate regulatory bodies can the environmental release of genetically engineered baculoviruses be allowed to proceed. The uncertainty of using baculoviruses with a phenotypic modification may be allayed by prior field experiments with viruses that have only minor changes in their genetic composition. In previous studies we used genetically marked viruses to mimic the field release of modified viruses.[12] Such an approach allows a model to be formulated of how altered baculoviruses may behave in the environment.

SUMMARY

The baculovirus, *Autographa californica* nuclear polyhedrosis virus, has been modified to incorporate the *Bacillus thuringiensis* δ-endotoxin gene or the *Heliothis virescens* juvenile hormone esterase gene. The biological effects of the expression of these proteins in insect larvae and the implications for the use of genetically engineered virus insecticides are considered.

REFERENCES

1. VOLKMAN, L. E. & B. A. KEDDIE. 1990. Nuclear polyhedrosis virus pathogenesis. Semin. Virol. **1:** 249–256.

2. BLISSARD, G. W. & G. F. ROHRMANN. 1990. Baculovirus diversity and molecular biology. Ann. Rev. Entomol. **35:** 127–155.

3. ENTWISTLE, P. F. & H. F. EVANS. 1985. Viral control. In: Comprehensive Insect Physiology, Biochemistry and Pharmacology. G. A. Kerkut & L. I. Gilbert, eds: 347–412. Pergamon Press. Oxford.

4. PODGWAITE, J. D. 1985. Strategies for field use of baculoviruses. *In* Viral Insecticides for Biological Control. K. Maramorosch & K. E. Sherman, eds.: 775–797. Academic Press. London & New York.

5. LUCKOW, V. A. & M. D. SUMMERS. 1988. Trends in the developments of baculovirus expression vectors. Bio/Technology **6:** 47–55.

6. MILLER, L. K. 1988. Baculoviruses as gene expression vectors. Ann. Rev. Microbiol. **42:** 177–199.

7. HÖFTE, H. & H. R. WHITELEY. 1989. Insecticidal proteins of *Bacillus thuringiensis.* Microbiol. Rev. **53:** 242–255.

8. HAMMOCK, B. D. 1985. *In* Comparative Insect Physiology Biochemistry and Pharmacology, **7:** 431–472. Pergamon Press. New York.

9. MERRYWEATHER, A. T., U. WEYER, M. P. G. HARRIS, M. HIRST, T. BOOTH & R. D.

POSSEE. 1990. Construction of genetically engineered baculovirus insecticides containing the *Bacillus thuringiensis* subsp. *kurstaki* HD-73 delta endotoxin. J. of Gen. Virol. **71:** 1535–1544.

10. HAMMOCK, B. D., B. C. BONNING, R. D. POSSEE, T. N. HANZLIK & S. MAEDA. 1990. Expression and effects of the juvenile hormone esterase in a baculovirus vector. Nature **344:** 458–461.

11. MARTENS, J. W. M., G. HONÉE, D. ZUIDEMA, J. W. M. VAN LENT, B. VISSER & J. M. VLAK. Insecticidal activity of a bacterial crystal protein expressed by a recombinant baculovirus in insect cells. Appl. Environ. Microbiol. **56:** 2764–2770.

12. BISHOP, D. H. L., P. F. ENTWISTLE, I. R. CAMERON, C. J. ALLEN & R. D. POSSEE. 1988. Field trials with genetically-engineered baculovirus insecticides. *In* The Release of Genetically-Engineered Micro-organisms. M. Sussman, C. H. Collins, F. A. Skinner & D. E. Stewart-Tull, eds.: 143–179. Academic Press. London.

Baculovirus Expression Vectors

A Review and Update

DONALD L. JARVIS

Department of Entomology and
Center for Advanced Invertebrate Molecular Sciences
Texas A&M University
College Station, Texas 77843

Baculoviruses have emerged as one of the most important vectors currently available for the expression of foreign gene products in a eukaryotic host. Relatively quick and simple methods have been devised for the expression of a chosen foreign gene product with these viruses. The product is usually expressed in exceedingly large amounts and participates in most of the protein-processing pathways found in mammalian cells. In addition, the viral infection shuts down the expression of host cell proteins which, coupled with the high levels of foreign gene expression, facilitates purification of the desired end-product. In this short overview, I review and update the ways in which baculoviruses can be used as vectors for foreign gene expression. The nature of baculovirus expression vectors, the methods used for their production, and the capabilities and limitations of this expression system will be briefly discussed. Finally, a new strategy of baculovirus-based foreign gene expression will be described. For more detailed information, the reader is referred to recent reviews by Kang,[1] Luckow and Summers,[2] Miller,[3] Vlak and Keus,[4] Blissard and Rohrmann,[5] and Jarvis and Summers.[6]

BACULOVIRUS EXPRESSION VECTORS

The subgroup A baculoviruses are enveloped, double-stranded DNA-containing viruses that infect insects. A distinguishing feature of these viruses is that they produce large particles called viral occlusions or polyhedra in the nuclei of infected cells. Polyhedra consist of progeny virions embedded within a protective crystalline matrix. The matrix consists mostly of a single viral-encoded protein, polyhedrin, that is produced in massive amounts in infected cells. The ability of baculoviruses to produce large amounts of polyhedrin depends on the overproduction of mRNA, which is synthesized under the control of an unusually strong transcriptional promoter in the polyhedrin gene. Polyhedrin is not required for the replication of baculoviruses in cell culture. Thus, it is possible to replace the polyhedrin coding sequence with the coding sequence for a foreign gene product and obtain viable recombinant viruses. When used to infect cultured insect cells, the recombinant virus will overexpress the foreign mRNA under the control of the polyhedrin promoter and the foreign protein product will usually accumulate in large amounts. The strength of the polyhedrin promoter is the most significant advantage that baculoviruses offer among eukaryotic expression vectors. In many cases, the recombinant protein will constitute over half the total protein in the infected cell by the very late phase of infection. Another advantage is that baculoviruses are infectious only for invertebrate hosts, so they are a relatively safe biological agent for laboratory manipulation. Finally, the baculovirus expression system has the advantage of

240

technical simplicity, as recombinant and wild-type viruses can quickly and easily be distinguished by differences in their plaque morphologies under a light microscope.

The construction and isolation of a baculovirus expression vector are accomplished by the procedures outlined in FIGURE 1. First, a cloned copy of the desired coding sequence is inserted into a "transfer vector" (FIG. 1, step 1). A variety of transfer vectors are available. Each consists of a bacterial plasmid containing the polyhedrin promoter, long 5' and 3' viral flanking sequences, and a multiple cloning

1. Construction

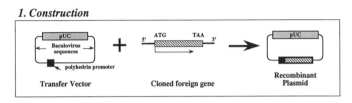

2. Cotransfection and visual selection

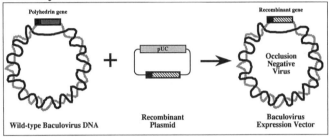

3. Foreign gene expression

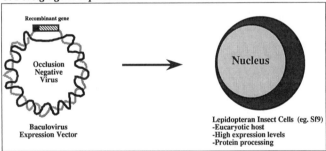

FIGURE 1. Preparation of a baculovirus expression vector. See text for detailed explanation.

site for insertion of the desired coding sequence. Almost all of the polyhedrin coding sequence has been deleted in these plasmids. The most effective transfer vectors are designed so that the foreign gene is inserted at a site downstream of the "A" residue from the polyhedrin initiation codon. Studies from several laboratories have shown that this will provide the highest levels of foreign gene expression, by preserving the transcriptional and translational control sequences located in the 5' noncoding region of the polyhedrin gene. Further details on these studies and information on many different transfer vectors are provided by Luckow and Summers[2] and Miller.[3]

After insertion of the foreign coding sequence, the next step is to incorporate the recombinant gene into the baculovirus genome (FIG. 1, step 2). This is necessary because the polyhedrin promoter cannot express anything by itself. It requires an unknown number of other viral functions, which must be expressed before the polyhedrin promoter becomes transcriptionally active. The recombinant gene is incorporated into the viral genome by cotransfecting insect cells with a mixture of wild-type viral DNA and transfer vector DNA. The flanking sequences in the transfer vector target the recombinant gene to the viral polyhedrin locus, where genetic exchange occurs by a process of homologous recombination. This generates a small percentage of recombinant viral progeny in which the wild-type polyhedrin gene has been replaced by the foreign gene. The mixture of recombinant and wild-type viral progeny can be resolved in a conventional plaque assay. The recombinants, which lack the polyhedrin coding sequence, produce occlusion-negative plaques, whereas wild-type viruses produce occlusion-positive plaques. As already mentioned, occlusion positive and negative plaques can be distinguished under the light microscope, and on this basis, recombinants can be visually identified, picked, and amplified. Once a working virus stock is obtained, it can be used to infect cultured insect cells or insects, and its ability to express the foreign gene product can be assessed (FIG. 1, step 3).

CAPABILITIES OF BACULOVIRUS EXPRESSION VECTORS

Generally, baculovirus expression vectors provide high levels of foreign gene expression, ranging from about 1–500 mg of recombinant product per liter of infected cells (2×10^9 cells). As a class, proteins that enter the secretory pathway seem to be expressed at lower levels than do nuclear or cytoplasmic proteins. However, this is not universally true, as there are several secretory pathway proteins that have been expressed at high levels in the baculovirus system (e.g., hepatitis B surface antigen; 90 mg/liter[7]). Moreover, it should be recognized that "low" levels of expression in the baculovirus system are often as high as the best levels of expression obtained in other eukaryotic systems. Nonetheless, most of the products that are relatively poorly expressed by baculoviruses are proteins that are processed within the cellular secretory pathway. The reason for this is unknown.

During the early stages of baculovirus expression vector development, questions arose regarding the protein-processing capabilities of lepidopteran insect cells, which were unfamiliar to most investigators. Few direct studies had been performed on the modification and transport of proteins in these cells, and in any case, most of the recombinant proteins that were being expressed in these cells were derived from vertebrates. Now that a large number of foreign proteins have been expressed in the baculovirus system, extensive information has accumulated on the types of protein processing that lepidopteran insect cells can carry out. This had led to one simple conclusion: these cells are qualitatively capable of almost every type of protein processing that has been examined, and the recombinant protein products invariably have had the antigenic and biological properties of the native product. More specifically, it has been shown that lepidopteran insect cells will cleave pre- and pro-sequences accurately from polypeptide precursors. N-glycosylation occurs, and although most evidence suggests that only high mannose or trimmed high mannose side chains are found in insect cell-derived N-glycoproteins,[8–10] the first demonstration of a complex sialic acid-containing side chain was recently reported.[11] Protein phosphorylation occurs in insect cells, as does acylation of proteins by the addition of either palmitic or myristic acid. Proteins are targeted to their correct subcellular destinations, as shown for cytoplasmic, nuclear, and plasma membrane-associated proteins, and secretory proteins are appropriately exported. The only protein processing pathway that appears to be missing in lepidopteran insect cells is the one that culminates with the alpha-amidation of a carboxy-terminal amino acid. This modification was not detected when human gastrin-releasing peptide was expressed in the Sf9 cell line derived from *Spodoptera frugiperda*.[12] However, indirect evidence exists for alpha-amidation of a *Manduca sexta* diuretic hormone expressed in baculovirus-infected silkworm larvae,[13] suggesting that the inability to alpha-amidate might be an Sf9 cell-specific limitation.

Important questions remain in the area of protein processing and targeting in the baculovirus expression system. For example, the kinetics and efficiencies of these processes are unknown. Subtle differences in the rates and efficiencies of protein processing and targeting might have a major influence on the overall levels of expression observed for a particular foreign protein product. The influence of the ongoing baculovirus infection on protein processing and targeting in lepidopteran insect cells also needs to be examined. It must be remembered that recombinant baculoviruses are essentially intact and are fully capable of a lytic infection. It has been shown that the ability of Sf9 cells to secrete human tissue plasminogen activator (t-PA) decreases dramatically with the time of infection.[14] This observation suggested that the ongoing virus infection might have an adverse effect on the host cell secretory pathway. This idea provided the rationale for the development of a new approach to baculovirus-based foreign gene expression, which was designed to avoid the potentially adverse effects of the viral infection.

A NEW METHOD OF BACULOVIRUS-BASED FOREIGN GENE EXPRESSION

This new strategy for baculovirus-based foreign gene expression is based on the use of promoters from viral genes that are expressed during the early phases of infection. Unlike the polyhedrin promoter, early viral promoters can be transcribed in uninfected insect cells. Recently, we demonstrated that the promoter from the immediate early 1 gene (IE1[15,16]) can be used to produce stably transformed Sf9 cell derivatives that will express a foreign gene product continuously.[17] First, we established that the IE1 promoter could induce continuous expression of the bacterial neomycin resistance (Neo) gene in uninfected Sf9 cells. This allowed us to use the standard cotransfection and antibiotic selection techniques developed in mammalian cell systems[18,19] to isolate transformed Sf9 cell derivatives that expressed a second gene, either *Escherichia coli* β-galactosidase (β-gal) or t-PA, under the control of the IE1 promoter. Biochemical and genetic analyses revealed that the transformed clones contained stably integrated copies of the plasmid DNAs and that these sequences were specifically and accurately transcribed from initiation sites within the IE1 promoter. A Western blot comparing expression of β-gal in recombinant baculovirus-infected cells and a transformed cell clone at different passage levels is shown in FIGURE 2. The cytoplasmic β-gal polypeptide was expressed continuously for over 50 passages in the transformed cells, but the level of expression was about 100–1,000-fold lower than that in baculovirus-infected cells. This comparison represents a worst-case example, as β-gal is very highly expressed in baculovirus-infected cells (about 200 mg/liter). Among several different transformed cell clones, there was a direct correlation between the levels of protein expression and the relative amounts of integrated DNA. This suggests that it should be possible to increase the levels of foreign gene expression in transformed Sf9 cells by manipulating transfection conditions in order to increase the number of plasmid copies integrated, as accomplished previously in the *Drosophila* expression system described by Rosenberg and coworkers.[20–22]

The secretory t-PA glycoprotein, which is relatively poorly expressed and secreted from baculovirus-infected cells (<1 mg/lit), was expressed and secreted at approximately equivalent levels by transformed Sf9 cells (FIG. 3). Our most exciting finding was that the transformed cells contained no detectable intracellular t-PA; virtually all of the product synthesized in these cells was secreted (FIG. 3). By contrast, about two thirds of the t-PA produced in infected cells was intracellular. We also found that t-PA secretion was faster in the transformed than in the infected cells (FIG. 4). These observations support and extend our previous hypothesis that

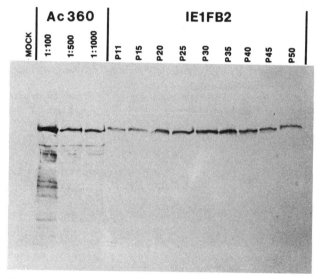

FIGURE 2. Expression of β-gal in infected and transformed insect cells. Extracts were prepared from a β-gal expressing Sf9 transformant (clone IE1FB2) at various passage levels after 2×10^6 cells had been seeded and grown for 48 hours. For comparison, extracts were prepared from 2×10^6 mock-infected or recombinant baculovirus (Ac360-βgal)-infected Sf9 cells (48 hours postinfection). The infected cell extract was diluted 1:100, 1:500, 1:1000, or 1:5000 for loading; all other extracts were undiluted. Analysis was by SDS-PAGE and western blotting, with monoclonal anti-β-gal as the primary antibody. Reproduced, with permission, from Jarvis *et al.*[13]

the host cell secretory pathway is compromised during the later phases of baculovirus infection.[14] Obviously, additional studies on the expression and processing of other foreign glycoproteins will be required to further test this hypothesis.

This work, with Sf9 cells,[17] and the work of Rosenberg's group, with *Drosophila* cells,[20–22] are the first examples of foreign gene expression using stably transformed insect cells. The transformed Sf9 cell expression system, in its present state, does not provide the high levels of expression that can be obtained for many gene products in baculovirus-infected Sf9 cells. However, the transformed cells represent a good alternative for the expression of a complex glycoprotein product that is relatively poorly expressed in infected cells. Under this condition, the ability of the transformed cells to provide continuous foreign gene expression and more efficient

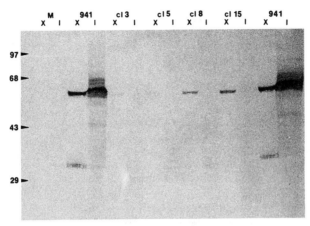

FIGURE 3. Expression of t-PA in infected or transformed insect cells. Extracellular (X) and intracellular (I) fractions were prepared from t-PA-expressing Sf9 transformants (cl 3, cl 5, cl 8, and cl 15) after 2×10^6 cells were seeded and grown for 48 hours. For comparison, samples were also prepared from 2×10^6 mock-infected (M) or recombinant baculovirus (941)-infected Sf9 cells (seeded for 24 hours and infected for an additional 24 hours). Fractions were analyzed by SDS-PAGE and western blotting, with polyclonal goat anti-human t-PA as the primary antibody. Reproduced, with permission, from Jarvis et al.[13]

protein processing become important advantages. A potential advantage of using transformed insect cells instead of mammalian cells for continuous production of a secretory glycoprotein relates to the inability of insect cells to process N-linked oligosaccharides to the complex form, as already noted.[8–10] The absence of carbohydrate can actually increase the *in vivo* half-life of certain therapeutic proteins, such as t-PA,[23] and it is tempting to speculate that the absence of complex N-linked

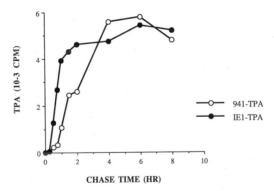

FIGURE 4. Processing of human t-PA in infected or transformed insect cells. 941-TPA-infected Sf9 cells or a t-PA expressing transformant (clone IE1TPA15) were grown for 24 hours and pulse-labeled for 15 minutes. The cells were chased immediately after labeling, and extracellular and intracellular fractions were harvested at the indicated times postchase, immunoprecipitated with polyclonal anti-t-PA, and separated by SDS-PAGE. The extracellular t-PA bands were excised, solubilized, and quantified by liquid scintillation spectroscopy. The figure shows the average amounts of extracellular t-PA from three separate pulse-chase experiments plotted against chase time. Reproduced, with permission, from Jarvis et al.[13]

oligosaccharides could have the same effect. Finally, a significant effort has been spent on the development of reactors and media for the scale-up of insect cell cultures, intended for use with traditional baculovirus expression vectors. This work has shown that, under the right conditions, insect cells are highly amenable to large scale culture and can be grown to densities as high as 5×10^6 cells/ml in airlift fermenters[24,25] or 1×10^8 cells/ml in hollow fiber bioreactors (B. Dale, personal communication). Cultures grown in the latter system can be maintained at high density for weeks in serum-free media. This technology could be transferred immediately and directly for use with transformed Sf9 cell cultures, resulting in a continuous source of recombinant product over a significant time period.

SUMMARY

Baculoviruses offer advantages over other vectors for the expression of foreign gene products in eukaryotic cells. The foremost of these is the ability to produce extremely large amounts of end-product based on the strength of the polyhedrin promoter, in a form that is relatively good starting material for subsequent purification. Although some differences in certain protein-processing pathways probably exist in mammalian and insect cells, recombinant proteins derived from insect cells appear to be structurally and biologically authentic. Future development of the baculovirus expression system should be directed at improving the expression of complex secretory and membrane-bound glycoproteins, which are often expressed in much lower amounts than are cytoplasmic and nuclear proteins. One possible reason for this is that infection with a recombinant baculovirus has an adverse effect on the host cell secretory pathway; however, this has not yet been definitively established. A new approach to baculovirus-mediated foreign gene expression is to use an immediate early viral promoter for continuous expression of a foreign gene product in uninfected, stably transformed insect cells. Currently, this approach provides lower levels of foreign gene expression than those generally obtained by using the lytic baculovirus expression vector system. However, expression is continuous, and a secretory glycoprotein was processed faster and more efficiently. Additional studies will be necessary to find ways to increase the levels of expression obtained in stably transformed Sf9 cells and to determine if glycoprotein processing is generally more efficient in transformed than it is in baculovirus-infected insect cells.

REFERENCES

1. KANG, C. Y. 1988. Adv. Virus Res. **35:** 177–192.
2. LUCKOW, V. L. & M. D. SUMMERS. 1988. Bio/Technology **6:** 47–55.
3. MILLER, L. K. 1988. Ann. Rev. Microbiol. **42:** 177–199.
4. VLAK, J. M. & R. J. A. KEUS. 1990. The baculovirus expression vector system for production of viral vaccines. *In* Viral Vaccines. A. Mizrahi, ed. Vol. 14. Alan R. Liss, Inc. New York, NY.
5. BLISSARD, G. W. & G. F. ROHRMANN. 1990. Ann. Rev. Entomol. **35:** 127–155.
6. JARVIS, D. L. & M. D. SUMMERS. 1992. Baculovirus expression vectors. *In* Recombinant DNA Vaccines: Rationale and Strategies. R. E. Isaacson, ed. Marcel Dekker, Inc. New York, NY.
7. LANFORD, R. E., V. LUCKOW, R. C. KENNEDY, G. R. DREESMAN, L. NOTVALL & M. D. SUMMERS. 1989. J. Virol. **63:** 1549–1557.
8. BUTTERS, T. D., R. C. HUGHES & P. VISCHER. 1981. Biochim. Biophys. Acta **640:** 672–686.
9. HSIEH, P. & P. W. ROBBINS. 1984. J. Biol. Chem. **259:** 2375–2382.

10. KURODA, K., H. GEYER, R. GEYER, W. DOERFLER & H.-D. KLENK. 1990. Virology **174:** 418–429.
11. DAVIDSON, D. J., M. J. FRASER & F. J. CASTELLINO. 1990. Biochemistry **29:** 5584–5590.
12. LEBACQ-VERHEYDEN, A.-M., P. G. KASPRZYK, M. G. RAUM, K. VAN WYKE COELINGH, J. A. LEBACQ & J. F. BATTEY. 1988. Mol. Cell. Biol. **8:** 3129–3135.
13. MAEDA, S. 1989. Biochem. Biophys. Res. Comm. **165:** 1177–1183.
14. JARVIS, D. L. & M. D. SUMMERS. 1989. Mol. Cell. Biol. **9:** 214–223.
15. GUARINO, L. A. & M. D. SUMMERS. 1986. J. Virol. **57:** 563–571.
16. GUARINO, L. A. & M. D. SUMMERS. 1987. J. Virol. **61:** 2091–2099.
17. JARVIS, D. L., J. G. W. FLEMING, G. R. KOVACS, M. D. SUMMERS & L. A. GUARINO. 1990. Bio/Technology **8:** 950–955.
18. WIGLER, M., M. PERUCHO, D. KURTZ, S. DANA, A. PELLICER, R. AXEL & S. SILVERSTEIN. 1980. Proc. Natl. Acad. Sci. USA **77:** 3567–3570.
19. SOUTHERN, P. J. & P. BERG. 1982. J. Mol. Appl. Gen. **1:** 327–341.
20. VAN DER STRATEN, A., H. JOHANSEN, R. SWEET & M. ROSENBERG. 1989. *In* Invertebrate Cell System Applications. J. Mitsuhashi, ed. Vol. 1: 183–195. CRC Press, Inc. Boca Raton, FL.
21. JOHANSEN, H., A. VAN DER STRATEN, R. SWEET, E. OTTO, G. MARONI & M. ROSENBERG. 1989. Genes & Development **3:** 882–889.
22. CULP, J. S., H. JOHANSEN, B. HELLMIG, J. BECK, T. J. MATTHEWS, A. DELERS & M. ROSENBERG. 1991. Bio/Technology **9:** 173–177.
23. LAU, D., G. KUZMA, C.-M. WEI, D. J. LIVINGSTON & N. HSIUNG. 1987. Bio/Technology **5:** 953–958.
24. MAIORELLA, B., D. INLOW, A. SHAUGER & D. HARANO. 1988. Bio/Technology **6:** 1406–1410.
25. MURHAMMER, D. W. & C. F. GOODHEE. 1988. Bio/Technology **6:** 1410–1418.

Engineering the Cardiovascular System

Blood Pressure Regulation

MARK E. STEINHELPER,[a] KAREN L. COCHRANE,[a] AND
LOREN J. FIELD[a]

Cold Spring Harbor Laboratory
Cold Spring Harbor, New York 11724

A decade ago the heart was found to act as an endocrine gland in addition to its well-documented mechanical function.[1] Atrial cardiomyocytes synthesize and secrete atrial natriuretic factor (ANF), a member of a family of natriuretic peptide hormones, in response to atrial distention resulting from increased cardiac preload. Acute biological effects of ANF include decreased arterial blood pressure and increased urine production and sodium excretion.[2] Also, ANF acutely affects the vasculature,[3] antagonizes aldosterone secretion from the adrenal cortex,[4] inhibits secretion of vasopressin from the pituitary,[5] and has direct effects within the kidney.[6,7] As with other peptide hormones, cell-surface receptors that bind the hormone with high affinity and specificity are expressed by target tissues. Recently, two high-affinity receptors for ANF were cloned.[8,9]

Given the large array of biological responses observed in the short-term studies just cited, we wished to investigate the role of ANF in the chronic regulation of the cardiovascular system. To this end we generated transgenic animals that exhibit elevated concentrations of ANF in the systemic circulation. In contrast to traditional infusion studies employing mechanical or osmotic pumps, our transgenic approach provides an unlimited temporal window in which to evaluate the biological responses to elevated hormone concentrations. Studies of the chronic effects of ANF using the traditional infusion systems have been restricted to at most 7 days and have resulted in contradictory observations.[10-19] The present report summarizes the initial characterization of transgenic mice expressing ANF in the liver.

METHODS

We attempted to target expression of the mouse ANF gene to the liver using transcriptional regulatory elements from the mouse transthyretin gene (TTR) and the rat cytosolic phosphoenolpyruvate carboxykinase gene (PEPCK), as summarized in TABLE 1. Briefly, the TTR promoter (cloned from a BALB/cCr library) consisted of sequences from −3 kb to +14 bp relative to the mRNA cap site.[20] The PEPCK promoter was the 620 bp *Bam*HI to *Bgl* II fragment previously described.[21] These regulatory elements were oriented using standard cloning protocols[22] so that fusion gene transcripts would encode pre-pro-ANF. Transgenic mice harboring the TTR-ANF and PEPCK-ANF fusion genes were generated and analyzed as previously described.[23]

[a]PRESENT ADDRESS: Indiana University School of Medicine, Krannert Institute of Cardiology, 1111 West 10th St., Indianapolis, IN 46202–2859.

RESULTS

As summarized in TABLE 1, the ANF fusion genes integrated into the germ line of mice and were transmitted to progeny, thereby establishing transgenic lineages. Representative mice from all lineages were examined periodically (at 1.5, 3, and 6 months of age) for evidence of fusion gene expression by Northern blot analysis of hepatic RNA. Of the 11 transgenic lines established from this series of microinjections, only 2 had detectable amounts of ANF transcripts present in liver, and these lineages were expanded for further study by backcrossing to C3HeB/FeJ inbred mice to yield mice hemizygous for the transgene. Suppression of fusion gene expression in the other lineages presumably resulted from positional effects arising from fusion gene integration.[24] As shown in FIGURE 1A, the male founder of the TTR-ANF-2 lineage was a germ-line chimera based on the low frequency (3%) of progeny inheriting the fusion gene. Transmission of the fusion gene in subsequent generations showed a pattern of inheritance consistent with integration at an autosomal locus. A similar segregation analysis in the TTR-ANF-4 lineage revealed a sex-linked pattern of inheritance (FIG. 1B), indicating that the fusion gene integrated on the X chromosome in this lineage.

TABLE 1. Summary of Transgenic Mice

Promoter	Lineages	Expressing Lineages	Sites of Expression[a]
TTR	4	2	Liver
PEPCK	7	0	. . .

[a]Expression was assessed by Northern blot analysis as described.[22,23]

The tissue specificity of fusion gene expression was determined in both lineages. The TTR-ANF fusion gene was expressed exclusively in the liver as determined by RNase protection assays.[23] Fusion gene expression was comparable between the two TTR-ANF lineages as estimated by Northern blot analysis.[23] To determine if the transgenic mice were translating the hepatic fusion gene transcripts, the concentration of immunoreactive ANF (irANF) was measured in extracts of transgenic and nontransgenic TTR-ANF-4 liver. Livers were perfused briefly through the portal vein with ice-cold phosphate-buffered saline solution to remove blood and circulating hormones from the hepatic vasculature. ANF was extracted from the liver by boiling the tissue in acetic acid; this was followed by chromatography on a reverse-phase resin. Transgenic liver contained detectable amounts of irANF (about 21 ng/g wet weight), whereas nontransgenic extracts contained essentially no detectable irANF (<1 ng/g wet weight). As expected, plasma irANF was elevated in the transgenic animals and correlated directly with the steady-state concentration of fusion gene transcripts detected by Northern blot analysis.[23]

To ascertain the chronic physiological effect of increased circulating irANF on blood pressure, cannulas were placed in the abdominal aorta, and pulsatile arterial pressure was recorded directly in conscious mice.[23] As shown in FIGURE 1, mice inheriting the TTR-ANF fusion genes had lower mean arterial blood pressure than did their nontransgenic siblings.

DISCUSSION

The present study demonstrates that increased steady-state concentrations of irANF in the systemic circulation caused a marked and persistent decrease in arterial blood pressure in transgenic animals irrespective of the etiology and potential

A

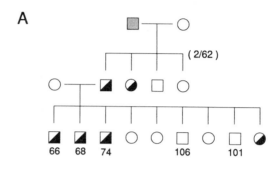

B

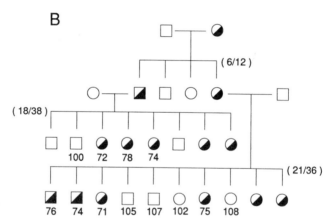

FIGURE 1. Partial pedigrees of TTR-ANF lineages that express the fusion gene. *Symbols:* Male (*squares*), female (*circles*), nontransgenic (*open*), transgenic (*half-closed*), and chimeric (*speckled*). Numbers below individual mice represent the mean arterial pressure measured as described.[23] Fractions in parentheses are ratio of transgenic to total progeny. **(A)** TTR-ANF-2 lineage. The male founder was chimeric and transmitted the fusion gene to only 2 of 62 progeny. **(B)** TTR-ANF-4 lineage. Transgenic males transmitted the fusion gene to all female offspring and never to male siblings. Transgenic females transmitted the fusion gene to both sexes at a frequency of approximately 50%.

down-regulation of ANF receptors. Whether the decreased mean arterial pressure observed in transgenic mice results from direct actions of ANF on the heart and vasculature or from indirect actions on other cardiovascular regulatory systems, such as the renin-angiotensin-aldosterone system or the sympathetic nervous system, is not clear at present. Additional studies to resolve issues related to kidney function have proved interesting.[25]

In addition, we cannot currently rule out any deleterious consequences of chronically elevated ANF concentrations on the health and survival of the transgenic mice. However, our observations over the last year all indicate that mice tolerate well the elevated ANF steady state. The ability of ANF to antagonize experimental manipulations designed to produce hypertension (i.e., DOCA salt and Goldblatt models) can be investigated using these transgenic mice and may prove informative. Also, the TTR-ANF transgenic mice should be useful for evaluating the consequences of chronic ANF overexpression on ANF receptors and on other regulatory aspects of the circulatory system. In summary, these transgenic models provide a unique experimental resource for investigating the chronic biological effects of ANF.

SUMMARY

We have generated several lineages of transgenic mice that exhibit chronic elevations in the steady-state concentration of atrial natriuretic factor (ANF) in the peripheral circulation. ANF, a peptide hormone synthesized primarily by atrial cardiomyocytes, is a potent natriuretic and diuretic. ANF also reduces blood pressure transiently when acutely administered. To address the potential role of ANF in chronic cardiovascular regulation, we generated transgenic mice that express the ANF gene in the liver. The fusion genes comprised either the mouse transthyretin (TTR) or rat phosphoenolpyruvate carboxykinase (PEPCK) promoters fused to the mouse ANF structural gene and were designed to target to the liver constitutive and inducible expression of pre-pro-ANF, respectively. Transgenic animals harboring the TTR-ANF fusion gene expressed chimeric ANF transcripts exclusively in the liver. In contrast, mice harboring the PEPCK-ANF fusion gene did not express detectable amounts of ANF mRNA in liver even after induction (24-hour fasting). In the TTR-ANF mice, hepatic and plasma immunoreactive ANF concentrations were proportional to the concentration of hepatic ANF transcripts. Moreover, mean arterial blood pressure recorded in conscious transgenic mice was inversely proportional to hepatic ANF expression. These transgenic models demonstrate that chronically elevated ANF concentration can induce sustained hypotension.

ACKNOWLEDGMENTS

The authors thank S. Teplin and P. Weinberg for technical assistance and Dr. M. McGrane for providing the PEPCK promoter.

REFERENCES

1. DeBold, A. J., H. B. Borenstein, A. T. Veress & H. Sonnenberg. 1981. A rapid and potent natriuretic response to intravenous injection of atrial myocardial extract in rats. Life Sci. **28:** 89–94.
2. Needleman, P., S. P. Adams, B. R. Cole, M. G. Currie, D. M. Geller, M. L. Michener, C. B. Saper, D. Schwartz & D. G. Standaert. 1985. Atriopeptins as cardiac hormones. Hypertension **7:** 469–482.
3. Currie, M. G., D. M. Geller, B. R. Cole, J. G. Boylan, W. Yusheng, S. W. Holmberg & P. Needleman. 1983. Bioactive cardiac substances: Potent vasorelaxant activity in the mammalian atria. Science **221:** 71–73.

4. ATARASHI, K., P. J. MULROW, R. FRANCO-SAENZ, R. SNAJDAR & J. RAPP. 1984. Inhibition of aldosterone production by an atrial extract. Science **224:** 992–994.
5. SAMSON, W. K. 1985. Atrial natriuretic factor inhibits dehydration and hemorrhage-induced vasopressin release. Neuroendocrinology **40:** 277–279.
6. SONNENBERG, H. 1988. Renal effects of atrial natriuretic factor. ISI Atlas of Science: Pharmacology 809–9083: 171–174.
7. MARIN-GREZ, M., J. T. FLEMING & M. STEINHAUSER. 1986. Atrial natriuretic peptide causes pre-glomerular vasodilation and post-glomerular vasoconstriction in rat kidney. Nature **324:** 473–476.
8. CHINKERS, M., D. L. GARBERS, M.-S. CHANG, D. G. LOWE, H. CHIN, D. V. GOEDDEL & S. SCHULZ. 1989. A membrane form of guanylate cyclase is an atrial natriuretic peptide receptor. Nature **338:** 78–83.
9. FULLER, F., J. G. PORTER, A. E. ARFSTEN, J. MILLER, J. W. SCHILLING, R. M. SCARBOROUGH, J. A. LEWICKI & D. B. SCHENK. 1988. Atrial natriuretic peptide clearance receptor: Complete sequence and functional expression of cDNA clones. J. Biol. Chem. **263:** 9395–9401.
10. GARCIA, R., G. THIBAULT, J. GUTKOWSKA, P. HAMET, M. CANTIN & J. GENEST. 1985. Effect of chronic infusion of synthetic atrial natriuretic factor (ANF8-33) in conscious two-kidney, one-clip hypertensive rats. Proc. Soc. Exp. Biol. Med. **178:** 155–159.
11. GARCIA, R., G. THIBAULT, J. GUTKOWSKA, K. HORKY, P. HAMET, M. CANTIN & J. GENEST. 1985. Chronic infusion of low doses of atrial natriuretic factor (ANF Arg 101-Tyr 126) reduces blood pressure in conscious SHR without apparent changes in sodium excretion. Proc. Soc. Exp. Biol. Med. **179:** 396–401.
12. GARCIA, R., M. CANTIN, J. GENEST, J. GUTKOWSKA & G. THIBAULT. 1987. Body fluids and plasma atrial peptide after its chronic infusion in hypertensive rats. Proc. Soc. Exp. Biol. Med. **185:** 352–358.
13. DEMEY, J. G., C. CUTHBERT, K. G. VON SZENDROI, G. VAN MALDER & J. ROBA. 1987. Smooth muscle relaxing, acute and long-term blood pressure lowering effects of atriopeptins: Structure-activity relationship. J. Pharmacol. Exp. Ther. **240:** 937–943.
14. KONDO, K., O. KIDA, A. SASAKI, J. KATO & J. TANAKA. 1986. Natriuretic effects of chronically administered human ANF in sodium depleted or repleted conscious SHR rats. Clin. Exp. Pharmacol. Physiol. **13:** 417–424.
15. SPOKAS, E. G., O. D. SULEYMANOV, S. E. BITTNER, J. G. CAMPOIN, R. J. GOREZYNSKI, A. LENAERS & G. M. WALSH. 1987. Cardiovascular effects of chronic high-dose atriopeptin III infusion in normotensive rats. Toxicol. Appl. Pharmacol. **91:** 305–314.
16. HOFBAUER, K. G., L. CRISCIONE, C. SONNENBERG, A. MUIR & S. C. MAH. 1986. Acute and chronic haemodynamic and natriuretic effects of AP II in conscious rats. J. Hypertension 4(suppl.): S41–S47.
17. NAGANO, M. & E. L. BRAVO. 1986. Impaired aldosterone production by chronic intravenous infusion of atrial natriuretic factor in rabbits. J. Hypertension 6(suppl.): S306–S316.
18. PARKES, D. G., J. P. COGHLAN, J. G. McDOUGALL, B. A. SCOGGINS & H. FLOREY. 1988. Long-term hemodynamic actions of atrial natriuretic factor in conscious sheep. Am. J. Physiol. **254:** H811–H815.
19. GRANGER, J. P., T. J. OPGENORTH, J. SALAZAR, J. C. ROMERO & J. C. BURNETT, JR. 1986. Long-term hypotensive and renal effects of atrial natiuretic factor. Hypertension 8(suppl. II): II-112–II-116.
20. COSTA, R. H., E. LAI & J. E. DARNELL, JR. 1986. Transcriptional control of the mouse prealbumin (transthyretin) gene: Both promoter sequences and a distinct enhancer are cell specific. Mol. Cell. Biol. **6:** 4697–4708.
21. McGRANE, M. M., J. DEVENTE, J. YUN, J. BLOOM, E. PARK, A. WYNSHAW-BORIS, T. WAGNER, F. M. ROTTMAN & R. W. HANSON. 1988. Tissue-specific expression and dietary regulation of a chimeric phosphoenolpyruvate carboxykinase/bovine growth hormone gene in transgenic mice. J. Biol. Chem. **263:** 11443–11451.
22. SAMBROOK, J., E. F. FRITSCH & T. MANIATIS. 1989. Molecular Cloning: A Laboratory Manual. 2nd ed. Cold Spring Harbor Laboratory Press. Cold Spring Harbor, NY.

23. STEINHELPER, M. E., K. L. COCHRANE & L. J. FIELD. 1990. Hypotension in transgenic mice expressing atrial natriuretic factor fusion genes. Hypertension **16:** 301–307.
24. STEIF, A., D. M. WINTER, W. H. STRATLING & A. E. SIPPEL. 1989. A DNA attachment element mediates elevated and position-independent gene activity. Nature **341:** 343–345.
25. FIELD, L. J., A. T. VERESS, M. E. STEINHELPER, K. COCHRANE & H. SONNENBERG. 1991. Kidney function in ANF-transgenic mice: Effect of blood volume expansion. Am. J. Physiol. **260:** R1–R5.

A Sugar-Inducible Excretion System for the Production of Recombinant Proteins with *Escherichia coli*

DIOGO ARDAILLON SIMOES, MALENE DAL JENSEN,
ERIC DREVETON, MARIE-ODILE LORET,
SYLVIE BLANCHIN-ROLAND,
JEAN-LOUIS URIBELARREA,
AND JEAN-MICHEL MASSON

I.N.S.A.
Laboratoire de Génie Biochimique et Alimentaire
UA 544 du CNRS
Avenue de Rangueil
F-31077 Toulouse CEDEX, France

The production of recombinant proteins from microorganisms requires that all the parameters be optimized for an industrial process to be designed and scaled up. As recombinant DNA technology brings its own problems as well as original solutions in designing a production process for a given protein, genetic and engineering expertise must be combined from the start to try and set up the most appropriate choices for a specific production. This should take into account the design of the strain as well as the production protocol, nutritional constraints, and downstream processing.

For such an approach to be practical, a sufficiently large set of technological solutions that can be combined adequately into a production process must be on hand. To expand the currently available production technologies with recombinant *Escherichia coli,* we have designed an expression vector that is especially suited for fermentation and allows for excretion of the recombinant protein in the culture medium, greatly simplifying the purification steps.

MATERIALS AND METHODS

Construction of the Vector

Restriction enzymes obtained from Pharmacia were used according to the manufacturer's specifications. All recombinant DNA techniques were as described by Maniatis *et al.*[1] The strain used in these experiments is XAC-1, *ara*Δ(*lac pro*), *gyr* A, *arg* E_{am}, *rpo B, thi*/F'(*lac* I_{373}*lac* Z_{u118am} *pro B*[+]). The fermentation plasmid is derived from pKK233-2 (Pharmacia), pING1 (Ingene), pGFIB1,[2] pINA901,[3] and pCTB113.[4] Site-directed mutagenesis was performed using Eckstein's method[5] to introduce relevant cloning sites when needed.

Minimal Medium

The medium contained (in g/L): KH_2PO_4 4.0, K_2HPO_4 4.0, Na_2HPO_4. $12H_2O$ 3.2, $(NH_4)_2HPO_4$ 3.5, NH_4Cl 0.2, $MnSO_4.H_2O$ 0.010, $CoCl_2.6H_2O$ 0.004, $ZnSO_4.7H_2O$

0.002, $Na_2MoO_4.2H_2O$ 0.002, $CuCl_2.2H_2O$ 0.001, H_3BO_3 0.0005, and $AlCl_3$ 0.001; the pH was adjusted to 6.5 with H_3PO_4. After autoclaving the following salts were added (in g/L): $MgSO_4.7H_2O$ 1.0, $FeSO_4.7H_2O$ 0.04, $CaCl_2.2H_2O$ 0.04, and thiamine 0.005.

Batch Fermentations

Inoculum was grown on 200 ml minimal medium (5 g/L glucose). Fermentations were done at 37°C in 1.7 L. The pH was controlled at 6.5 with a 28% aqueous solution of NH_3. Partial oxygen tension (pO_2) was measured with an oxygen sensor (Ingold). Initial concentration of glycerol was 15 g/L; when all the glycerol was consumed (shown as a fast increase in the pO_2), another injection of 15 g/L was given. The biomass is expressed in grams of dry weight per liter.

α-Amylase Assay

Samples were centrifuged, and extracellular α-amylase was measured directly on the supernatant. The cell pellet was resuspended in the α-amylase assay buffer (0.05 M KH_2PO_4, 1 mM $CaCl_2$, pH 6.0 with NaOH) to an approximate cell concentration of 4 g/L. Cellular α-amylase was measured after gentle agitation of 2 mL of cell suspension with three drops of toluene for 30 minutes at 37°C.

α-amylase was measured according to the method of Wilson and Ingedew[6] with slight modifications. Commercial α-amylase from *Bacillus licheniformis* (Sigma) was used as standard. One unit hydrolyzes 1.0 mg of maltose from starch in 3 minutes at pH 6.9 at 20°C.

RESULTS AND DISCUSSION

The cytoplasmic accumulation of a heterologous gene product in a recombinant strain of *E. coli* is generally recognized to hinder bacterial growth as well as the genetic stability of the population. In addition, further purification of such an intracellular product will require more complicated steps than if it is previously excreted to the culture medium. Excretion of the gene product to the abiotic phase is then the central feature of the expression vector pCTBfer2 to be described. The steps followed in its construction are summarized in FIGURE 1 and its final configuration is shown in FIGURE 2.

Heterologous Protein

The α-amylase of *B. licheniformis* was chosen as a model protein to be constitutively expressed in our system. The cloned gene from pINA901 also contains the natural signal sequence from *B. licheniformis,* and the expressed α-amylase was already shown to accumulate in the periplasmic space of *E. coli.*[4]

Excretion System and Its Control

Although *E. coli* is not generally considered a host for protein excretion, it does naturally secrete a few proteins in its culture medium, including colicins. In the case

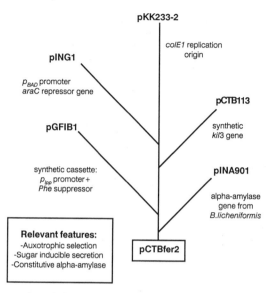

FIGURE 1. Construction of the fermentation vector. The *araC* gene was cloned onto a modified pKK233-2, as a *Bam*HI/*Pvu*II fragment. An *Ava*I cut followed by *Bal*31 digestion shortened the gene by approximately 600 base pairs at its 3′ end. The p_{BAD} promoter was then modified by site-directed mutagenesis and cloned as a *Bam*HI/*Nco*I (*filled in*) fragment. The next step was to introduce the suppressor cassette as a *Pvu*II fragment at the *Pvu*II site of the vector. The *kil3* gene from pCTB113 was cloned as a *Hind*III/*Bam*HI fragment in the *Hind*III/*Bgl*II sites of the new polylinker next to the p_{BAD} promoter. Finally, the α-amylase gene was taken as an *Eco*RV fragment from pINA901 and cloned at the *Hinc*II site of the fermentation vector.

of colicin E1, excretion is mediated by a small peptide that is coded by the *kil* gene on the colicin-producing plasmid and modifies the inner and outer membrane permeability. We have shown that the expression of a synthetic *kil* gene, under the control of the *lac* promoter, efficiently promotes the semispecific release of recombinant periplasmic proteins into the medium.[4] The leakiness of the *lac* promoter, however, implies that the *kil* gene is continuously expressed, albeit at a low level, even in the absence of IPTG. This hinders the control of the excretion of the foreign protein and

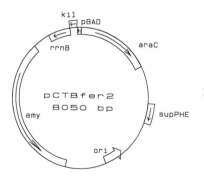

FIGURE 2. The fermentation plasmid pCTBfer2.

may also affect the growth of the host strain, as the *kil* peptide is known to be toxic to *E. coli* cells.

More efficient control is obtained with the *araBAD* promoter. In *E. coli,* expression of the *araB* gene is tightly regulated in a rather sophisticated manner: it is subjected to catabolite repression (through cAMP-CAP interaction) and inducer exclusion by glucose as well as to the specific repression by the *araC* protein.[7] Both the *araBAD* promoter and the *araC* gene were subcloned from the plasmid pING1; several levels of expression (up to a thousandfold the repressed level) can be achieved with this system, if advantage is taken of the regulation phenomena by using different types of carbon sources or their combinations.[8]

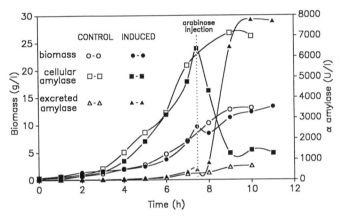

FIGURE 3. Evolution of biomass concentration, excreted α-amylase, and cellular (periplasmic) α-amylase for an induced (*full symbols*) and a control experiment (*open symbols*). All enzymatic activities are expressed in units per liter of culture.

Auxotrophic Selection

For protein excretion to be truly advantageous the composition of the culture medium must be as simple as possible. Auxotrophy suppression by a suppressor tRNA is the method of choice to ensure plasmid stability on minimal medium. This was achieved by cloning the small *Pvu*II fragment of vector pGFIB1 which encompasses a suppressor tRNA expression cassette,[2] harboring a synthetic *tRNA*[Phe] gene.[9] In an *argE*$_{am}$ strain like XAC-1, this small plasmidic insert provides the selective pressure for plasmid maintenance.

Bench Scale Experiments

A preliminary investigation was performed in a 2-L fermentor on minimal medium with glycerol, a substrate that does not elicit catabolite repression or inducer exclusion. The results of an experiment in which induction of the excretion system is achieved by a single injection of arabinose are compared to those of a control, noninduced culture in FIGURE 3. Before induction of the excretion system, the α-amylase has almost completely been accumulated into the cells, no significant activity being found in the supernatant. Excretion of the recombinant protein to the culture medium starts immediately after the injection of 0.15 g/L of arabinose in the induced experiment. It dramatically changes the ratio of excreted to cellular activity of α-amylase. Up to 80% of the total α-amylase activity is excreted at the end of the induced culture, compared to roughly 10% for the control experiment. No major effects on growth were observed, as the dry weight curves of both cultures continued to be parallel after induction. Absence of cell lysis was confirmed in another experiment in which the supernatant activity of a typically cytoplasmic enzyme (β-galactosidase) showed no variation after induction of the excretion system based on the *kil* gene.[10]

CONCLUSION

The expression vector presented herein was designed to meet some of the requirements of an industrial process for recombinant proteins. It does not require that the gene coding for the heterologous protein be modified, provided the protein is expressed in the bacterial periplasm. Plasmid maintenance is ensured with a nonantibiotic selective pressure that takes advantage of the choice of a minimal medium. Induction of excretion is independent of the production of the heterologous protein, thus providing great flexibility in the design of process operation. Very good yields of excretion are achieved, and this advantage, from the downstream processing standpoint, will be fully exploited in high cell density cultures of the recombinant strain[10] to improve even further the supernatant concentration of the heterologous protein.

REFERENCES

1. MANIATIS, T., E. F. FRITSCH & J. SAMBROOK. 1982. Molecular cloning: A laboratory manual. Cold Spring Harbor Laboratory Press. Cold Spring Harbor, NY.
2. MASSON, J. M. & J. H. MILLER. 1986. Expression of synthetic suppressor genes under the control of a synthetic promoter. Gene **47:** 179–183.
3. DECLERCK, N., P. JOYET, C. GAILLARDIN & J. M. MASSON. 1990. Use of amber suppressors to investivate the thermostability of *Bacillus licheniformis* alpha-amylase. Amino acid replacements at six histidine residues reveal the importance of position 133. J. Biol. Chem. **265:** 15481–15488.
4. BLANCHIN-ROLAND, S. & J. M. MASSON. 1989. Protein secretion controlled by a synthetic gene in *Escherichia coli.* Protein Eng. **2:** 473–480.
5. TAYLOR, J. W., J. OTT & F. ECKSTEIN. 1985. The rapid generation of oligonucleotide-directed mutations at high frequency using phosphorothioate-modified DNA. Nucl. Acids Res. **13:** 8765–8785.
6. WILSON, J. J. & W. M. INGEDEW. 1982. Isolation and characterization of *Schwanniomyces alluvius.* Amylolytic enzymes. Appl. Environ. Microbiol. **44:** 301–307.
7. SCHLEIF, R. 1987. The L-arabinose operon. *In Escherichia coli* and *Salmonella typhimurium.* F. C. Niedhardt, ed.: 1473–1481. American Society for Microbiology. Washington, DC.
8. CAGNON, C., V. VALVERDE & J. M. MASSON. 1991. A new family of sugar inducible expression vectors for *Escherichia coli.* Protein Eng. **4:** 843–847.
9. NORMANLY, J., J. M. MASSON, L. G. KLEINA, J. ABELSON & J. H. MILLER. 1986. Construction of two *Escherichia coli* ambre suppressor genes: *tRNA^phe* and *tRNA^cys*. Proc. Natl. Acad. Sci. USA **83:** 6548–6552.
10. Manuscript in preparation.

Study of High Density *Escherichia coli* Fermentation for Production of Porcine Somatotropin Protein

LING-LING CHANG, LIH-YUEH HWANG,
CHIN-FA HWANG, AND DUEN-GANG MOU

Development Center for Biotechnology
Taipei, Taiwan, Republic of China

Porcine somatotropin (PST) is a protein of 191 amino acids that is synthesized in the pituitary gland of swine. Through the application of genetic engineering techniques, the PST gene was cloned into *Escherichia coli* and produced by fermentation.[1-5]

Genetic engineering methods for producing proteins by means of fast-growing microorganisms are already in commercial use. The economical use of recombinant microorganisms that form intracellular products requires a fermentation process that results in both a high intracellular level of product and a high cell concentration in the fermentor. In this case, because of its ability to maintain ideal substrate concentration for optimal growth and production, fed-batch culture often becomes the method of choice.

Different feeding strategies have been proposed for supplying the substrates and nutrients necessary for fermentation. They include strategies based on respiratory quotient,[6] mass balancing,[7] dissolved oxygen concentration,[1,8] or carbon source limitation.[2-5] All of these strategies can be used to control the substrate feeding. In this study, feed rate was manipulated manually or by computer to maintain as low a residual glucose concentration as possible. This was done to avoid the crabtree effect or the accumulation of an inhibitory level of acetic acid. We further demonstrated that the computer control of feed rate could be used in *E. coli* fermentation to facilitate experimental process development and to achieve high cell density and a high level of gene expression.

Fermentation parameters studied in this work included oxygen enrichment, yeast extract (YE) in feed, optimal specific growth rate to initiate PST expression, and feed strategies. To suppress possible protease activity for better expression,[9] the effect of YE on cell density and expression level was studied in great detail.

Because *E. coli* cells of three different genetic constructs were made available to us during the course of this study, observations on the effects of possible host-vector interaction on PST gene product expression and necessary process modification were presented and discussed.

MATERIALS AND METHODS

Strain

The *E. coli* host strains cl-C (temperature-sensitive repressor gene cl on chromosome) and cl-P (temperature-sensitive repressor gene cl on plasmid) in combination with plasmids p_LAR-PGH and pGem-408N-PGH, respectively, were used. In the plasmid p_LAR-PGH and host cl-C construct, plasmid containing ampicillin- or kanamycin-resistant genes were used in building the first two producers. With

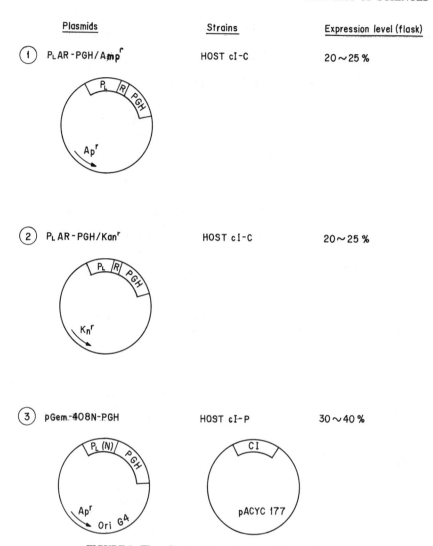

FIGURE 1. Three host-vector systems used in experiments.

plasmid pGem-408N-PGH and host cl-P, ampicillin-resistant construct was used. It was necessary to adjust culture temperature from a growth temperature of 30°C to 42°C to express the temperature-sensitive p_L promoter gene. Please refer to FIGURE 1 for an outline of the three host-vector systems used.

Preculture Procedures

One milliliter of a frozen glycerol stock culture was aseptically transferred into 25 ml of Luria broth medium with 1.25 mg of antibiotic contained in a 250 ml baffled

TABLE 1. Composition of Basal Medium

Component	Concentration
Na_2HPO_4	6 g/L
KH_2PO_4	3 g/L
NaCl	0.5 g/L
Casamino acid	5 g/L
NH_4Cl	1 g/L
$MgSO_4 \cdot 7H_2O$	1M $MgSO_4 \cdot 7H_2O$ 4 ml/L
$CaCl_2$	0.01 M $CaCl_2$ 10 ml/L
Glucose	50% (w/v) glucose 10 ml/L

shaker flask. This culture was grown overnight at 30°C at 200 rpm in an orbital shaker (stroke 70 mm, Hotech Inc. Corp., Taiwan). Two milliliters of this culture were then used to inoculate 200 ml of basal medium (composition shown in TABLE 1) contained in a 500 ml baffled shaker flask. This culture was also grown overnight at 30°C at 200 rpm on the same shaker. The entire culture was used to inoculate 3 L of medium in a 5-L glass jar fermentor. In the fed-batch process, feed began manually or automatically when residual glucose concentration reached zero. This was detected by a digital blood glucose meter. The results of seven runs of host cl-C (p_LAR-PGH/Amp^r) fed-batch fermentation (the feed medium shown in TABLE 2) based on the foregoing method are shown in TABLE 3. They demonstrate good experimental consistency in our fermentation and SDS-page methodology.

Feed Control by Computer

From mass balance, the change of cell mass X, substrate concentration S, and broth volume can be expressed as:

$$\frac{dX}{dt} = \left(\mu - \frac{F}{V}\right) X \tag{1}$$

TABLE 2. Media Used in Fed-Batch Fermentation

Component	Composition	
	Fermentation	Feed
Na_2HPO_4	6 g	18 g
KH_2PO_4	3 g	9 g
NaCl	0.5 g	
Casamino acid	5 g	180 g
NH_4Cl	1 g	36 g
Yeast extract	5 g	
$MgSO_4 \cdot 7H_2O$	1 M $MgSO_4 \cdot 7H_2O$ 4 ml	17.748 g
$CaCl_2$	0.01 M $CaCl_2$ 10 ml	0.4 g
Glucose	50% (w/v) glucose 10 ml	180 g
Antibiotic 5% w/v	1 ml	
Trace metal solution[a]	2 ml	10 ml
Antifoam DL-2000	0.134 ml	
Distilled water	1 L	0.8 L

[a]Fieschko and Ritch.[2]

TABLE 3. Results of PST Fed-Batch Fermentation[a]

Run No.	Run Time (hr)	OD	Expression
1	27	5.73	19%
2	27	5.96	22%
3	29	5.27	21%
4	29	5.24	20%
5	28	6.90	21%
6	28	5.22	19%
7	25	5.26	20%
Average	27.6	5.65	20%
Root mean square error	1.29	0.61	1%

[a]Fermentation medium used in these seven runs was the basal medium.

$$\frac{dS}{dt} = \frac{F}{V}(S_0 - S) - \left(\frac{\mu}{Y_{x/S}} + m\right)X \tag{2}$$

$$\frac{dV}{dt} = F \tag{3}$$

where μ, F, S_0, $Y_{x/S}$, and m are the specific growth rate (hr^{-1}), flow rate of feed medium (L/hr), feed substrate concentration (g/L), yield coefficient (OD-L/g), and maintenance coefficient (g/OD-L-hr), respectively.

The objective of control was to keep broth glucose concentration near zero to avoid the accumulation of toxic levels of organic acids[10] and to achieve high cell density. The desired feed rate can be derived from Equation 2 as:

$$F^* = \frac{\left(m + \frac{\mu}{Y_{x/S}}\right)VX}{S_0 - S} \tag{4}$$

In Equation 4, V is integrated from Eq. 3; μ, X are estimated on-line from exhaust carbon dioxide concentration;[7] and $Y_{x/S}$, m are determined from experimental measurement to be 1 OD-L/g and 0.2 g/OD-L-hr, respectively. Low residual glucose in computer-controlled runs were verified by independent glucose analysis, and its concentrations were consistently below 1 g/L.

Apparatus

The fermentor used was a 5-L Mituwa Model KMJ-5A (Mituwa Rikagaku Kogyo Co. Ltd., Osaka, Japan) glass jar fermentor with dissolved oxygen concentration and temperature controllers. The jar fermentor had a six-bladed disc-turbine impeller and four baffles. A CO_2 gas analyzer model IR-703 D (Infrared Ind. Inc., California, USA) was used to measure the carbon dioxide concentration in exhaust gas, and an IBM compatible PC-XT was used for on-line calculations of cell concentration and instantaneous specific growth rate and on-line control of feed. Temperature was maintained at 30°C during growth and switched to 42°C for gene expression. The pH value was kept at 6.7–7.0 by adding NH_4OH (25%).

Analysis

Culture turbidity was measured at 600 nm using a Beckman DU-62 spectropho-tometer, and glucose concentration was measured using a micro processor digital blood glucose meter (model 5000, Metertech Inc., Taiwan).

For PST assays, cell pellets were diluted to 1 OD unit, centrifuged, and resuspended in 1 ml of the sample buffer (100 mM Tris-HCl, pH 6.8, 2% SDS, 20% 2-mercaptoethanol, 20% glycerol, 4 mM EDTA, and 0.01% bromophenol blue). A portion of the sample (about 10–20 μl) was used for sodium dodecyl sulphate-12.5% polyacrylamide gel electrophoresis (SDS-page). To determine the level of PST expression, the gel was stained with Coomassie blue and scanned by a LKB 2222-020 UltroScan laser densitometer (LKB Produkter AB, Sweden).

RESULTS AND DISCUSSION

Plasmid Stability: Amp^r vs Kan^r

Plasmid retention by host cl-C (p_LAR-PGH/Amp^r) and host cl-C (p_LAR-PGH/ Kan^r) in LB, basal, and basal + 0.5% YE media are shown in FIGURES 2 and 3, respectively. FIGURE 2 shows that the plasmid Amp^r was not stable, especially in the basal medium. FIGURES 2 and 3 demonstrate that the Kan^r plasmid was more stable than was the Amp^r one in host cl-C cells harboring the p_LAR plasmids. Because the mechanism of ampicillin resistance involves an antibiotic-degrading enzyme that is periplasmic in nature, the added drug might be readily degraded upon its addition. Hence, its role of killing the plasmid-negative and preserving the plasmid-plus cells could not be fulfilled.

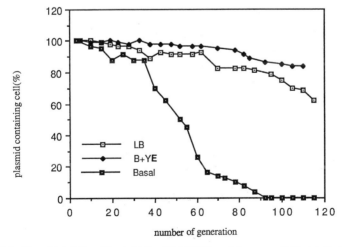

FIGURE 2. Plasmid stability of host cl-C(p_LAR-PGH/Amp^r) in sequential batch flask fermen-tations in different media. Sequential transfers into fresh media were made with 1% seed and every 24 hours. ☐ Luria broth medium, ◆ basal medium + 0.5% yeast extract, ■ basal medium.

Oxygen Enrichment

The fermentation and feed media used in this study are shown in TABLE 2. The feed medium was manually added to the fermentor after 5–6 hours when glucose concentration became depleted. The effects of oxygen supply on the fermentation of host cl-C (p_LAR-PGH/Kan^r) are shown in FIGURE 4. Cell turbidity and PST expression reached, respectively, 20 OD_{600} and 20% when house air was supplied at a rate of 1 L/L-min. Cell turbidity and PST expression reached, respectively, 40 OD_{600} and 15% when dissolved oxygen concentration was controlled at 50% air saturation by using oxygen-enriched air. Hence, oxygen enrichment was later adopted as a standard operating condition.

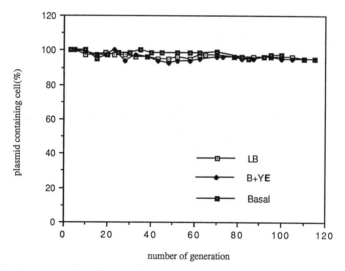

FIGURE 3. Plasmid stability of host cl-C(p_LAR-PGH/Kan^r) in sequential batch flask fermentations in different fermentation media. Sequential transfers into fresh media were made with 1% seed and every 24 hours. □ Luria broth medium, ◆ basal medium + 0.5% yeast extract, ■ basal medium.

Effect of Cell Growth on PST Expression

Experiments testing the effect of cell growth on PST expression by host cl-C(p_LAR-PGH/Amp^r) in fed-batch basal medium fermentations were performed by varying medium strength and timing of heat shock. TABLE 4 gives the experimental results and shows that PST expression is inversely correlated to cell density for host cl-C (p_LAR-PGH/Amp^r).

Effects of Yeast Extract

The house air control experiment was repeated without the 5 g/L YE in the fermentation medium. Results of four duplicated runs using host cl-C with the Amp^r

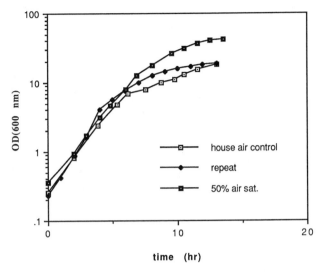

time (hr)

FIGURE 4. Effect of oxygen enrichment on fed-batch fermentation of host cl-C(p_LAR-PGH/Kan^r) using basal medium plus 0.5% yeast extract. Feed began manually at 5–6 hours when glucose was near depletion. Temperature shift occurred roughly at 10–12 hours. □, ◆ are duplicate house air control runs at 1 L/L-min and ■ is the oxygen-enriched run with dissolved oxygen controlled at 50% of air saturation.

plasmid are shown in FIGURE 5. When compared with the experiment with YE (FIG. 4), it was found that the YE increased μ_{max} from 0.15 to 0.57 hr^{-1}. Hence, the growth phase can be shortened significantly with the use of YE.

To reach an even higher culture density and sooner, the effect of supplementary YE feed was studied for the strain host cl-C (p_LAR-PGH/Kan^r) and host cl-P (pGem-408N-PGH) under an oxygen-enriched fed-batch condition. For host cl-C, 30 g of YE were added to 0.8 L of the feed medium (composition shown in TABLE 2), and the results are shown in FIGURE 6. Supplementary YE feed increased cell turbidity from 40 to 50 OD_{600}, but reduced PST expression from 16 to 10%. The decrease in PST expression with increased OD is again consistent with our earlier observation.

Further experiments tested the effect of four different concentrations of YE feed (0, 1.875, 3.75, and 11.25% before heat shock; YE-free feed after heat shock) on the PST fed-batch fermentation using host cl-P. FIGURE 7 shows that high μ could be

TABLE 4. Effect of Cell Density on PST Expression in Fed-Batch Fermentations

Fermentation Medium	Feed Medium	OD before 4 Hours of Heat Shock	OD after 4 Hours of Heat Shock	Inclusion Body $(-,+)^a$	Run Time (hr)
Basal	Basal	2.5	5.7	++++	28
Basal	Basal	4.0	11.1	+++	32
Basal	Basal	5.5	18.7	++	38
Basal	Basal	15.09	24.72	+	35

^{a}Estimated by microscopic observation.

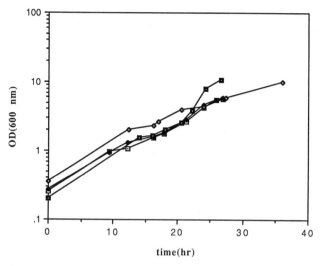

FIGURE 5. Growth curves of host cl-C(p_LAR-PGH/Amp^r) fed-batch fermentation using the basal medium. House air was used for aeration.

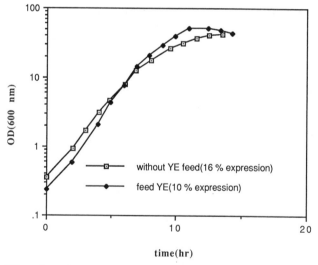

FIGURE 6. Effect of supplementary yeast extract feed on the growth of host cl-C(p_LAR-PGH/Kan^r) in fed-batch fermentation under oxygen enrichment conditions. Feed began manually at 5–6 hours when glucose was near depletion. □: feed without yeast extract; ◆: feed with yeast extract.

maintained longer at higher cell density with YE supplementation exceeding 3.75% in the feed. With time and subsequent heat shock, final cell density in all four runs reached 43–45 OD_{600}. Despite the high OD, FIGURE 8 shows that host cl-P, unlike host cl-C, had PST expression relatively insensitive to YE supplementation and/or high cell density (TABLE 5). All four runs produced a PST expression level of 35–40%. This result has obvious practical significance, but we do not have a good explanation at this point.

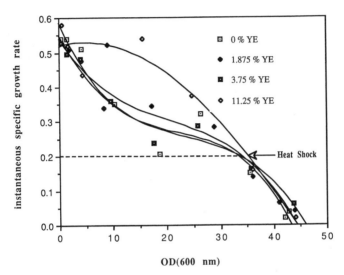

FIGURE 7. Effect of supplementary yeast extract feed (supplementation ended at heat shock) on the specific growth rate of host cl-P (pGem-408N-PGH) at a different cell density under oxygen-enrichment conditions. Feed began manually when residual glucose concentration reached zero at 5–6 hours. Temperature shift started when specific growth rate dropped to 0.2 hr^{-1}. $\square$: 0% yeast extract in feed; $\blacklozenge$: 1.875% yeast extract in feed; $\blacksquare$: 3.75% yeast extract in feed; $\lozenge$: 11.25% yeast extract in feed.

Specific Growth Rate to Switch on PST Expression

Because it was desired that temperature shift be computer controlled, an index was required to enable the computer to determine the right time for heat shock. Here, we selected an instantaneous specific growth rate as the state variable for control action. Fed-batch experiments were carried out using host cl-P and the media in TABLE 2 under oxygen-enrichment conditions, and heat shock (42°C) initiated each at instantaneous specific growth rates of 0.4, 0.2, and 0.1 hr^{-1}. Results in FIGURES 9 and 10 indicate that heat shock at the highest instantaneous specific growth rate (0.4 hr^{-1}) produced the highest PST expression (40%), but the lowest cell density (20 OD_{600}). However, at a modest instantaneous specific growth rate of 0.2 $hour^{-1}$, cell turbidity was more than doubled to an OD_{600} of 43, whereas PST expression reduced by roughly one third to 28%. The foregoing results indicate that heat shock at an instantaneous specific growth rate of roughly one third the μ_{max} can produce the maximum amount of PST per liter of broth for our cl-P host/vector

1 2 3 4 5 6 7 8 9

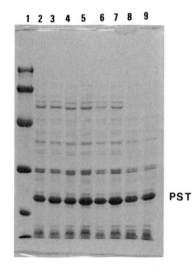

FIGURE 8. Effect of different levels of yeast extract (YE) supplementation on PST expression in fed-batch fermentation of host cl-P(pGem-408N-PGH) as shown by SDS-page. *Lane 1:* MW markers; *lane 2:* 0% YE 2 hours after heat shock, 28% expression; *lane 3:* 0% YE 3 hours after heat shock, 36% expression; *lane 4:* 1.875% YE 2 hours after heat shock, 30% expression; *lane 5:* 1.875% YE 3 hours after heat shock, 35% expression; *lane 6:* 3.75% YE 2 hours after heat shock, 25% expression; *lane 7:* 3.75% YE 3 hours after heat shock, 38% expression; *lane 8:* 11.25% YE 2 hours after heat shock, 32% expression; *lane 9:* 11.25% YE 3 hours after heat shock, 37% expression.

PST

system. Hence, the instantaneous specific growth rate of 0.2 hr^{-1} was later adopted as a control index for the temperature shift toward gene expression.

Computer Control

With process and control parameters properly identified, two computer-controlled experiments using host cl-P under an oxygen-enriched fed-batch condition were carried out. In one, no YE was added to the feed medium. The results are shown in FIGURES 11 and 12. In the other, 11.25% YE was added to the feed medium. The results are shown in FIGURES 13 and 14. The on-line computer-calculated cell density and instantaneous specific growth rate were determined by using real-time exhaust carbon dioxide concentration data. The supplementary feed rate was then controlled by computer to maintain a near zero residual glucose concentration using on-line mass balancing using Eq. 4. When the instantaneous specific growth rate dropped to 0.2 hr^{-1}, the temperature set point was manually shifted to 42°C for PST expression. Computer-controlled feed continued to regulate near zero residual glucose concentration. FIGURES 11 and 13 show that the on-line estimated OD of growth was comparable to that of the off-line measured ones. Cell turbidity and PST expression reached, respectively, 40 OD$_{600}$ and 30% when YE was not added to the feed. They reached, respectively, 55 OD$_{600}$ and 35%, when 11.25% YE was added to the feed.

TABLE 5. Effect of Cell Density, YE Supplementation on PST Expression from Host cl-C and Host cl-P in Fed-Batch Fermentations

Strain Condition	Host cl-C		Host cl-P	
	OD$_{600}$	Expression	OD$_{600}$	Expression
House air control	20	20%		
Oxygen enrichment	30	15%	40	30%–40%
YE added in feed	40	10%	55	30%–40%

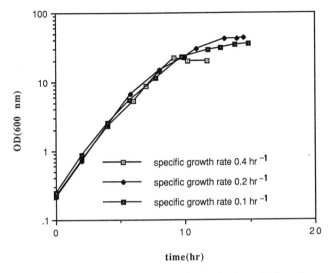

FIGURE 9. Effect of different specific growth rates in starting heat shock on the growth curves of host cl-P(pGem-408N-PGH) in oxygen-enriched fed-batch fermentations. Feed began manually at 5–6 hours when glucose was near depletion. □: 0.4 hr^{-1}; ◆: 0.2 hr^{-1}; ■: 0.1 hr^{-1}.

The foregoing results show that the addition of YE to the feed after heat shock can further increase the final culture cell density, and the goal of using computer control to achieve high cell density together with a high PST expression level in a recombinant *E. coli* fermentation was met.

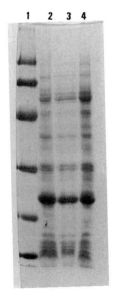

FIGURE 10. Effect of different specific growth rates in starting heat shock on PST expression (2 hours after heat shock) in fed-batch fermentation of host cl-P(pGem-408N-PGH) as shown by SDS-page. *Lane 1:* MW markers; *lane 2:* 0.4 hr^{-1}, 40% expression; *lane 3:* 0.2 hr^{-1}, 28% expression; *lane 4:* 0.1 hr^{-1}, 30% expression.

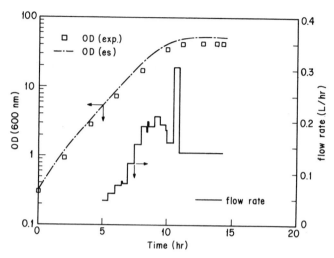

FIGURE 11. Computer-controlled growth of host cl-P pGem-408N-PGH) in fed-batch fermentation without yeast extract supplementation. □: measured OD; —··—: computer-calculated OD; —: feed rate by computer.

SUMMARY

Recombinant *E. coli* strains and culture conditions were studied for the fermentation expression of porcine somatotropin (PST) inclusion bodies under the control of a p_L promoter. Our objective was to achieve high cell density together with a high level of recombinant protein expression. Improved fermentation conditions included oxygen enrichment, yeast extract (YE) effect, optimal specific growth to switch on

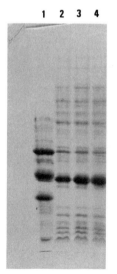

FIGURE 12. PST expression in computer-controlled experiment from FIGURE 11 (without YE feed) as shown by SDS-page. *Lane 1:* MW markers; *lane 2:* 23% 2 hours after heat shock; *lane 3:* 26% 3 hours after heat shock; *lane 4:* 30% 3.5 hours after heat shock.

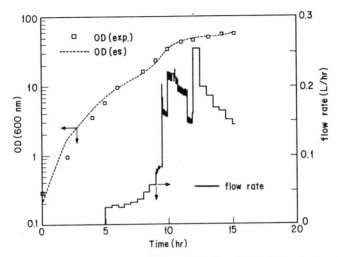

FIGURE 13. Computer-controlled growth of host cl-P (pGem-408N-PGH) in fed-batch fermentation with yeast extract (11.25%) in feed supplementation throughout the run. □: measured OD; ······: computer-calculated OD; —: feed rate by computer.

gene expression, and feeding strategies. To maintain a low residual glucose concentration, a medium feed rate was controlled on a real-time basis by using cell density information estimated from on-line carbon dioxide monitoring of a fermentor's exhaust gas. The optimal specific growth rate required to initiate a temperature shift in our system was found to be around 0.2 hr^{-1}. The cell density and PST expression level could reach 55 OD$_{600}$ and 35%, respectively, after 16 hours of cultivation under optimal conditions by applying computer-controlled nutrient feed. In our recombi-

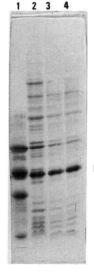

FIGURE 14. PST expression in computer-controlled experiment from FIGURE 13 (with YE feed) as shown by SDS-page. *Lane 1:* MW markers; *lane 2:* 22% 1 hour after heat shock; *lane 3:* 30% 2 hours after heat shock; *lane 4:* 35% 3 hours after heat shock.

nant host/vector system, the location of cl gene appears to affect gene expression under YE-supplemented and/or a high cell density culture condition. With cl gene placed on plasmid, our *E. coli* host no longer showed sensitivity toward YE in PST gene expression.

ACKNOWLEDGMENTS

We thank Dr. Lih-Hwa Hwang of our Molecular Biology Division for contructing the strains used in this study.

REFERENCES

1. CUTAYAR, J. M. & D. POILLON. 1989. High cell density culture of *E. coli* in fed-batch system with dissolved oxygen as substrate feed indicator. Biotechnol. Lett. **11:** 155–160.
2. FIESCHKO, J. & T. RITCH. 1986. Production of human alpha consensus interferon in recombinant *Escherichia coli*. Chem. Eng. Commun. **45:** 229–240.
3. ZABRISKIE, D. W., D. A. WAREHEIM & M. J. POLANKY. 1987. Effects of fermentation feeding strategies prior to induction of expression of a recombinant malaria antigen in *Escherichia coli*. J. Ind. Microb. **2:** 87–95.
4. RINAS, U., H. ANDREAS, K. HELM & K. SCHIIGERL. 1989. Glucose as a substrate in recombinant strain fermentation technology. Appl. Microb. Biotechnol. **31:** 163–167.
5. CALCOTT, P. H., J. F. KANE, G. G. KRICI & G. BOGOSIAN. 1988. Parameters affecting production of bovine somatotropin in *Escherichia coli* fermentation. Dev. Ind. Microb. **29:** 257–266.
6. WANG, H. Y., C. L. COONEY & D. I. C. WANG. 1977. Computer-aided baker's yeast fermentation. Biotechnol. Bioeng. **19:** 69–81.
7. MOU, D. G. & C. L. COONEY. 1983. Growth monitoring and control in complex medium: A case study employing fed-batch penicillin fermentation and computer-aided on-line mass balancing. Biotechnol. Bioeng. **25:** 257–269.
8. YANO, T., T. KOBAYASHI & S. SHIMIZU. 1985. High concentration cultivation of *Candida brassicae* in a fed-batch system. J. Ferment. Technol. **63:** 415–418.
9. TSAI, L. B., M. MANN, F. MORRIS, C. ROTGERS & C. FENTON. 1987. The effect of organic nitrogen and glucose on the production of recombinant human insulin-like growth factor in high cell density *Escherichia coli* fermentation. J. Ind. Microbiol. **2:** 181–187.
10. MACDONALD, H. L. & J. O. NEWAY. 1990. Effects of medium quality on the expression of human interleukin-2 at high cell density in fermentor cultures of *Escherichia coli* K-12. Appl. Environ. Microb. **56:** 640–645.

Optimization of Continuous Two-Stage Fermentations of Recombinant Bacteria

Modeling and Experimental Analysis

FUDU MIAO AND DHINAKAR S. KOMPALA

Department of Chemical Engineering
University of Colorado
Boulder, Colorado 80309–0424

The optimal expression of foreign DNA is the common objective in all recombinant bacterial fermentations. This objective may be accomplished through two complementary approaches, namely, genetic engineering and optimal induction. The genetic engineering approach is mainly aimed at designing an efficiently expressed vehicle and a proper host organism. The expression plasmid can be elaborately constructed at the molecular level to enhance transcription and translation. A promoter gene is usually cloned into the plasmid to speed the initiating of the transcription of the foreign DNA to the corresponding messenger RNA. The initiation efficiency depends on the promoter strength, which is related to the feature of Pribnow boxes. The mutation or deletion of the nucleotide sequence spacing from -35 to -10 should be avoided because it weakens the binding strength of the RNA polymerase to the promoter.[1] The translational efficiency can be facilitated by strengthening the ribosome binding site (RBS) or the Shine-Dalgarno sequence. This purine-rich 6–8 nucleotide sequence should be centered on the RBS. Any spacing variation will dramatically reduce the ribosome binding strength, thereby lowering the translational efficiency.[2]

The recombinant protein production rates will also be increased by a greater concentration of genes. The rate-determining step is usually the transcription of the mRNA, which is roughly proportional to gene content. Therefore, one way to increase the foreign protein formation rate is to choose a fast-replicating plasmid to maintain high gene copies per cell. The most common method of inducing high plasmid copy number is the use of a plasmid containing a temperature-sensitive promoter, such as λP_R.[3] Elevating culture temperature deactivates the *cI*857 repressor; thus, the promoter initiates the replication events at the *ori,* which then amplifies plasmid copy number. However, when the plasmid copy number is high enough, the rate-determining step for foreign protein synthesis will no longer be dominated by plasmid copy number. Therefore, the foreign protein production rate does not always increase in proportion to the increase in plasmid content.

In addition to the genetic approaches, induction strategy is also ultimately important for optimal gene expression. For a plasmid carrying an efficient controllable promoter, inducing foreign protein synthesis will change the cell growth kinetics. Overexpression of the heterogolous protein can greatly reduce the growth rate of the recombinant cells. In such a case, outgrowing of the plasmid-free cells to the plasmid-bearing cells will cause plasmid instability. An appropriate induction strategy should therefore be employed to achieve a high-level gene product of interest.

This paper analyzes the optimal induction strategies in both batch and two-stage continuous fermentations of highly expressed recombinant *Escherichia coli.* The plasmid used in this study contains a strong T7 promoter; a large amount of foreign protein is expected to be produced when the recombinant cells are induced.[4] For batch cultures, proper selection of the induction time and inducer concentration can improve plasmid stability. For continuous cultures, however, growth rate reduction of the recombinant cells will cause severe plasmid instability; hence, the two-stage fermentation strategy is used. The present study briefly elucidates the optimal induction strategy for this highly efficient expression system with the aid of a structured kinetic model.

METABOLIC MODEL

The structured kinetic model, developed by Bentley and Kompala,[5] for analyzing recombinant cell metabolism was employed in the present study. This model divides the internal composition of the recombinant cells into eight constituent pools. The rate equations for these components are described using the saturation terms and the maximum synthesis terms, which account for the mass flow from the substrate through the intermediates to the macromolecular products, and the information flow for directing product synthesis. The unique attribute of this model framework is that the instantaneous specific growth rates of both the recombinant cells and the plasmid-free cells are directly calculated by the simple addition of the synthesis rate expressions of all constituents. This model contains only two adjustable parameters, μ_4 and μ_5, that govern the maximum rates of foreign protein synthesis and plasmid replication, respectively. The value of μ_5 only depends on the plasmid/host system; therefore, it usually does not vary with operating conditions, except in plasmid amplification. The value of μ_4 represents the induction strength associated with both the transcriptional and the translational efficiency. Therefore, it increases with an increase in inducer concentration until saturation. The growth rate reduction resulting from foreign protein overexpression can be simulated by varying the value of μ_4.

HOST/PLASMID EXPRESSION SYSTEM

Both the plasmid and the host were kindly provided by Green and Gold (MCD Biology Department, University of Colorado at Boulder). The plasmid shown in FIGURE 1 was transformed into the host BL21(DE3)[4] using the calcium chloride method described by Maniatis et al.[6] This recombinant cell can continuously secrete β-lactamase to degrade the antibiotic ampicillin used as selective pressure. Therefore, it is difficult and also expensive to stabilize the plasmid by adding antibiotics.

The plasmid was the derivative of pBC26 containing the origins of replication of both ColE1 and phage f1.[7] A T7 promoter was cloned into the plasmid to control the target gene expression. This phage promoter is recognized exclusively by T7 RNA polymerase rather than bacterial RNA polymerases which are used in most other host/vector systems. The native bacterial chromosome usually does not contain the T7 RNA polymerase gene, so that the T7 promoter is dormant in these microorganisms. The method used in this system to deliver the T7 RNA polymerase is cloning its gene into the bacterial chromosome. The T7 RNA polymerase produced by the host will direct the T7 promoter to transcribe the DNA. To control the expression of this gene, a commonly used bacterial promoter, lacUV5, was also cloned into the bacterial chromosome. This promoter can easily be induced by a chemical inducer, such as isopropyl-β-D-thiogalactopyranoside (IPTG). The inducer turns on the synthesis of the T7 RNA polymerase, which in turn initiates the expression of the product gene on the plasmid. Although the *lac*UV5 promoter is relatively not strong, a low level of the T7 RNA polymerase is enough to drive the functioning of the strong T7 promoter. In this way, the target gene expression is delicately regulated by the inducer.

To enhance the translational efficiency of mRNA, a single-stranded DNA sequence was cloned at the 5′ end of the *lac*Z gene and right downstream of the T7 promoter gene. The effect of the positioning of the Shine-Dalgarno sequence on the translation was studied in detail by Gold *et al.*[2,7] The plasmid used in this study contains an optimal spacing of the Shine-Dalgarno sequence for maximizing the translational efficiency.

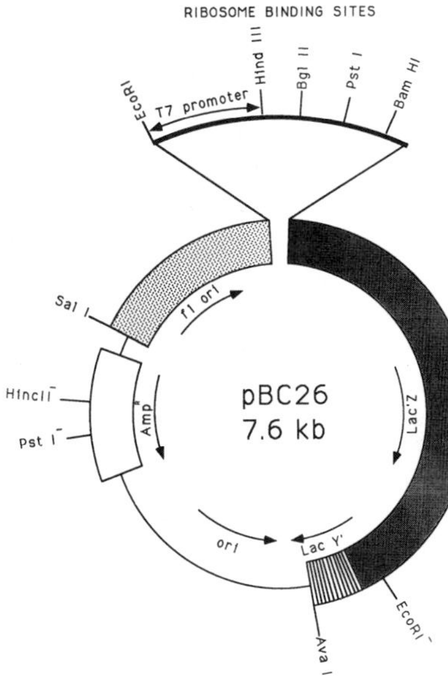

FIGURE 1. Plasmid used in this study. The plasmid contains a T7 promoter to control gene expression. A ribosome binding site was also cloned into the notch between *Bgl* II and *Pst* I sites of pBC26.

METABOLIC BURDEN

The variations in cell growth kinetics resulting from the overexpression of heterologous protein are shown in FIGURE 2. In this batch experiment, the inducer was added at the onset of the culture, and 100 μg/ml ampicillin was introduced to each shaker flask every 2 hours to suppress the occurrence of plasmid-free cells. The specific growth rates of the recombinant cells at different inducer levels can be calculated from the slopes of the curves in this semilogarithmic plot. It is clearly indicated that cell growth rates greatly differ after induction. However, the distinct decline in growth rate occurs only a couple of hours after introduction of the inducer, being highly dependent on inducer concentration. This gradual induction may be caused by the active transport of the inducer, which has been studied by others,[8] and the transport kinetics has been described in a lumped form by Ray *et al.*[9] Another likely reason is that the inducer transported from the bulk solution needs to react with the repressor which has accumulated to a certain level, and then the T7 RNA polymerase will be produced to induce the T7 promoter to synthesize the foreign protein. These processes take some time and delay the induction response. Consequently, the use of a higher level of inducer results in a faster response in growth rate reduction because of higher mass transport or reaction rates. At inducer levels higher than 0.05 g/L, the growth rate pattern does not change further (data not shown here), probably because the transport capacity of the inducer is limited by the carrier (transferase) embedded in the plasma membrane. It is interesting to observe from FIGURE 2 that at the high bulk inducer concentration (0.05 g/L), the cell growth rate recovers to a lower level after passing the declining period. This surprising phenomenon has been predicted by the metabolic model, and the model simulation results are presented in FIGURE 3,[10] which shows that the growth rate shock becomes more apparent with higher induction strength.

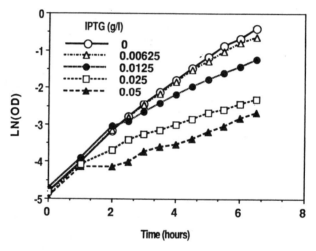

FIGURE 2. Effect of induction on growth rates of recombinant *E. coli.* Cells were induced at the beginning of the cultures, and 100 μg/ml of ampicillin was added to each shaker flask culture to kill the plasmid-free cells. This figure illustrates the dynamic reduction in cell growth rates caused by induction.

OPTIMAL INDUCTION OF BATCH CULTURES

As discussed earlier, induction of foreign protein synthesis will dramatically decrease growth rates for an efficient expression system. Hence, the induction strategy will be of importance for maximizing productivity in batch cultures. This concept is well demonstrated in FIGURE 4, which shows the dependence of the *final* β-galactosidase levels on inducer concentration and induction time, defined as the time from the culture inoculation to the addition of the inducer. An optimal induction time exists for each inducer level. Moreover, the optimal induction time becomes longer with increasing inducer concentration. At high inducer levels, earlier induction yielded less gene product, because the culture was overtaken by the

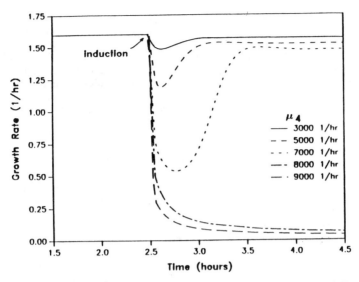

FIGURE 3. Simulation of induced batch culture, reprinted from Bentley and Kompala.[10] Simulation was performed by varying the induction strength, μ_4, which is related to inducer concentration.

plasmid-free cells, whereas the low yield was also observed if the induction was too late simply because not enough nutrient was left for the recombinant cells to synthesize foreign protein. Therefore, induction should be placed at an intermediate exponential phase.[13] At low inducer concentration, however, induction should be as early as possible, because plasmid instability was not a serious problem in this case.

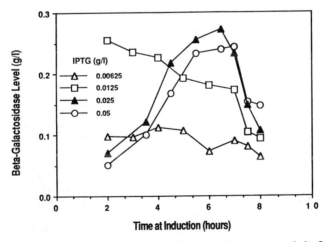

FIGURE 4. Optimal induction in batch cultures. Each symbol represents a shake flask culture. Final β-galactosidase levels were measured at the harvesting time of 15 hours. Cultures were induced at the different inducer concentrations and different induction times.

Earlier induction would prolong the time for foreign protein synthesis. But if the inducer level was too low (e.g., 0.00625 g/L), the final yields were always low whether the induction was early or late, because of the slow foreign protein synthesis rates. This optimal induction behavior was predicted by the metabolic burden model.[10] The model suggests that strong induction greatly reduces cell growth rates and therefore may not produce high levels of product, even with stable plasmid, unless the culture time is long enough. In the present case, the low protein level is due not only to the slow growth rates of the recombinant cells, but also to the fast-growing plasmid-free cells, consuming nutrients.

TWO-STAGE FERMENTATION MODEL

Two-stage continuous fermentations are conducted by operating two fermentors in series, as shown in FIGURE 5. The first stage is dedicated to cell growth only, with the gene expression *un*induced, and the produced recombinant cells are continuously pumped into the second stage, where foreign gene expression is induced.[11] When the plasmid-bearing cells enter the second stage, they are exposed to a

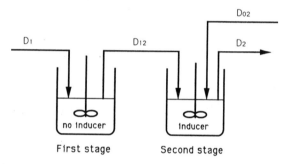

FIGURE 5. Schematic of two-stage continuous culture. The first stage is for cell growth and the second stage is for gene expression.

different nutrient environment containing the inducer and experience a dynamic change in their metabolism. As the intracellular inducer accumulates, the synthesis rate of foreign protein goes up, whereas the cell growth rate slows down until full induction. Because the cells in the second stage have different induction times corresponding to their residence time distribution, they exhibit different growth rates and different foreign protein synthesis rates. The observed productivity of the foreign protein and the cell growth rates are contributed virtually by the various cells with different residence times. Therefore, both the residence times of the plasmid-bearing cells and the dynamics of cell growth and protein formation should be considered in modeling the two-stage fermentations. These dynamic variations may play a dominant role in plasmid stability and overall foreign protein productivity.

The mathematical model for describing cell growth and substrate consumption, based on these characteristics in two-stage cultures, is developed in detail elsewhere.[12] For the first stage (denoted by subscript 1), the material balance equations are:

$$\frac{dX_1^+}{dt} = (1 - \nu_1)\mu_1^+ X_1^+ - D_1 X_1^+ \tag{1}$$

$$\frac{dX_1^-}{dt} = \mu_1^- X_1^- + \nu_1 \mu_1^+ X_1^+ - D_1 X_1^- \tag{2}$$

$$\frac{dS_1}{dt} = D_1(S_0 - S_1) - \frac{1}{Y_{x/s}}(\mu_1^+ X_1^+ + \mu_1^- X_1^-) \tag{3}$$

where μ_1^+ and μ_1^- are the specific growth rates of the plasmid-containing and plasmid-free cells, respectively, and X_1^+ and X_1^- are the corresponding cell mass concentrations. S_0 and S are the limiting nutrient concentrations in the feed and the first stage reactor. D_1 is the dilution rate of the first stage. ν_1 is the segregation coefficient from the plasmid-bearing cells to the plasmid-free cells. $Y_{x/s}$ is the yield coefficient from the nutrient to the cell mass.

For the second stage (denoted by subscript 2), the rate equations are expressed in terms of the cells with different residence times:

$$\frac{dX_2^+}{dt} = D_{12}X_1^+ - D_2 X_2^+ + (1 - \nu_2)D_{12} \int_0^t X_1^+ \hat{\mu}_2^+(\alpha) e^{\int_0^\alpha [\hat{\mu}_2^+(\alpha)(1-\nu_2) - D_2]d\alpha} \, d\alpha \tag{4}$$

$$\frac{dX_2^-}{dt} = D_{12}X_1^- - D_2 X_2^- + \mu_2^- X_2^- + \nu_2 D_{12} \int_0^t X_1^+ \hat{\mu}_2^+(\alpha) e^{\int_0^\alpha [\hat{\mu}_2^+(\alpha)(1-\nu_2) - D_2]d\alpha} \, d\alpha \tag{5}$$

$$\frac{dS_2}{dt} = D_{12}S_1 + D_{02}S_0 - D_2 S_2$$

$$- \frac{1}{Y_{x/s}} D_{12} \int_0^t X_1^+ \hat{\mu}_2^+(\alpha) e^{\int_0^\alpha [\hat{\mu}_2^+(\alpha)(1-\nu_2) - D_2]d\alpha} \, d\alpha - \frac{1}{Y_{x/s}} \mu_2^- X_2^- \tag{6}$$

Product formation rate can be expressed as:

$$\frac{dP_{f2}}{dt} = \frac{D_{12}X_1^+ P_{f1}}{X_2^+} - \frac{P_{f2}}{X_2^+}\frac{dX_2^+}{dt} - D_2 P_{f2}$$

$$+ \frac{(1 - \nu_2)D_{12}}{X_2^+} \int_0^t X_1^+ \hat{\mu}_2^+(\alpha) \hat{P}_{f2}(\alpha) e^{\int_0^\alpha [\hat{\mu}_2^+(\alpha)(1-\nu_2) - D_2]d\alpha} \, d\alpha \tag{7}$$

where $\hat{P}_{f2}(\alpha)$ is the mass fraction of the foreign protein in the plasmid-bearing cells with residence time α, and P_{f2} is the average mass fraction. $\hat{\mu}_2^+(\alpha)$ is the specific growth rate of the recombinant cells with the residence time α. S_2 is the limiting substrate concentration in the second stage. D_{12} and D_{02} are dilution rates from the first stage to the second stage and the nutrient supply for the second stage, respectively, and D_2 is obviously equal to D_{12} plus D_{02}. The integral terms in Equations 4–7 represent the contribution of the various cells with different residence times in the second stage. The overall productivity of the two-stage fermentation, P, is defined by the equation:

$$P = \frac{D_2 X_2^+ P_{f2}}{1 + \beta} \tag{8}$$

where β is the volume ratio of the first bioreactor to the second.

OPTIMAL INDUCTION OF TWO-STAGE CULTURES

For two-stage continuous cultures, dilution rates and inducer levels are two key factors influencing foreign protein productivity. The effect of inducer concentration on productivity was investigated by maintaining all dilution rates constant. Two sets of experiments were conducted, one with a lower dilution rate and the other higher. The plasmid was maintained stably in the first stage throughout the duration of fermentation by choosing appropriate dilution rates. The productivities of the foreign protein were calculated from the data measured at 50 hours of fermentation, which could be considered a pseudo-steady state, because the protein levels were almost constant. The experimental results are presented in FIGURE 6. The inducer concentration was varied from 0.00625 g/L to 0.05 g/L, because both cell growth rates and foreign protein production rates change significantly over this range of inducer concentration in batch cultures (FIG. 3). It is very surprising to note, however, that inducer concentration does not significantly affect foreign protein production in two-stage continuous cultures as it does in batch cultures.

The productivity is almost constant at the lower dilution rate ($D_2 = 0.2 \ hr^{-1}$) for all inducer concentrations used. Although the cultures at very low inducer levels (less than 0.00625 g/L) are not investigated here, we can predict that the productivity will be lower because they approach a noninduction condition, whereas at the higher dilution rate ($D_2 = 0.5 \ hr^{-1}$), the data show an optimal inducer concentration that maximizes productivity. However, the optimal behavior is not as distinct as predicted by the metabolic model, which is shown in FIGURE 7. The reason is that the growth rates of the recombinant cells gradually decrease rather than recover to a new level, after entering the second stage. At high dilution rates, the average residence time of the cells is short, so that most of the recombinant cells are virtually growing within the period of dynamic reduction. Thus, the actual growth rates of the recombinant cells are not low. Another reason in this T7 expression system is that the growth rates

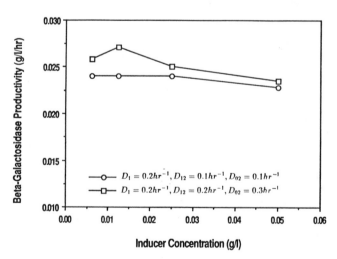

FIGURE 6. Effect of inducer concentration on foreign protein productivity in two-stage fermentations. Productivity, P, was calculated from experimental measurements at the pseudo-steady state condition using Equation 8.

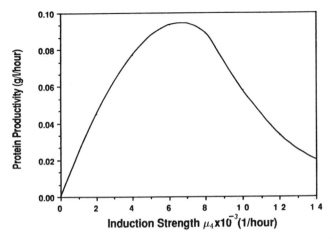

FIGURE 7. Effect of induction strength on foreign protein productivity. Simulation was performed by using Bentley's structured kinetic model and the material balance equations in this paper. Here, μ_4 is equal to zero in the first stage and varies in the second stage. Productivity for the second stage was calculated from the equation: $P = D_2 X_2^+ P_{f2}$. Dilution rates were: $D_1 = 0.3 \text{ hr}^{-1}$, $D_{12} = 0.2 \text{ hr}^{-1}$, and $D_{02} = 0.3 \text{ hr}^{-1}$.

of the plasmid-free cells also slightly decrease in the presence of the inducer.[13] This reduction decreases the growth rate difference between the two types of cells. At lower dilution rates, the growth rate difference is smaller because of the lower cell growth rates. The fraction of plasmid-bearing cells will not be very low despite the long average cell residence time. Therefore, plasmid stability has been compensated in both sets of experimental conditions. However, the metabolic model assumes that the recombinant cells are suddenly induced, so that it predicts that induction strength can more strongly affect foreign protein productivity.

The effect of dilution rate, D_2, on foreign protein productivity is illustrated in FIGURE 8. In this experiment, dilution rates D_1 and D_{12} and inducer concentration are kept constant while varying D_{02}. Maximum productivity is obtained for D_2 at about 0.4 h^{-1} ($D_{02} = 0.2 \text{ hr}^{-1}$). We can predict that the productivity will be low at a very low D_{02} due to the lack of the nutrient. This phenomenon suggests that high dilution rates would not produce high productivity but rather waste the nutrient. At high D_2 (when D_{12} is fixed), the lower productivity is due to the lower fraction of plasmid-bearing cells, lower plasmid copy number, and lower cell density. As the dilution rate is decreased, higher productivity results because of higher copy number, cell density, and plasmid stability. But a too low dilution rate will also lead to lower productivity, because there is not enough nutrient for recombinant cells to produce foreign protein. We also observed that productivity of two-stage fermentation is higher than that of batch cultures. In batch culture, the maximum average productivity (final protein level divided by culture time) is only 0.017 g/L per hour, whereas it is about 0.028 g/L per hour in the two-stage continuous cultures. Therefore, it may be advisable to use two-stage fermentation to achieve a higher foreign protein productivity with this recombinant bacterial system.

CONCLUSIONS

Production of a large amount of heterologous protein in recombinant *E. coli* will significantly reduce cell growth rate and cause plasmid instability. Hence, the recombinant cells should be optimally induced to maximize the cloned-gene product. In batch cultures, induction is optimal in mid-exponential phase at high inducer concentrations, whereas earlier induction will produce higher levels of the foreign protein at low inducer concentration. In two-stage continuous fermentations, the recombinant cells experience a dynamic reduction process in growth rate after entering the second stage. Therefore, both the growth rate dynamics and cell residence time distribution have to be considered in evaluating fermentation performance. The experimental results show an optimal inducer concentration and an optimal dilution rate, which are consistent with the model predictions. However, the optimal inducer concentration is less distinct than that of the model predicted,

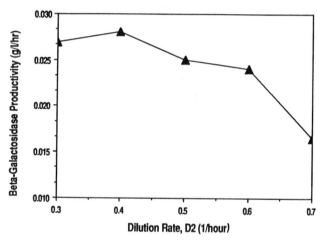

FIGURE 8. Effect of dilution rate on foreign protein productivity. The inducer concentration in the second stage was kept at 0.025 g/L. Dilution rates were: $D_1 = 0.2 \text{ hr}^{-1}$ and $D_{12} = 0.2 \text{ hr}^{-1}$.

because the recombinant cells cannot suddenly be induced by a chemical inducer. Still there is a clear advantage for two-stage continuous fermentations over batch cultures, because the uninduced cell growth in the first stage improves the plasmid stability in the induced second stage.

Nomenclature

D   Dilution rate (hr^{-1})
P   Productivity of foreign protein (g/L/hr)
P_f Mass fraction of foreign protein (g protein/g cell mass)
$\hat{P}_f(\alpha)$ Mass fraction of foreign protein with residence time α (g protein/g cell mass)
S   Substrate concentration (g/l)
t   Time (hr)

X   Cell mass concentration (g/l)
$Y_{x/s}$ Yield coefficient (g cell/g glucose)
α   Residence time in the second stage (hr)
μ   Specific cell growth rate (hr^{-1})
$\hat{\mu}(\alpha)$ Specific cell growth rate with residence time α (hr^{-1})
ν   Segregation coefficient

Subscripts

0   Feed
02   New feed to second stage
1   First stage
12   From first stage to second stage
2   Second stage

Superscripts

\+   Plasmid-bearing cells
−   Plasmid-free cells

REFERENCES

1. BROSIUS, J., M. ERFLE & J. STRORELLA. 1985. Spacing of the −10 and −35 regions in the tac promoter. J. Biol. Chem. **200**: 3539–3541.
2. GOLD, L. & G. D. STORMO. 1990. High-level translation initiation. Methods Enzymol. **185**: 89–94.
3. BETENBAUGH, M. J., C. BEATY & P. DHURJATI. 1988. Effects of plasmid amplification and recombinant gene expression on the growth kinetics of recombinant *E. coli*. Biotechnol. Bioeng. **33**: 1425–1436.
4. STUDIER, F. W. & B. A. MOFFATT. 1986. Use of bacteriophage T7 RNA polymerase to direct selective high level expression of cloned genes. *J. Mol. Biol. 189: 113–130.*
5. BENTLEY, W. E. & D. S. KOMPALA. 1987. A novel structured kinetic modeling approach for the analysis of plasmid instability in recombinant bacterial cultures. Biotechnol. Bioeng. **33**: 49–61.
6. MANIATIS, T., E. F. FRITSCH & J. SAMBROOK. 1982. Molecular Cloning.: 250–251. Cold Spring Harbor Laboratory Press. Cold Spring Harbor, New York.
7. CHILDS, J., K. VILLANUEBA, D. BARRICK, T. D. SCHNEIDER, G. D. STORMO, L. GOLD, M. LEITHER & M. CARUTHERS. 1985. Ribosome binding site sequences and function. UCLA Symposia on Molecular and Cellular Biology. Vol. 30: 341–350. Alan R. Liss, Inc. New York.
8. HENGGE, R. & W. BOOS. 1983. Maltose and lactose transport in *Escherichia coli*. Biochim. Biophys. Acta **737**: 443–478.
9. RAY, N. G., W. R. VIETH & K. VENKATASUBRAMANIATIAN. 1986. Regulation of *lac* operon expression in mixed sugar chemostat cultures. Biotechnol. Bioeng. **29**: 1003–1014.
10. BENTLEY, W. E. & D. S. KOMPALA. 1990. Optimal induction of foreign protein expression in recombinant bacterial cultures. Ann. N.Y. Acad. Sci. **589**: 121–138.
11. LEE, S. B., D. D. Y. RYU, R. SIEGEL & S. H. PARK. 1985. Performance of recombinant fermentation and evaluation of gene expression efficiency for gene product in two-stage continuous culture system. Biotechnol. Bioeng. **31**: 805–820.
12. MIAO, F. & D. S. KOMPALA. Analysis of two-stage recombinant bacterial fermentations using a structured kinetic model. Bioprocess Eng. In press.
13. MIAO, F. & D. S. KOMPALA. Overexpression of cloned genes using recombinant *Escherichia coli* regulated by a T7 promoter. Biotechnol. Bioeng. In press.

Structured Modeling of Bioreactors

MATTHIAS REUSS

Institute for Biochemical Engineering
University of Stuttgart
Stuttgart, Germany

In contrast to engineers, who are used to scaling-up bioreactors with the aid of averaged values, microorganisms are unable to recognize such abstract entities. The tiny little yeast cell, schematically illustrated in FIGURE 1, can only take notice and therefore will dynamically respond to local values of concentrations. These concentrations will change depending on the interaction between transport phenomena and biological reactions. This report addresses this problem with the aid of a selected model system—baker's yeast *Saccharomyces cerevisiae* growing in a stirred tank bioreactor. Some new ideas and concepts regarding mathematical modeling will be presented and discussed. The ultimate goal of this exercise is to predict the operating conditions of the bioreactor based on a knowledge of the yeast cell traveling through the different regions of the reactor and recognizing changing concentrations of glucose and oxygen.

It is well known that dynamically changing environmental conditions may result in drastic changes in metabolism and consequently the final outcome of the process. Although well known in classical fermentation processes, these problems may have even more serious consequences when dealing with recombinant microorganisms as, for example, during scale-up of high-density cultures. The long-term mathematical description of these phenomena requires flexible tools that can easily be adapted to different systems that integrate the process and the reactor. Flexible tools require the design of conceptual instruments for structuring both the abiotic and biotic phases of the system of interest.

By applying the concept of black-box and gray-box modeling, a descriptive term for problems of bioreactor modeling suggested by Kossen,[1] the demand for designing appropriate structures for both components and coupling the two may be graphically visualized, as in FIGURE 2.

On the basis of the ratio between time constants for the intracellular response and the corresponding time constants for changes in the extracellular environment,[2] problem-specific structures for the two components with balanced complexity are required. Unfortunately, the present state of the art of modeling such a system may be characterized by: (a) a group of models incorporating more and more complexity and structure into the description of the abiotic phase (gas and liquid), thereby neglecting important aspects of the dynamics of the biological phase (such as two-phase turbulence models), and (b) a class of models combining enormous complexity regarding the biological phase including regulation phenomena at metabolic, genetic, and epigenetic levels, thereby neglecting the aspects of transport phenomena in the abiotic phases in most points. These single-cell models are obviously inappropriate for scaling up.

STRUCTURED MODELS FOR THE ABIOTIC PHASES OF STIRRED BIOREACTORS

To identify bottlenecks in the present state of the art of modeling the different components of the system, the tools for describing the transport phenomena in the abiotic phases are much better developed than are the tools for developing and designing intracellular network kinetics at the time scale of mixing. What follows is a brief survey of some structures that have been suggested for the extracellular environment (gas and liquid). Structured models describing the transport phenomena in and between the abiotic phases of a stirred tank reactor can roughly be classified into three groups: (1) turbulence models (single- and two-phase flow); because of the many weak points of the two-phase turbulent models that still exist (particularly the boundary conditions in the impeller regions), we are far from being able to couple these complex fluid dynamic models to complex intracellular metabolic reactions; (2) recirculation time models; and (3) compartment models.

FIGURE 3 is an example of a recirculation/time-distribution model suggested for

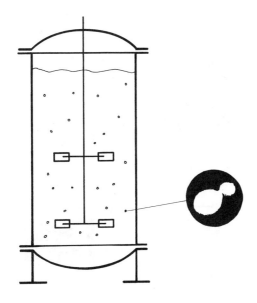

ABIOTIC PHASE   BIOTIC PHASE

FIGURE 1. A budding yeast cell in the environment of a stirred tank reactor.

single and multiple impeller systems by Bajpai and Reuss.[3-5] According to the model structure, illustrated more simply in FIGURE 4 with a single impeller, the vessel is divided into two regions, a micromixer in the immediate vicinity of the impeller and a macromixer consisting of the rest of the tank volume. Flow pattern through the macromixer is defined by the residence time distribution of the various elements in it, that is, the frequency with which the elements reenter the micromixer has a distribution. All the fluid elements passing through the micromixer lose their identity because of the micromixing therein, and they enter the macromixer as a fresh, new element. This model structure has successfully been applied to a quantitative explanation of the influence of mixing on the Crabtree effect in baker's yeast production[3] as well as the oxygen supply in the highly viscous non-Newtonian fermentation broth.[4,5]

A tremendous number of compartment models have been suggested, but only a few examples will be discussed. FIGURE 5 shows the well-known two-compartment model of Sinclair and Brown.[6] This model was proposed as a continuously stirred tank reactor to explain the influence of the intensity of mixing. The model consists of two well-mixed compartments that interact through exchange flow. The solution of the material balance equations for biomass and substrate illustrates the strong effect of the intensity of mixing on the exit cell concentration in a CSTR.

FIGURE 6 shows the five-compartment model of Oosterhuis and Kossen.[7] This

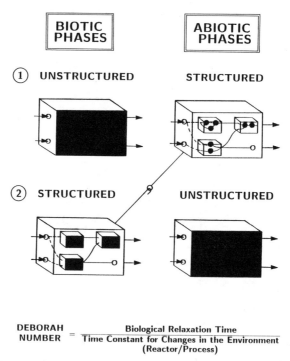

$$\text{DEBORAH NUMBER} = \frac{\text{Biological Relaxation Time}}{\text{Time Constant for Changes in the Environment}}$$
(Reactor/Process)

FIGURE 2. Black-box and gray-box modeling[1] applied to the biotic and abiotic phases of a bioreactor.

model consists of two impeller compartments and three compartments that characterize the flow conditions in the rest of the tank. For estimating the volumetric mass transfer coefficient gas/liquid it was assumed that in the impeller compartment, van't Riet's correlation[8] for noncoalescing systems can be applied. The oxygen transfer coefficients in the rest of the tank are estimated from an equation for bubble columns.[9]

FIGURE 7 shows another approach, the multiturbine impeller model of Bader.[10] The model separates the reactor into a series of mixing zones. Mixing of the liquid phase is represented by liquid flow between the compartments and is related to the

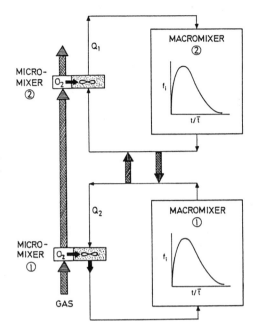

FIGURE 3. Two-environment circulation model for mixing in a stirred bioreactor with two impellers.

pumping capacity of the impellers. Backmixing of the gas phase is neglected in this model.

FIGURE 8 illustrates the multicompartment model of Singh and coworkers.[11] The model consists of a series of well-mixed single-phase compartments with a recycling unit. The number of compartments per impeller stage is assumed to be a function of the turbulence intensity. A similar structure has been suggested by Bajpai and Sohn.[12] The model (FIG. 9) consists of a cascade of well-mixed vessels with a recycle and exchange flow.

The compartment models discussed so far have two serious limitations. First, they do not account for the influence of back mixing of the gas phase. Secondly, in most models the number of compartments is related to the intensity of back mixing or turbulent intensity in the liquid phase. In attempting to simulate the dynamic behavior of a microorganism traveling through different regions of the bioreactor, serious problems may arise if the compartment structure causes discontinuities in the

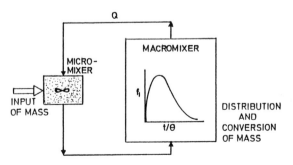

FIGURE 4. Micro-macromixer model[3-5] for a single impeller.

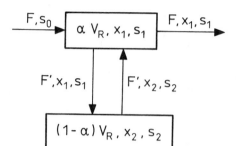

FIGURE 5. Two-region mixing model of Sinclair and Brown.[6]

extracellular concentrations. These discontinuities may cause systems-specific dynamic responses of organisms that are quite different from those in a structure with smoother changes.

The following approach (FIG. 10) introduces a new concept that incorporates the mixing of both phases, gas and liquid, as well as mass transfer and kinetics. For simplicity the discussion is restricted to two-phase compartments (FIG. 10b). The model proposed by Ragot and Reuss[13] is based on an appropriate aggregation of well-mixed multiphase compartments including mass transfer between the phases. FIGURE 10b schematically illustrates a single two-phase compartment that consists of a gas and a liquid phase. The ratio of the two is estimated from the gas holdup.

Aggregation is then achieved by connecting the gas and liquid fractions of the compartments through circulation streams and backflow (FIG. 11). Thus, the model structure accounts for different mixing intensities of the gas and liquid in the tank. The most significant advance in this model structure is that the number of compartments is independent of the intensity of mixing. This is an essential prerequisite for coupling the two phases, gas and liquid having different mixing intensities. Choosing the number of compartments is similar to selecting an appropriate step size in a numerical integration procedure.

If the oxygen consumption of suspended microorganisms in the liquid phase is considered, the material balance equations for oxygen in the liquid and gas phase are

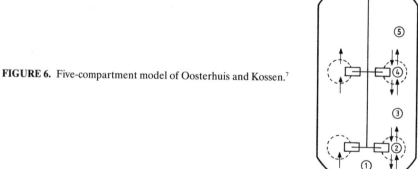

FIGURE 6. Five-compartment model of Oosterhuis and Kossen.[7]

given by

$$V_{L_{i,j}} \frac{dC_{L_{i,j}}}{dt} = \ldots$$

$$\underbrace{- (1 - \omega) f C_{L_{i,j+1}} + f C_{L_{i,j-1}}}_{\text{liquid circulation}} + \underbrace{k_L a_j V_{L_{i,j}} \left(\frac{y G_{i,j} P_i}{H'} - C_{L_{i,j}} \right)}_{\text{mass transfer}}$$

$$\underbrace{+ r_{O_{2,ij}} (C_{X_{i,j}}, C_{S_{i,j}}, \ldots) V_{L_{i,j}}}_{\text{reaction}}$$

$$V_{G_{i,j}} \frac{dy_{G_{i,j}}}{dt} = \ldots$$

$$\underbrace{+ \alpha \dot{V}_{G_{i,j-1}} y_{G_{i,j-1}} + \beta \dot{V}_{G_{i+1,j}} y_{G_{i+1,j}}}_{\text{gas circulation}} - \underbrace{k_L a_j V_{L_{i,j}} \frac{RT}{P_i} \left(\frac{y G_{i,j} P_i}{H'} - C_{L_{i,j}} \right)}_{\text{mass transfer}}$$

where f and ω are the circulation and backmixing parameters in the liquid phase, and α and β denote the corresponding properties of the gas phase.

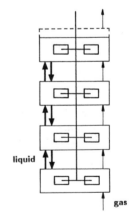

FIGURE 7. Mixing-cell model for a multiturbine fermentor of Bader.[10]

The model can only be applied if, in addition to gas holdup and mass transfer coefficients, reliable data for f, ω, α, and β are available. To demonstrate the application of the model the structure in FIGURE 12 was used to measure the response to a fluorescent dye injection in the liquid phase of a 100-liter vessel. Parameters f and ω were identified by comparison of the dynamic response of the model and the measured data. Estimations were conducted with the help of Nelder and Mead's simplex algorithm. The two parameters characterizing the mixing of the gas phase can be estimated from the measured residence time distributions of the gas

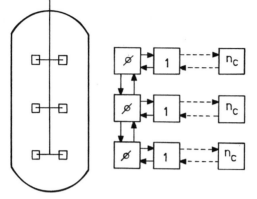

FIGURE 8. Recycle-backmix circulation model of Singh *et al.*[11]

phase. FIGURE 13a and b show the results of these measurements using methane as a tracer. Measurements of the response to step change (a) and pulse injection (b) were performed in a 30-liter vessel with the aid of a flame ionization detector that was placed in the exhaust gas line. Parameters α and β were again estimated from the solution of the material balance equations and application of Nelder and Mead's simplex algorithm.

Once the parameters of mixing in the gas and liquid phase are identified, the combined model can be used to calculate the performance of the stirred tank as a bioreactor. FIGURE 14 shows the distribution of oxygen in the gas and liquid phases, incorporating a Monod type of kinetics for oxygen consumption. It is easy to see that the influence of the two additional parameters, backmixing of the gas phase and hydrostatic pressure, become all the more important with increasing size of the operation. One important consequence of these distributions of oxygen in the tank is that classical scale-up rules, such as $P/V =$ idem, may result in the wrong conclusions. In FIGURE 15 the average rate of oxygen uptake predicted from the distribution in FIGURE 14 is plotted against the power input and compared with the results from a well-mixed system.

To summarize this part of the model structure and applications, the model includes a mixing of gas and liquid phase, oxygen depletion of the gas phase,

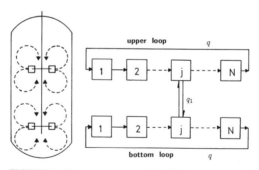

FIGURE 9. Compartment model of Bajpai and Sohn.[12]

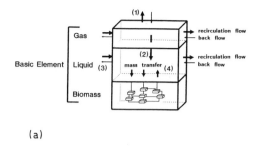

(a)

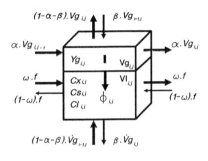

(b)

FIGURE 10. Basic element for the three-phase (a) and two-phase (b) multicompartment model of Ragot and Reuss.[13]

influence of hydrostatic pressure and mass transfer between the phases as well as reaction in the liquid phase.

We apparently do have some appropriate tools to tackle some of the problems in the abiotic phases of the bioreactor. Assuming that we are able to quantify the effect of mixing and mass transfer in the two abiotic phases, we are then faced with the problem of designing an appropriate problem-specific metabolic model to take care of those intracellular dynamics that are beyond a Monod type of kinetics.

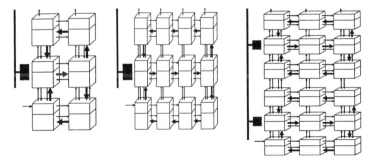

FIGURE 11. Aggregation of two-phase basic elements to multicompartment models for a stirred tank with single and double impellers.

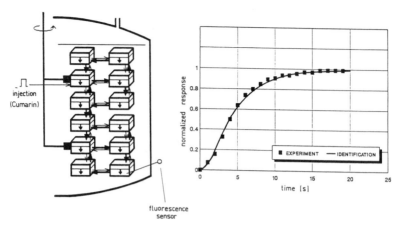

FIGURE 12. Estimation of the mixing parameters for the liquid phase from fluorescence measurements.

STRUCTURED MODEL FOR THE BIOTIC PHASE

Baker's yeast, *Saccharomyces cerevisiae*, has been chosen as a model system to illustrate the importance of structured enzymatic modeling of catabolic metabolism. The problems of interest with this yeast are the dynamic responses to local oxygen limitation, known as the Pasteur effect, and to local glucose excess, known as the Crabtree effect. Many publications have been devoted to the explanation of these two effects in baker's yeast fermentation. Application of the available information to study the influence of local concentration gradients on the dynamic response of the yeast is, however, rather limited. The reason for this limitation is, first, the tendency to most often restrict the observations of these phenomena to steady-state conditions of continuous cultures or to unsteady-state conditions in a range of relaxation times

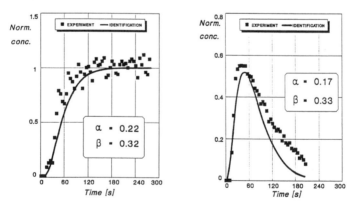

FIGURE 13. Estimation of mixing parameters for the gas phase from experimental observations of the dynamic response to a step change (methane) at the gas inlet in a 30-liter fermentor: (a) step change, and (b) pulse injection.

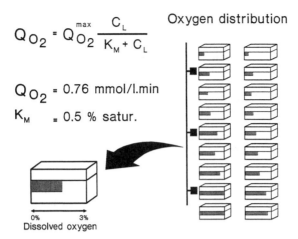

$$Q_{O_2} = Q_{O_2}^{max} \frac{C_L}{K_M + C_L}$$

Oxygen distribution

$$Q_{O_2} = 0.76 \text{ mmol/l.min}$$

$$K_M = 0.5 \text{ \% satur.}$$

0%   3%
Dissolved oxygen

FIGURE 14. Computed distribution of oxygen profiles in a 30 m³ tank. Oxygen consumption is predicted with the aid of a Monod kinetic.

that are far from the time scale of mixing. Second, the guiding work of the groups of Hess and Rapapport[15] in the 1960s and 1970s and also of Liao and Lightfoot[16] is unfortunately restricted to resting cells and/or crude extract of yeasts and cannot be applied to growing cells without some difficulty. Current research is aimed at filling this gap through the development of a dynamic mechanistic model and its verification through experimental observations in the range of relaxation times of interest (in seconds).

The two phenomena mentioned, the Pasteur effect and the Crabtree effect, as far as the time scale of interest is concerned, are related to the catabolism of the cell. Therefore, if we are designing a structured model for the biophase, anabolic reactions such as synthesis of protein, RNA, DNA, lipid, and the like can be considered as frozen in the initial state.

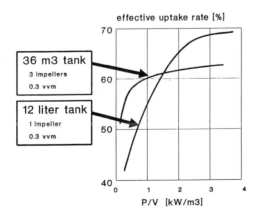

effective uptake rate [%]

36 m3 tank
3 impellers
0.3 vvm

12 liter tank
1 impeller
0.3 vvm

P/V [kW/m3]

FIGURE 15. Oxygen uptake as a function of power input. Comparison between the results for a well-mixed system and the two-phase multicompartment model.

An approach to a simplified model of catabolism is schematically summarized in FIGURE 16. The model incorporates the well-known key reactions of gycolysis, balances of the cometabolites, and the production and excretion of glycerol, ethanol, and acetic acid. The communication between the intracellular and extracellular environment is restricted to glucose and oxygen consumption and the metabolites that are excreted from the cells.

This contribution is not aimed at discussing all of the individual reactions of the model in detail.[17,18] A few remarks, however, are in order to explain the overall strategy.

Transport of glucose through the cell membrane is aided by a reversible transport kinetic including inhibition through G6P. Reaction 2, the phosphorylation of glucose by hexokinase, is modeled as a reversible double substrate-double product kinetic. For reaction 3, the phosphoglucose isomerase, a near equilibrium reaction is assumed. For the subsequent reaction 4, the well-known key enzyme of glycolysis, phosphofructokinase, we have applied a reaction kinetic based on a four-state allosteric enzyme model, including AMP and ADP as activators and ATP as an inhibitor.[19] All the reactions to and/or between fructose-di-phosphate and glycerinaldehyde-3-phosphate and dihydroxyaceton-phosphate are assumed to be in near equilibrium. They have been introduced for glycerol production that has been experimentally observed during the Crabtree effect in the time range of interest. The reaction catalyzed by pyruvate kinase is assumed to be represented by a double substrate-double product kinetic with a concerted allosteric regulation, as suggested by Hess et al.[20]

Pyruvate is considered the key metabolite for both the Crabtree and the Pasteur effect. First, it is assumed that pyruvate dehydrogenase can be modeled by a special rate equation for a multienzyme complex. It is further assumed that the rate of production of reducing equivalents via TCA is equivalent to this rate multiplied by a stoichiometric coefficient. All reducing equivalents are collected as substrate for the respiratory chain which is mathematically simplified with the aid of a double-substrate Michaelis-Menten kinetic for the cytochrome oxidase. The second substrate in this reaction is oxygen. It is easy to see that a local oxygen limitation results in an accumulation of NADH which will be reduced in the remaining pathway and results in ethanol production. This pathway, in turn, consists of an allosteric reaction mechanism for pyruvate decarboxylase, as suggested by Boiteux and Hess,[21] and reversible Michaelis-Menten kinetics for the remaining enzyme.

This is an appropriate place to comment critically on the complexity of the model and the number of parameters involved. It must be emphasized that we are exclusively applying intrinsic rate equations for the individual enzymes and fixing the parameters in these kinetics to values that have already been reported. Therefore, although the complexity and the number of parameters are high, the degree of freedom for manipulations is reduced.

The initial attempt to use the model for simulation was made by introducing the concentrations of intracellular metabolites published in the available literature. These concentrations are used as initial conditions to estimate the maximum rates from the given kinetic equations.

In the following, one example of the comparison between model predictions and experimental observations will be presented. Because the interest is the fast response in a time window that is in the same order of magnitude as are the recirculation times, special experiments were performed in a continuous culture. Starting from a steady-state observation of S. cerevisiae, the dynamic response of the population to pulse injections of glucose to a level of 1 g/L was measured[18] (FIG. 17).

Experimentally observed and predicted by the model is a very fast and sharp

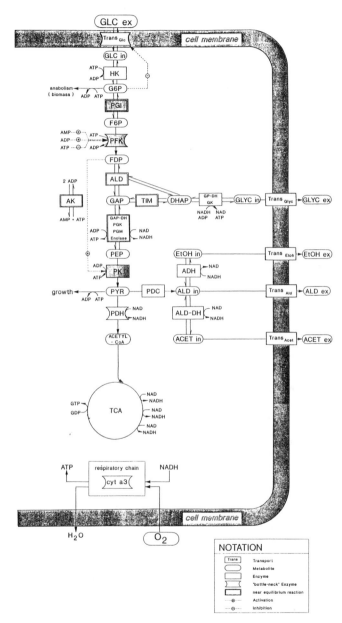

FIGURE 16. Structured enzymatic model for simulating the fast dynamic effects of the Crabtree and the Pasteur effect.

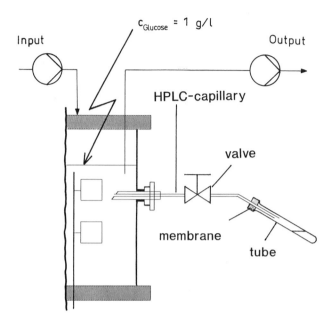

- vaccum-tight precooled (-25°C) tubes

- perchloric acid extraction (35%, -25°C)

FIGURE 17. Chemostate culture of *S. cerevisiae* with pulse injection of glucose and fast sampling device.

decrease of ATP after puls injection of glucose (FIG. 18). Measured and predicted time courses of the extracellular concentrations of gycerol and acetic acid show reasonable agreement, as illustrated in FIGURE 19. The kinetics of ethanol formation need to be improved.

A critical assessment of these initial comparisons between model predictions and

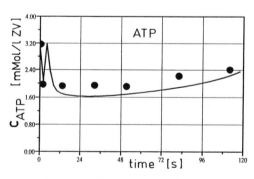

FIGURE 18. Comparison of model predictions and experimental observations of intracellular ATP after glucose injection.

experiments leads to the conclusion that the model must be further improved. From different simulations it can be concluded that this improvement cannot be achieved by just tuning the parameters in the rate equations. Additionally, some of the assumptions in the model, for example, some near-equilibrium conditions, must be critically examined. These investigations are accompanied by an increasing number of intracellular state variables that are experimentally determined in the ongoing work.

The next step towards the complete structured model of the bioreactor is schematically illustrated in FIGURE 20, which shows the extension of the two-phase

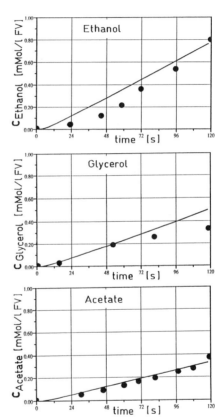

FIGURE 19. Comparison of model predictions and experimental observations of extracellular ethanol, glycerol, and acetic acid after glucose injection.

model to include the dynamics of the biophase. Local concentrations of glucose that depend on the mixing process determine the local flux into the cell. The local oxygen concentrations depend on mixing and mass transfer gas/liquid and influence the rate of respiration via cytochrome oxidase which results in changes in NADH and excreted metabolites.

Much work remains to be done to verify the proposed model structure. Both experimental and theoretical work on the suggested structures for the abiotic and biotic phases is under active investigation. Work is also being done to develop new concepts for model reduction that can have practical application.

SUMMARY

This contribution addresses the advances in mathematical modeling of stirred tank bioreactors involving a new method for structuring the abiotic phases and comprehensive enzymatic network kinetics for the biotic phase. The concept for modeling the effects of mixing in the gas and liquid phase is based on an aggregation of well-mixed multiphase compartments including mass transfer between the phases. Aggregation is performed by connecting the gas and liquid fractions of the compartments through circulation streams and backflow. As a model system for the biological phase, the influence of the lack of homogeneity of glucose (Crabtree effect) and/or oxygen (Pasteur effect) on the dynamics of cell metabolism of *S. cerevisiae* was chosen. To study the influence of local gradients in glucose and oxygen concentrations during circulation of the yeast cells in the bioreactor, a mechanistic model was

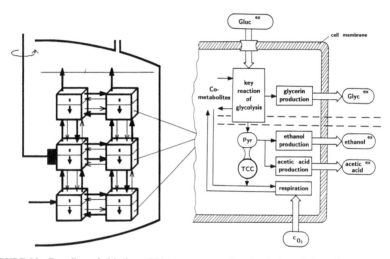

FIGURE 20. Coupling of abiotic and biotic structures for simulation of the influence of mass transfer and mixing on the dynamics of the Crabtree and the Pasteur effect.

developed using key reactions of the glycolytic pathway, incorporating their intrinsic rate equations and regulations. This intracellular structured model is coupled with the structured multiphase compartment model for the extracellular environment.

ACKNOWLEDGMENTS

The author expresses his deep appreciation to the following individuals who have contributed to this work: M. Baltes, F. Ragot, M. Rizzi, and U. Theobald.

REFERENCES

1. KOSSEN, N. W. F. & N. M. G. OOSTERHUIS. 1985. Modelling and scaling up of bioreactors. *In* Biotechnology. H.-J. Rehm & G. Reed, eds. Vol. **2:** 572–605. VCH Verlagsgesell-schaft. Weinheim-Deerfield Beach/Florida-Basel.

2. ROELS, J. A. 1983. Energetics and Kinetics in Biotechnology. Elsevier Biochemical Press. Amsterdam, New York, Oxford.

3. BAJPAI, R. K. & M. REUSS. 1982. Coupling of mixing and microbial kinetics for evaluation of the performance of bioreactors. Can. J. Chem. Eng. **60:** 834–392.

4. REUSS, M. 1983. Mathematical models for coupled oxygen transfer and microbial kinetics in bioreactors. *In* Modelling and Control of Biochemical Processes. A. Halme, ed.: 33–45. Pergamon Press. Oxford.

5. REUSS, M. & R. K. BAJPAI. 1991. Stirred tank models. *In* Biotechnology. H.-J. Rehm & G. Reed, eds. Vol. 4. VCH Verlagsgesellschaft. Weinheim-Deerfield Beach/Florida-Basel.

6. SINCLAIR, C. G. & D. E. BROWN. 1970. Effect of incomplete mixing on the analysis of the static behavior of continuous cultures. Biotechnol. Bioeng. **12:** 1001–1017.

7. OOSTERHUIS, N. M. G. & N. W. F. KOSSEN. 1984. Dissolved oxygen concentration profiles in a production-scale bioreactor. Biotechnol. Bioeng. **26:** 546–550.

8. VAN'T RIET, K. 1979. Review of measuring methods and results in nonviscous gas-liquid mass transfer in stirred vessels. Ind. Eng. Chem. Proc. Des. **18:** 367–375.

9. HEIJNEN, J. J. & K. VAN'T RIET. 1982. Mass transfer, mixing and heat transfer phenomena in low-viscous bubble column reactors. Proc. IVth Eur. Conf. on Mixing, April 27–29th. Noordwijkerhout. The Netherlands.

10. BADER, F. G. 1987. Modelling, mass transfer and agitator performance in multiturbine fermentors. Biotechnol. Bioeng. **30:** 37–51.

11. SINGH, V., R. FUCHS & A. CONSTANTINIDES. 1987. A new method for fermentor scale-up incorporating both mixing and mass transfer effects. I. Theoretical basis. *In* Biotechnology Processes, Scale up and Mixing. C. S. Ho & J. Y. Oldshue. eds.: 200–214. American Institution of Chemical Engineers. New York.

12. BAJPAI, R. K. & P. U. SOHN. 1987. Stage models for mixing in stirred bioreactors. *In* Biotechnology Processes, Scale up and Mixing. C. S. Ho & J. Y. Oldshue, eds.: 13–21. American Institution of Chemical Engineers. New York.

13. RAGOT, F. & M. REUSS. 1991. A multi-phase compartment model for stirred bioreactors incorporating mass transfer and mixing. *In* Biochemical Engineering—Stuttgart. M. Reuss, H. Chmiel, H.-J. Knackmuss & E. D. Gilles, eds. Gustav Fischer Verlag. Stuttgart, New York.

14. JURY, W., G. SCHNEIDER & A. MOSER. 1988. Modelling approach to industrial bioreactors. Proc. VIth Europ. Conf. on Mixing, May 24–26th. Pavia, Italy.: 451–456.

15. CHANCE, B., E. K. PYE, A. K. GOSH & B. HESS. 1973. Biological and Biochemical Oscillators. Academic Press. New York, London.

16. LIAO, J. C., E. N. LIGHTFOOT, JR., S. O. JOLLY & G. K. JACOBSON. 1988. Application of characteristic reaction paths: Rate-limiting capability of phosphofructokinase in yeast fermentation. Biotechnol. Bioeng. **31:** 855–868.

17. RIZZI, M., U. THEOBALD, M. BALTES & M. REUSS. 1991. Structured modelling of bioreactor systems. *In* Biochemical Engineering—Stuttgart. M. Reuss, H. Chmiel, H.-J. Knackmuss & E. D. Gilles, eds. Gustav Fischer Verlag. Stuttgart, New York.

18. THEOBALD, U., M. BALTES, M. RIZZI & M. REUSS. 1991. Structured metabolic modelling applied to dynamic simulations of the Crabtree- and Pasteur-effect in baker's yeast. *In* Biochemical Engineering—Stuttgart. M. Reuss, H. Chmiel, H.-J. Knackmuss & E. D. Gilles, eds. Gustav Fischer Verlag. Stuttgart, New York.

19. HOFMANN, E. & G. KOPPERSCHLÄGER. 1982. Phosphofructokinase from yeast. *In* Methods in Enzymology. W. A. Wood, ed. Vol 90: 49–60. Academic Press. New York.

20. HESS, B. 1971. Allgemeine Prinzipien der Regulation der Glykolyse. Ergeb. Exp. Med. **9:** 66–87.

21. BOITEUX, A. & B. HESS. 1970. Allosteric properties of yeast pyruvate decarboxylase. FEBS Letts. **9:** 293–296.

High Cell Density Fermentation of Recombinant *Escherichia coli* with Computer-Controlled Optimal Growth Rate

W. A. KNORRE,[a] W.-D. DECKWER,[b] D. KORZ,[b]
H.-D. POHL,[a] D. RIESENBERG,[a] A. ROSS,[b] E. SANDERS,[b]
AND V. SCHULZ[a]

[a] *Zentralinstitut für Mikrobiologie und experimentelle Therapie (ZIMET)*
0 6900 Jena, Thüringen, Germany

[b] *Gesellschaft für Biotechnologische Forschung mbH (GBF)*
W 3300 Braunschweig Niedersachsen, Germany

In recent years interest in the production of recombinant DNA products, such as enzymes and pharmaceuticals, has been growing. High cell density cultivation has been one of the most effective ways to increase cell as well as product yield. Basic work in this field was carried out by Bauer and coworkers (1974–1981) with nonrecombinant strains of *Escherichia coli*.[1,2]

A variety of process strategies have been developed in fed-batch fermentation of *E. coli*.[3] Common goals are to control the oxygen demand within the oxygen transfer capabilities of the fermentor and to avoid the accumulation of acetate and ethanol.[4] These goals can be met by feeding the carbon source to achieve C-limited growth, reducing temperature to decrease the growth rate, or increasing the oxygen transfer capability of the fermentor using oxygen as the sparging gas. Cell densities of about 110 g/L have been reported for a process-controlled fed-batch plus oxygen mode for cultivating nonrecombinant *E. coli* in a 2.5-L fermentor.[5] An impressive application of rDNA products is the production of alpha-consensus interferon by recombinant *E. coli*. Fieschko and Ritch[6] reported product concentrations of 5.5 g/L from 65 g/L cell dry mass.

ADVANTAGES OF HIGH CELL DENSITY FERMENTATION

In general, the main advantages of high-density cultivation are: reduced fermentor and closed system volume, improved space time yield (volumetric productivity), reduced medium costs, reduced volume in primary downstream processing, frequent omission of concentration steps, and reduced plant and operating costs.

HOST STRAIN *E. COLI*

For decades, *E. coli* has played an important role in molecular biological work, which explains its use as the host strain for the majority of protein productions from rDNA. For example, insulin, hGH, alpha$_2$-interferon, alpha$_{2b}$-interferon, and rennin are available from cultivations of recombinant *E. coli*. For none of them has a high cell density process been used.

Of great interest is the control of growth rate because it strongly influences the formation of both products and inhibitory metabolites. Acetate formation is reported to increase drastically when the growth rate exceeds $0.35 \ h^{-1}$ (defined medium) and $0.2 \ h^{-1}$ (complex medium).[7]

Seo and Bailey[8] showed the existence of an optimum dilution rate (growth rate) for β-lactamase production in continuous cultivation. This finding was not confirmed in batch experiments when growth rate control was changed by altering growth medium. Despite this, Riesenberg *et al.*[9] reported an optimum growth rate in batch production of alpha$_1$-interferon.[9]

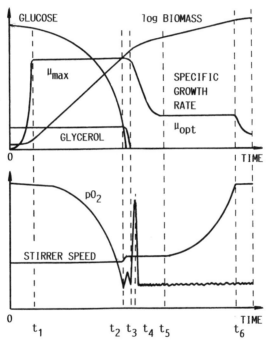

FIGURE 1. Scheme of the ZIMET HDF 30/450: t_1 = end of lag-phase; t_2 = start of pO_2-control via stirrer speed; t_3 = exhaustion of glycerol; t_4 = start of glucose feeding; t_5 = start of exponential shift in stirrer speed; and t_6 = end of exponential shift in stirrer speed.

ZIMET HIGH CELL DENSITY FERMENTATION 30/450

In the ZIMET a high cell density fermentation (HDF) process for a glucose/mineral salt medium allows growth of a recombinant *E. coli* strain (TG 1, pBB 210) up to a cell density of 60 g/L in a 30- and 450-L Chemap fermentor.[9] Except for the feeding of glucose as a carbon source and of aqueous ammonia for pH control, there was no need for the feeding of other nutrients and for the supply of oxygen-enriched air. FIGURE 1 schematically illustrates this process with a batch phase with glucose consumption and a feed-batch phase with glucose feeding. In the fed-batch phase the pO_2 was kept at 20% of saturation via closed-loop controls with two variables, namely, stirrer speed and feeding rate of glucose. The fed-batch mode prevented significant accumulation of acetate and other metabolic byproducts. The recombinant *E. coli* expressed alpha$_1$-interferon constitutively with a higher efficiency at a lower specific growth rate ($\mu_{opt} = 0.17 \ h^{-1}$) than at the maximal specific growth rate ($\mu_{max} = 0.45 \ h^{-1}$) (FIG. 2). Therefore, after reaching a suitable cell density with growth at μ_{max}, the culture was forced to grow at the optimal specific growth rate, μ_{opt}, by open-loop control for agitation directing the input of oxygen and hence the supply of glucose. The stirrer speed was increased according to an e-function profile.

ZIMET/GBF-HIGH CELL DENSITY FERMENTATION 70/1500

To overcome the well-known disadvantages of high density processes, GBF and ZIMET in 1989 developed a special HDF process for *E. coli* that produces more than

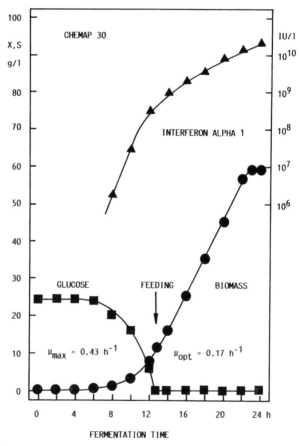

FIGURE 2. ZIMET HDF 30 process: Kinetics of glucose, biomass, and alpha$_1$-interferon.

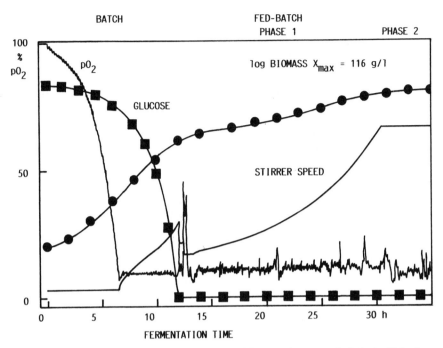

FIGURE 3. ZIMET/GBF HDF process. Kinetics of biomass, glucose, and pO_2 in the 70-L pilot fermentor.

100 g/L cell dry mass. The special cultivation strategy prevents oxygen limitation and hence the accumulation of inhibitory metabolites such as acetate and ethanol. (See FIG. 5 for acetate kinetics.)

FIGURE 3 shows a typical time course of the HDF process with two exponential growth phases. The first is characterized as a batch phase under maximum growth rate corresponding to the media and process parameter, that is, temperature and pH. The subsequent fed-batch operation can be subdivided into μ-controlled phase 1, implemented as a dissolved oxygen control loop via nutrient feed and a μ control loop via agitation rate, and a second period (phase 2) with decreasing growth rate at the maximum attainable oxygen transfer rate.

In the example shown, growth rate was switched from $\mu_{max} = 0.45$ h^{-1} after 12 hours to a lower constant value of $\mu = 0.11$ h^{-1}, which corresponded to an optimal growth rate for rDNA product formation. Over more than 18 hours, respectively, of three doublings of biomass (12–95 g/L), growth rate was kept constant (FIG. 4).

The special advantages of the ZIMET/GBF HDF process (FIG. 5) include: defined nutrient medium without turbidity; the simple composition of the one-feed medium; a lowered risk of contamination; aeration with air instead of oxygen; if desired, N, P or other limitations are possible; a controlled growth rate over a long period of time; a low concentration of inhibitory metabolites; and no peptides from medium components in downstream processing.

In contrast to the work of Seo and Bailey,[8] it is possible to study product formation and plasmid content of the cells at different growth rates without changing the medium. This enables the researchers to determine the optimum growth rate in

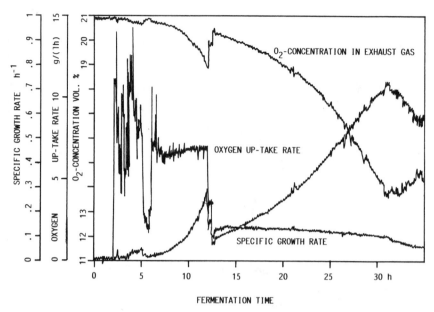

FIGURE 4. ZIMET/GBF HDF process. Kinetics of oxygen uptake and concentration in the exhaust gas. Time course of the specific growth rate.

batch experiments instead of the chemostat operation which may lead to conclusions that are not always easily applicable to subsequent batch cultivations.

SCALE-UP

Process development was carried out in a three-stage bioreactor cascade of 70, 300, and 1,500 L. The reactors are equipped with multichannel control microcomputers (MICON P200) that act as standard controllers for basic parameters such as temperature, pH, gas flow, and pressure. Furthermore, a second P200 calculates the actual growth rate from exhaust gas analysis and controls it by appropriate alteration of the oxygen transfer to the culture.

Any operation of the user such as input of setpoints and switching between cultivation phases and therefore different modes of the controllers can be performed at a terminal connected to a process computer, which provides for data transfer between P200 microcomputers, exhaust gas analysis, and host computer link.

With data from the 70-L cultivation experiments it was no problem to scale-up the process to 1,500 L. The only critical problem that arose with increasing scale was the removal of heat.

SCALE-DOWN

Further investigations on HDF and its applications on rDNA protein production are under way. Actually, variations of HDF are being analyzed and appropriate host/vector systems are being prepared. This work can be carried out in laboratory scale fermentors, so that scale-down to 5-L working volume was performed in an early phase of process development. The major difference from the original scale (≥ 70 L) HDF is operation at atmospheric pressure with the use of glass fermentors. In this case the only way to ensure sufficient oxygen transfer into the broth is to aerate with oxygen-enriched air.

In 5-L scale a slightly modified HDF process was successfully applied to production of recombinant β-galactosidase using a chemically inducible system. Specific activities (U/g) obtained in standard cultivations were also achieved under dense cultivation conditions at a cell mass 10–50 times higher, that is, up to 50 times higher volumetric activity.

FUTURE PROSPECTS

In the last decade, most scientific attention has focused on genetics-based solutions to productivity issues, and progress has been impressive. Ultimately, the returns from a purely genetics approach will diminish, and further optimization will depend on understanding the relationship between the microbial environment and synthesis of the desired protein. More recently, Plückthun[10] reported an expression system with which fully functional antibody F_v or F_{ab} fragments can be expressed in *E. coli.* Both chains are co-expressed and co-secreted into the periplasm of *E. coli* with correct signal processing, disulfide formation, and chain association. Such expression systems should also be suitable for other similar proteins. Therefore, it is necessary

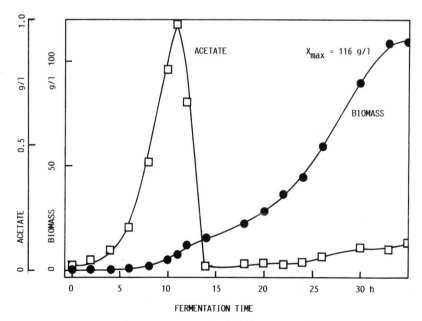

FIGURE 5. ZIMET/GBF HDF process. Typical time course of biomass and acetate.

to develop HDF processes for *E. coli* with such an expression system and to optimize the accumulated periplasmatic proteins.

The research and commercial forces that are guiding the development of recombinant DNA technology will certainly motivate the growing effort in HDF process research devoted to recombinant microorganisms, especially as products move from research into development.

SUMMARY

In recent years recombinant DNA technology has enabled us to produce various proteins of therapeutic importance with microorganisms. As an appropriate host organism, *E. coli* plays a dominant role. Yields of *E. coli* dry cell mass in shaker flask culture range from 1–2 g/L, whereas in fermentors up to 10 g dry cells/L can be achieved. ZIMET and GBF have developed a high cell density fermentation process that produces *E. coli* (on a glucose/mineral salt medium) up to more than 100 g dry cells/L in a special fed-batch mode. This cultivation strategy prevents oxygen limitation and hence the accumulation of acetate and other metabolic byproducts. The specific growth rate can be adjusted so that product formation reaches its optimum value. An example of the production of alpha$_1$- interferon is presented. The high cell density fermentations were realized in 30- and 450-L Chemap fermentors (ZIMET) and in a three-stage bioreactor scale-up system (72, 300, and 1,500 L) developed in cooperation with GBF and B. Braun Melsungen AG. Multiloop controllers were used to control the process variables.

REFERENCES

1. BAUER, S. & J. SHILOACH. 1974. Maximal exponential growth rate and yield of *E. coli* obtainable in a bench-scale fermentor. Biotechnol. Bioeng. **16:** 933–941.
2. GLEISER, I. E. & S. BAUER. 1981. Growth of *E. coli* W to high cell concentration by oxygen level linked control of carbon source concentration. Biotechnol. Bioeng. **23:** 1015–1021.
3. ZABRISKIE, D. W. & E. J. ARCURI. 1986. Factors influencing productivity of fermentations employing recombinant microorganisms. Enzyme Microbiol. Technol. **8:** 706–717.
4. PAN, J. G., J. S. RHEE & J. M. LEBEAULT. 1984. Physiological constraints in increasing biomass concentration of *E. coli* B in fed batch culture. Biotechnol. Lett. **9:** 89–94.
5. EPPSTEIN, L., J. SHEVITZ, X.-M. YANG & S. WEISS. 1989. Increased biomass production in a benchtop fermentor. Bio/Technology **7:** 1178–1181.
6. FIESCHKO, J. & T. RITCH. 1986. Production of human alpha consensus interferon in recombinant *E. coli*. Chem. Eng. Commun. **45:** 229–240.
7. MEYER, H.-P., C. LEIST & A. FIECHTER. 1984. Acetate formation in continuous culture of *E. coli* K12 D1 on defined and complex media. J. Biotechnol. **1:** 355–358.
8. SEO, J.-H. & J. E. BAILEY. 1986. Continuous cultivation of recombinant *E. coli:* Existence of an optimum dilution rate for maximum plasmid and gene product concentration. Biotechnol. Bioeng. **28:** 1590–1594.
9. RIESENBERG, D., K. MENZEL, V. SCHULZ, K. SCHUMANN, G. VEITH, G. ZUBER & W. A. KNORRE. 1991. High-cell-density-fermentation of recombinant *E. coli* expressing human interferon alpha-1. Appl. Microbiol. Biotechnol. **34:** 77–82.
10. PLÜCKTHUN, A. 1990. Recombinant antigen binding fragment of an antibody expressed in *E. coli:* folding *in vivo,* properties and catalytic activity. Engineering Foundation Conference: Progress in Recombinant DNA Technology and Application. June 3–8, Potosi, Missouri.

Application of Biochemical Engineering Principles to Develop a Recovery Process for Protein Inclusion Bodies

E. KESHAVARZ-MOORE, R. OLBRICH, M. HOARE, AND
P. DUNNILL

SERC Centre for Biochemical Engineering
Department of Chemical and Biochemical Engineering
University College London
Torrington Place
London WC1E 7JE, England

The accumulation of non-native proteins of commercial interest as insoluble aggregates (also known as inclusion bodies) in the protoplasm of genetically modified microorganisms has created a new challenge in the downstream processing and purification of these proteins. The unit operations traditionally used to recover intracellular products require reexamination and reassessment of their applicability. Although the inactive insoluble state of inclusion bodies may be beneficial, firstly, because of potentially lowered susceptibility to proteolytic attack[1] and, secondly, because of particle characteristics that differentiate them from other cellular matter,[2] these very properties create their own problems. The recovery of active recombinant protein involves solubilization and refolding steps that impose an added level of complexity to the process. Furthermore, the particle characteristics of inclusion bodies are system specific.[3] In the early stages of process development, the relatively limited quantities of process material available favor small scale operations that could directly predict the industrial scale production. In practice, this has been difficult to achieve given the volumes required to operate pilot equipment. In this paper, ways in which a process scale high pressure homogenizer and disc stack centrifuge may be scaled down to recover prochymosin inclusion bodies from recombinant *Escherichia coli* cells are described. The impact of choices of operating conditions in the fermentor and in the high pressure homogenizer on the properties of cell debris, inclusion bodies, and their effect on the subsequent continuous centrifugal recovery process are reported.

CONCEPTS IN THE DESIGN OF A RECOVERY PROCESS FOR INCLUSION BODIES

The initial downstream processing steps in the recovery of inclusion bodies from host cells consist of cell harvesting, followed by cell disruption, particle-particle separation, and solubilization/renaturation (FIG. 1). Traditionally, the design of a production process is based on the performance optimization of individual unit operations in isolation. As will be shown, the recovery of inclusion bodies exemplifies the importance and relevance of an integrated approach to process design.

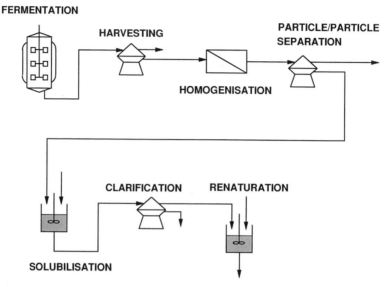

FIGURE 1. Schematic diagram of recovery of prochymosin inclusion bodies produced by *E. coli* HB101 pCT70.

RELEASE OF INCLUSION BODIES

The release of inclusion bodies from harvested washed cells was carried out in a high pressure homogenizer. The details of this operation are given subsequently herein. The host strain *E. coli* HB101 was shown to be more difficult to disrupt compared with the recombinant strain; this difference is reflected in the homogenization rate constant K (TABLE 1). Furthermore, freshly harvested cells were more resilient to cell breakage than were previously frozen cells. Maximal cell disruption, as measured by soluble protein release, was achieved after three to five passes at 55 MPa. However, the aim of the operation is not merely to disrupt the cells but to release and dissociate the inclusion bodies from cell debris. Repeated homogenization resulted in a decrease in cell debris size for each of the first five passes beyond which no further size reduction was achieved. The size of the inclusion bodies, on the other hand, remained unchanged throughout the process and was distinctly larger than that of the cell debris, as indicated by the respective size distributions (FIG. 2).

SCALE DOWN OF HOMOGENIZATION PROCESS

Recent studies[4] on the mechanism of cell disruption in a high pressure homogenizer have identified the most important operating parameters that determine

TABLE 1. Effect of Strain and Biomass Storage on Disruption Rate

Strain	Disruption Rate	Storage
Host cells	0.413	Stored at −70°C
Recombinant strain	0.445	Fresh cells
Recombinant strain	0.590	Stored at −70°C

performance. It is evidently necessary to maintain the valve geometry and fluid hydrodynamics, and consequently scale down is best achieved using a commercial valve geometry operating at normal throughput but only for short periods of time. The smallest high pressure homogenizers suitable for industrial use generally have flow rates above 40 L/h. It is however possible to adapt a pilot/production scale homogenizer to process small quantities (e.g., well below 10 L) of cellular suspension. At University College London, a 30-CD APV-Gaulin high pressure homogenizer with a maximum flow rate of 113 L/h was used. Ancillary equipment was designed and constructed to achieve a flexible integrated homogenization system that would allow disruption on a discrete pass basis using a total suspension volume as small as 5 L. Given the very high pressures achieved in this equipment (up to 105 MPa) it was also essential to design for the cooling of the homogenized fluid. FIGURE 3 is a line diagram of the integrated homogenization system. Several specific features accommodated the scale-down operation. One feature consisted of three hoppers (8.3 L total volume, 5 L working volume), two of which were arranged in parallel to

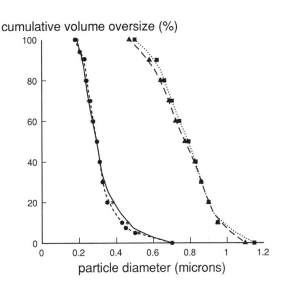

FIGURE 2. Particle size analysis of inclusion bodies and cell debris after homogenization. Cell debris after 5 passes, —*—; after 7 passes, -●-. Inclusion bodies after 5 passes, ··■··; after 7 passes, —▲—.

allow interchangeability and to alleviate the necessity for rapid discharge of the vessels and associated aerosol formation. These hoppers were jacket cooled and contained an overhead stirrer with variable speed control, a temperature indicator, and a removable baffle at the exit orifice. A conductivity probe connected to an alarm would alert the operator to the low fluid level in the vessel. All connections were hard piped. The total length of piping was minimized to reduce hold-up volume in the system. Fast acting, pneumatically driven valves manually operated from a central control panel facilitated efficient disruption under a discrete pass regimen.

SCALE DOWN OF THE CENTRIFUGATION PROCESS

In this study an intermittent-discharge disc stack centrifuge was used (Westfalia BSB7). The feed occurs via a rotating pipe into the settling space of the centrifuge. A

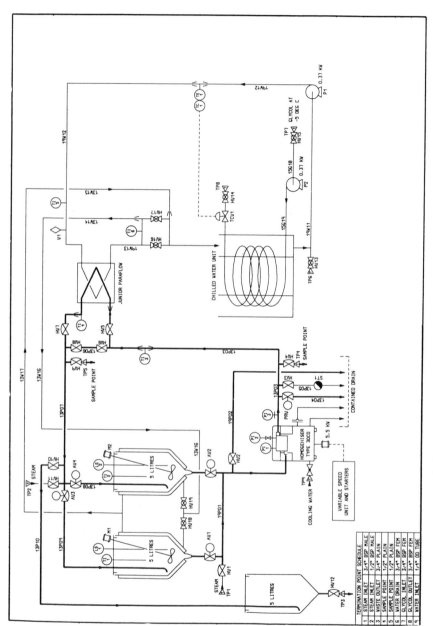

FIGURE 3. Line diagram of the integrated homogenization system.

set of truncated, conical discs ensures that the distance the solids must traverse before setting on a surface is minimized. The solids are intermittently discharged by means of a hydraulically operated piston. The supernatant leaves via the top of the centrifuge. To evaluate the applicability of such a centrifuge on an industrial scale it is necessary to use equipment capable of operating at high hydraulic throughputs of greater than 1,000 L/h. The problem arises when a relatively small amount of process fluid is available. This necessitates the scale down of the centrifuge. This has been achieved at UCL by reducing the active proportion of disc stacks available for separation to 10% of the total available separation area.[5] It is thus possible to use feed volumes of 50–100 liters. FIGURE 4 indicates that, with careful choice of the location of active discs, the efficiency of particle separation using a scale-down

grade efficiency

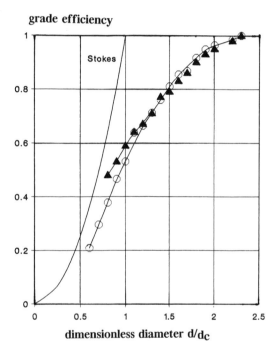

FIGURE 4. Comparison of grade efficiency curves of a disc stack with reduced sedimentation area, ○, and a full set of discs of an industrial centrifuge (Westfalia Separator), ▲. d = diameter of solids; d_c = critical diameter of solids for a disc stack centrifuge.

dimensionless diameter d/d_C

centrifuge, as expressed by grade efficiency curves, compares well with that of a full stack, especially in the region of full particle recovery (d/d_c greater than 2). When a high proportion of particles appears in the overflow, then the scale down results in an underestimation of performance in terms of particle recovery. This is important in the recovery of cell debris from inclusion bodies, where it is required to avoid the sedimentation of cell debris. In this case an overestimation of performance is obtained and care is needed in translation to full scale operation.

APPLICATION OF LABORATORY TECHNIQUES

In the recovery of inclusion bodies, it is desirable to predict the separation of these particles from cell debris under different operating conditions in order to optimize the overall process operation. This is achievable using a grade efficiency curve. However, particle and carrier fluid properties are needed. Laboratory-based techniques such as centrifugal sedimentation (using equipment such as a Joyce-Loebl photosedimentometer) and particle size measurement (using an electrical sensing zone particle sizer) provide the necessary parameters. FIGURE 5 illustrates the sedimentation characteristics of inclusion bodies, cell debris, and total homogenate suspension. These data in conjunction with particle size distribution of inclusion bodies and cell debris (FIG. 2) were used with grade efficiency profiles (FIG. 4) to examine the effect of flow rate on recovery of inclusion bodies.

PREDICTION OF THE SEPARATION PERFORMANCE

A prediction is made using experimental data obtained from size distribution of inclusion bodies and cell debris and experimental grade efficiency curves. The

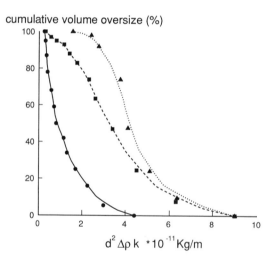

FIGURE 5. Sedimentation characteristics of homogenate suspension, inclusion bodies, cell debris. -■-, homogenate; —●—, cell debris; ···▲···, inclusion bodies. d = diameter of solids; $\Delta\rho$ = density difference between solids and liquid phase; k = shape factor.

portion of the grade efficiency curve to be used is defined by the flow rate, centrifuge operating conditions, the particle characteristics such as density and shape, and suspension viscosity.

FIGURE 6 summarizes the performance of the centrifuge. At low throughputs there is good recovery of the inclusion bodies as solids, but poor removal of cell debris that also sediment out. At high flow rates, the inclusion body recovery is low as the particles are carried over in the overflow which also has a high content of cell debris. It is therefore evident that a compromise must be made between yield and separation of inclusion body. Under optimal operating conditions for this process (e.g., a flow rate of 200 L/h), it is possible to achieve 95% recovery of inclusion bodies with as much as 75% removal of cell debris. The most significant result of such an analysis is that it enables the process development engineer to predict which operating window to examine and to explore the effect of different operating parameters on the recovery of inclusion bodies from cell debris.

recovery or removal, %

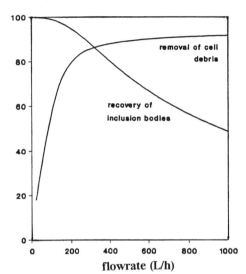

FIGURE 6. Simulated recovery of prochymosin inclusion bodies in an industrial disc stack centrifuge (Westfalia Separator, BSB 7) showing the separation of cell debris from the inclusion bodies.

flowrate (L/h)

EFFECT OF CHANGING CONDITIONS ON RECOVERY

The recovery of inclusion bodies is governed by operating conditions adopted at each unit operation stage, with results from each stage affecting subsequent processing. This is best illustrated in FIGURE 7 which shows the effect of inoculum size and antibiotic concentration on particle size distribution of inclusion bodies. A low inoculum concentration (0.2% of the original inoculum) resulted in larger inclusion bodies ($D_{50} = 0.84$ μm compared to $D_{50} = 0.75$ μm for original inoculum of 1.7% v/v). Although an eightfold decrease in ampicillin concentration did not affect the

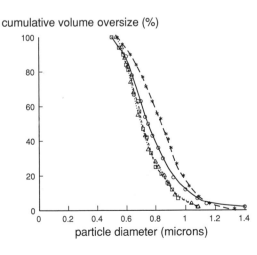

cumulative volume oversize (%)

particle diameter (microns)

FIGURE 7. Effect of fermentation conditions on inclusion body particle size distribution. —O— = 0.05 g/L ampicillin; --□--, 0.40 g/L ampicillin; ··△··, 1.50 g/L ampicillin; —*— = low inoculum.

particle size distribution, a 30-fold decrease did result in a small but discernable increase (over 4%) in the D_{50} value of the inclusion bodies obtained. Given the importance of size distribution in centrifugation processes, it is evident that tight control of fermentation is imperative in process design and operation.

It must be noted that whereas the harvesting of the cells from the fermentor could lead to process variation such as cell disruption discharge from the harvesting stage, such variations are minimal compared to the effects of subsequent homogenization.

CONCLUSIONS

This study has revealed the importance of several factors in the design of a biochemical process for inclusion body release and recovery. Firstly, the importance of integrating unit operations was made clear by illustrating the effect of operating conditions of each process on the subsequent step, including the inoculation stage in fermentation. Secondly, given small process volumes, it is possible to modify pilot scale equipment with no loss of information. Finally, the performance of large scale operations may be simulated and predicted by using laboratory techniques together with data obtained from experimentation with scaled down equipment.

REFERENCES

1. SCHEIN, C. H. 1989. Production of soluble recombinant proteins in bacteria. Bio/technology 7: 1142–1148.
2. MITRAKI, A. & J. KING. 1989. Protein unfolding intermediates and inclusion body formation. Bio/Technology 7: 689–697.
3. DUNNILL, P. 1990. Integrated bioprocessing for recombinant proteins. Proceedings of Bioseparations '90. Cambridge, September 27–28.
4. KESHAVARZ MOORE, E., M. HOARE & P. DUNNILL. 1990. Disruption of bakers' yeast in a high pressure homogenizer: New evidence on mechanism. Enzyme Microb. Technol. 12: 764–769.
5. MANNWEILER, K. 1990. University of London PhD thesis.

Engineering Metal Binding Sites into Recombinant Proteins for Facile Purification

MICHELE C. SMITH

Lilly Research Laboratories
Eli Lilly and Company
Lilly Corporate Center
Indianapolis, Indiana 46285–0444

Protein purification involves differentiating the unique physicochemical properties of a desired protein from other proteins in a crude mixture with similar properties. With the advent of recombinant DNA technology, it became possible to attach a "purification handle" to a protein, which imparts a useful physicochemical property and aids in the purification of the protein. A powerful and economical method for purifying recombinant proteins, by engineering an accessible metal binding site into the protein, is presented following a brief review of protein purification methodology.

A few physicochemical properties of proteins and the corresponding protein purification techniques based on those properties are listed in TABLE 1. For instance, the molecular weight of a protein can be calculated from its amino acid composition. An appropriate size exclusion chromatographic medium with the desired molecular weight range can then be selected and used to separate the protein from other proteins with different molecular weights. The overall charge of a protein can be determined from the sum of the charges of the individual amino acids with ionizable side chains. With this information at hand, an anion or cation exchange resin can be chosen to effect the desired separation. The presence of hydrophobic residues in proteins allows them to be purified using techniques based on hydrophobic interactions, such as reverse-phase high performance liquid chromatography (HPLC). In addition to the properties based on the amino acid composition, methods of affinity chromatography are based on the three-dimensional structure of proteins and their ability to recognize substrates, inhibitors, cofactors, and antibodies immobilized on hydrophilic supports.

The one characteristic all these techniques have in common is their predictability, especially for recombinant proteins whose amino acid compositions and sequences are known. Reasonable estimates of size, charge, and hydrophobicity can be made and used to design a purification scheme using ion exchange chromatography, size exclusion chromatography, or HPLC. Affinity chromatography is a powerful technique that can be used once a substrate, cofactor, or antibody is immobilized.

An often neglected property some proteins share is the ability to bind transition metals. Nevertheless, over one third of all known proteins contain metals in their active sites, and many more contain surface binding sites.[1] Immobilized metal ion affinity chromatography, or IMAC, takes advantage of this property to purify proteins based on their differential affinity for transition metals.[2,3] Unfortunately, the metal binding affinities of proteins are not easily predicted and must be determined

TABLE 1. Protein Purification

Property	Technique	Biotech	Reference
Size	Gel filtration	. . .	. . .
Charge	Ion exchange	Polyarginine	10
Hydrophobicity	RP-HPLC	. . .	. . .
Shape	Affinity chromatography	β-Galactosidase	6–9
		IgG binding sites	11–12
		D-Y-K-D$_4$-K	14
		Maltose binding protein	15
		Glutathione-S-transferase	16
Metal binding	Immobilized metal ion affinity chromatography	Chelating peptides	17,18,20,21,23,28
		Polyhistidine	24–27

experimentally. Consequently, IMAC has not been widely used to purify proteins, because their metal binding affinities are largely unknown.

Early work on IMAC proposed that the affinity of a given protein correlated with the number of histidine residues.[2] This idea soon gave way to the concept that the affinity depended on the number of *exposed* histidine residues.[4,5] Unfortunately, a detailed description of the three-dimensional structure of a protein is required to determine this number. Few proteins have been characterized crystallographically compared to the total number of known proteins, so it is virtually impossible to predict the number of exposed histidine residues for most proteins. TABLE 2 lists some proteins whose elution behavior on a Ni(II) IMAC column has been examined. The proteins were eluted with a pH gradient, and the pH required for elution is listed. A lower elution pH indicates a higher affinity for immobilized Ni(II). The number of histidine residues is also listed for those proteins whose compositions are known. The lack of correlation between the number of histidine residues and affinity is clear, because the protein with the highest affinity is porcine insulin with only two histidines. Human carbonic anhydrase, with 11 histidines, has no affinity for immobilized nickel, whereas bovine carbonic anhydrase, which has the same number of histidines, has a high affinity, reflected in an elution pH of 4.95. This lack of predictability has hampered the widespread use of IMAC.

Recombinant DNA technology provides unique opportunities for improving the efficiency of protein purification, as the protein itself can be modified to facilitate that process. This can be done by extending the gene of the protein to code for additional amino acids which will confer a desired and useful property to the recombinant protein. Such a modification distinguishes it from normal proteins present and makes the subsequent purification easier. The same modification, or purification handle, can then be included in the expression system of any new recombinant protein, and the same purification method can be used to obtain pure protein. This aspect alone promises to eliminate the need to devise new purification schemes for each new protein and make purification handles the method of choice.

A few examples of proteins engineered to facilitate their purification are also listed in TABLE I. The first published report by Bastia and coworkers[6] in 1983 described the purification of an initiator protein, R6K, fused to β-galactosidase using a β-galactosidase affinity column. The following year the method was refined and subsequently employed by others.[7–9] A fusion protein was constructed consisting of R6K connected to β-galactosidase through a short collagen sequence. Following elution from the β-galactosidase affinity column, the collagen site was cleaved enzymatically with collagenase, generating mature R6K.[7,9] Early that year, Sassenfeld and Brewer[10] reported the addition of a polyarginine tail to the COOH-terminus of urogastrone. The additional positively charged amino acids led to tight binding to a cation exchange column, which required higher salt concentrations for elution than did other proteins in the crude *Escherichia coli* lysate. After purification was achieved, carboxypeptidase B was used to remove the polyarginine tail from urogastrone. A second pass of the cleaved material over a cation exchange column resulted in further purification, because the contaminating proteins still bound the cation exchange column tightly but the cleaved urogastrone eluted early in the salt gradient.

TABLE 2. Elution pHs of Proteins on a Ni(II) IMAC Column

Protein	Ni(II) IMAC Elution pH[a]	Histidine Number
Alcohol dehydrogenase (yeast)	[b]	10
Alcohol dehydrogenase (horse)	...	...
Bovine serum albumin	6.05	17
Carbonic anhydrase (bovine)	4.95	11
Carbonic anhydrase (human)	...	11
Carboxypeptidase A (bovine)	...	8
Cathepsin C (bovine)	5.65	...
Ceruloplasmin (bovine)	4.7–5.7[c]	...
Ceruloplasmin (human)	5.6	41
Chymotrypsinogen A (bovine)	...	2
Conalbumin (chicken)	...	12
Cytochrome c (horse)	...	3
Cytochrome c (tuna)	...	2
Ferritin (horse)	7.5, 6.1[d]	6
Y globulins (rabbit)	5.5	...
Met-Growth hormone (human)	6.05	3
Immunoglobulin G (human)	5.35	...
Insulin (porcine)	3.65	2
Insulin A chain (S-CM)	...	0
Insulin B chain (S-CM)	5.05	2
Lactalbumin (bovine)	...	3
Lysozyme (chicken)	...	1
Myoglobin (horse)	5.25	11
Myoglobin (sperm whale)	4.85	12
Ovalbumin (chicken)	...	7
Pancreatic trypsin inhibitor (bovine)	...	0
Proinsulin (human)	4.9	2
Proinsulin S-sulfonate	[b]	2
Ribonuclease A (bovine)	7.5, 5.95[d]	4
Superoxide dismutase (bovine)	7.5, 4.95[d]	8
Thyroglobulin (bovine)	7.5, 6.0[d]	31
Transferrin (human)	5.05	19
Tyrosinase (mushroom)	6.0	10

[a]Proteins, obtained from Sigma, were applied to a Ni(II) Chelating Sepharose 6B column as described.[21] Entries without an elution pH, eluted at pH 7.5 during the wash of unbound protein from the column.

[b]These proteins bind to the column but precipitate at acidic pH, so that the exact elution pH could not be determined. Proinsulin S-sulfonate was eluted with EDTA.

[c]Protein elutes as three overlapping peaks with the reported elution pH range.

[d]Two protein peaks elute with approximately a 60/40 distribution, so both elution pH values are reported.

Protein A fusion proteins were purified using IgG affinity columns around the same time and then further developed and improved.[11,12]

Other adaptations of traditional affinity chromatography to purifying recombinant proteins with purification handles were reported more recently.[13] A small octapeptide on the end of a recombinant protein allows the protein to be purified using immobilized monoclonal antibodies raised against the peptide. The octapeptide can be cleaved with the enzyme enterokinase to liberate the desired protein.[14] Chimeric proteins fused to a maltose binding protein have been purified over a cross-linked amylose column.[15] Glutathione affinity chromatography has been used to purify glutathione S-transferase fusion proteins.[16] Although all these techniques can be used to purify recombinant proteins, there are some limitations. A major disadvantage is the use of expensive reagents for the purification, such as sophisticated substrates, inhibitors, or monoclonal antibodies. In addition to the cost, these reagents are often specialty items and not readily available. The expression of fusion proteins can be inefficient when the purification handle comprises a large part of the expression product. Using metal binding peptides as purification handles eliminates these disadvantages, as the peptides are small and the materials for carrying out the IMAC are readily available and inexpensive. In addition, IMAC can be carried out under a wide variety of conditions, which also increases its utility.

In developing a strategy to prove our hypothesis that a metal binding peptide site on the end of a recombinant protein could be used to purify that protein using IMAC, we considered the minimal requirements for a chelating peptide.[17,18] Commercially available Chelating Sepharose Fast Flow (Pharmacia) immobilizes metal ions through iminodiacetic acid, a tridentate ligand that occupies three coordination sites about the metal. For metal ions that form six coordinate complexes, this leaves three coordination sites available for binding an incoming protein or peptide. A review of the literature on metal-peptide complexes revealed that a simple dipeptide could occupy three coordination sites about a metal ion through the NH_2-terminal amine, the COOH-terminal carboxylate, and a deprotonated amide nitrogen.[19] Peptides or amino acids with coordinating side chains, such as histidine, cysteine, and aspartic acid, form stable complexes with six or five membered rings made up of the metal ion, amine, and side chain donor atom. Our initial survey of di- and tripeptides containing histidine, lysine, and aspartic acid revealed a wide range of affinities for immobilized metal ions.[20] Three peptides, His-Trp, His-Tyr-NH_2, and His-Gly-His, of approximately 50 that were examined had unusually high affinities for Co(II), Ni(II), and Cu(II) IMAC columns and were identified as potential chelating peptide purification handles. This study also revealed that Ni(II) was well suited for resolving small differences in affinities, compared to Cu(II), and was therefore used for subsequent purifications.

The next step in the development of chelating peptide-immobilized metal ion affinity chromatography, or CP-IMAC, was to show that the metal ion affinity of a small peptide could be transferred to a larger polypeptide. This was accomplished with luteinizing hormone-releasing hormone (LHRH) analogs that contained the chelating peptide sequence His-Trp.[21] The sequence of 2-10 LHRH (His-Trp-Ser-Tyr-Gly-Leu-Arg-Pro-Gly-NH_2) contains the chelating peptide sequence His-Trp, whereas the 4-10 LHRH analog (Ser-Tyr-Gly-Leu-Arg-Pro-Gly-NH_2) lacks the chelating peptide. When these polypeptides were chromatographed over a Ni(II) IMAC column, only the 2-10 LHRH analog bound Ni(II) and required the same low pH of 4.4 for elution as did the peptide His-Trp alone. These experiments were key in demonstrating that the high affinity of a small dipeptide could be transferred to a polypeptide that has no affinity for immobilized metal ions, allowing the CP-peptide to be purified.

Human proinsulin was chosen as a model recombinant protein to be modified with a chelating peptide for the next step in the development of CP-IMAC. As shown in TABLE 2, proinsulin itself binds immobilized Ni(II). The effect of the addition of His-Trp to the NH$_2$-terminus on the immobilized metal ion affinity of proinsulin was therefore a stringent test of our hypothesis. The chromatographic behavior of proinsulin and His-Trp-proinsulin S-sulfonates on a Ni(II) column was examined. The His-Trp–proinsulin eluted from the column with a lower pH, by about 1.5 units, than did proinsulin itself. This model system demonstrated the ability to improve on the inherent metal binding characteristics of a protein by simply attaching a high affinity dipeptide sequence to the NH$_2$-terminus.[21]

The purification of recombinant proteins using CP-IMAC has been accomplished with a number of different proteins and chelating peptide sequences. Our laboratory has purified human insulin-like growth factor II (IGF-II) by expressing it with a chelating peptide sequence, Met-His-Trp-His, on the NH$_2$-terminus followed by an enzymatic cleavage site, the ompA signal peptide.[22] The expression product, Met-His-Trp-His-ompA signal peptide-IGF-II, was purified over a Cu(II) IMAC column. The purified protein, or bound fraction, was digested with signal peptidase to release IGF-II.[23] The digestion mixture was then chromatographed over a Cu(II) IMAC column, and the unbound fraction, which contained purified cleaved IGF-II, was collected.

Hochuli and coworkers[24] purified mouse dihydrofolate reductase over a Ni(II) IMAC column with a tetradentate-immobilizing ligand, nitrilotriacetic acid. A polyhistidine chelating peptide with either 2, 3, 4, 5, or 6 histidines was attached to the NH$_2$-terminus or COOH-terminus. The COOH-terminal chelating peptide could also be removed using limited digestion with carboxypeptidase A. A number of HIV viral proteins have been purified using CP-IMAC, including Rev, Tat, and reverse transcriptase with the His-His-His-His-His-His chelating peptide sequence attached to the NH$_2$-terminus.[25-27] The presence of the chelating peptide did not interfere with the biological activity of these proteins; therefore, the fusion proteins were used directly. Nilsson and coworkers[28] investigated the use of multimers of a chelating peptide, (Ala-His-Gly-His-Arg-Pro)$_n$ where n is either 2, 4, or 8, to purify two recombinant proteins. The chelating peptide sequences were attached to the COOH-terminus of the ZZ protein, derived from protein A, and to the NH$_2$-terminus of β-galactosidase and purified over a Zn(II) IMAC column when four or eight copies of the peptide were used. The Zn(II)-immobilized ZZ-His$_4$ fusion protein was also used as an affinity column to bind IgG.[28]

The longer chelating peptides described recently seem to have a high affinity for iminodiacetic acid (IDA) immobilized metal ions, because Zn(II) can be used for the separation.[28] Zn(II), compared to Cu(II) or Ni(II), has a rather low affinity for peptides and proteins.[1,20] The binding behavior observed by Ljungquist and coworkers[28] would only be expected if the affinity of the His-Gly-His multimers for Zn(II) were greater than those for the dipeptides, inasmuch as the latter did not bind to Zn(II) under similar conditions.[20] This increase in affinity is most likely due to the dynamic nature of coordinate covalent bonds. The rate of dissociation of a coordination complex with a large formation constant is very small but nevertheless finite, so that at any given time a small fraction of the complex is dissociated. With the longer chelating peptide, the probability of encountering another histidine residue and reforming the same or a similar complex is greater than it is with a simple dipeptide with only one histidine residue.

Another advantage CP-IMAC offers is the versatility of chromatographic conditions that can be used to carry out a separation. Recombinant proteins are expressed as either soluble or insoluble products and IMAC can be used in either case as the first step. Supernatants of lysed cells can be applied directly to an IMAC column with or without high salt concentrations. Denaturants, such as detergents, urea, guanidine hydrochloride, and organic solvents, can be used in the IMAC chromatographic separation when the expression product is insoluble. The bound proteins can be eluted with a pH gradient or with increasing concentrations of a displacing ligand, such as imidazole, in cases in which the protein of interest cannot tolerate a low pH. IMAC columns are also resistant to microbial contamination because of the high metal content.

The only restrictions for carrying out an IMAC separation are based on the need to provide the correct chemical environment to form a coordination complex. Proteins must be applied to the IMAC columns at neutral or basic pH values where the imidazole ring of histidine is deprotonated and available for coordinating an immobilized metal ion. The buffers cannot contain large amounts of chelating agents or amines that will coordinate metals, including those immobilized on the column, thereby removing them from the column. This can be detected easily, as the columns are colored when metal ions other than zinc are bound and opaque when free of transition metals. During pH gradient elutions, pH values less than 3.5 should be avoided, as the carboxylic acid groups of IDA will become protonated, releasing metal ions and contaminating the purified protein. Gradient elutions with imidazole eliminate this concern, as metal leaching is not a problem. These unfavorable conditions can easily be avoided.

The increasing use of CP-IMAC to purify recombinant proteins points to its utility compared to other types of purification handles. The small size required to achieve an affinity purification often does not interfere with the biological activity of the protein, which then eliminates the need for removing the chelating peptide. The efficiency of expression is optimal because the purification handle is tiny compared to that of most fusion proteins whose purification handle is as large or larger than the protein of interest. Together with the ready availability of the chromatographic resin, these aspects bode well for the future use of this technology.

ACKNOWLEDGMENTS

I would like to thank T. Furman, J. Cook, C. Pidgeon, T. Ingolia, H. Hsiung, and W. Muth, who have been involved in the development of CP-IMAC.

REFERENCES

1. COTTON, F. A. & G. WILKINSON. 1980. Advanced Inorganic Chemistry, 4th Ed. Chapt. 20, 28, & 31. Wiley. New York.
2. PORATH, J., J. CARLSSON, I. OLSSON & G. BELFRAGE. 1975. Nature **258:** 598–599.
3. SULKOWSKI, E. 1989. BioEssays **10:** 170–175.
4. SULKOWSKI, E., K. VASTOLA, D. OLESZEK & W. VONMUENCHHAUSEN. 1982. *In* Affinity Chromatography and Related Techniques. T. C. J. Gribnau, J. Viser & R. J. F. Nivard, eds.: 313–322. Elsevier Scientific Publishing Company. Amsterdam.
5. HEMDAN, E. S., Y.-J. ZHAO, E. SULKOWSKI & J. PORATH. 1989. Proc. Natl. Acad. Sci. USA **86:** 1811–1815.
6. GERMINO, J., J. GRAY, H. CHARBONNEAU, T. VANAMAN & D. BASTIA. 1983. Proc. Natl. Acad. Sci. USA **80:** 6848–6852.
7. GERMINO, J. & D. BASTIA. 1984. Proc. Natl. Acad. Sci. USA **84:** 4692–4696.

8. ULLMANN, A. 1984. Gene **29**: 27–31.
9. SCHOLTISSEK, S. & F. GROSSE. 1988. Gene **62**: 55–64.
10. SASSENFELD, H. M. & S. J. BREWER. 1984. Bio/Technology **2**: 76–81.
11. NILSSON, B. & L. ABRAHMSÉN. 1990. Methods Enzymol. **185**: 144–161.
12. MOKS, T., L. ABRAHMSEN, E. HOLMGREN, M. BILICH, A. OLSSON, M. UHLÉN, G. POHL, C. STERKY, H. HULTBERG, S. JOSEPHSON, A. HOLMGREN, H. JÖRNVALL & B. NILSSON. 1987. Biochemistry **26**: 5239–5244.
13. UHLÉN, M. & T. MOKS. 1990. Methods Enzymol. **185**: 129–143.
14. HOPP, T. P., K. S. PRICKETT, V. L. PRICE, R. T. LIBBY, C. J. MARCH, D. P. CERRETTI, D. L. URDAL & P. J. CONLON. 1988. Bio/Technology **6**: 1204–1210.
15. DI GUAN, C., P. LI, P. D. RIGGS & H. INOUYE. 1988. Gene **67**: 21–30.
16. SMITH, D. B. & K. S. JOHNSON. 1988. Gene **67**: 31–40.
17. SMITH, M. C. & C. PIDGEON. February 11, 1986. U. S. Patent 4,569,794. U. S. Patent Office. Washington, DC.
18. SMITH, M. C. & C. PIDGEON. June 6, 1986. European Patent Application 184355. European Patent Office, Munich, Germany.
19. FREEMAN, H. C. 1967. Adv. Protein Chem. **22**: 257–437.
20. SMITH, M. C., T. C. FURMAN & C. PIDGEON. 1987. Inorg. Chem. **26**: 1965–1969.
21. SMITH, M. C., T. C. FURMAN, T. D. INGOLIA & C. PIDGEON. 1988. J. Biol. Chem. **263**: 7211–7215.
22. SMITH, M. C., J. A. COOK, T. C. FURMAN, P. D. GESELLCHEN, D. P. SMITH & H. HSIUNG. 1990. *In* Protein Purification: From Molecular Mechanisms to Large-Scale Processes. M. R. Ladish, R. C. Willson, C. C. Painton & S. E. Builder, eds. ACS Symposium Series **427**: 168–180. American Chemical Society. Washington, DC.
23. WICKNER, W., K. MOORE, N. DIBB, D. GEISSERT & M. RICE. 1987. J. Bacteriol. **169**: 3821–3822.
24. HOCHULI, E., W. BANNWARTH, H. DÖBELI, R. GENTZ & D. STÜBER. 1988. Bio/Technology **6**: 1321–1325.
25. COCHRANE, A. W., C.-H. CHEN, R. KRAMER, L. TOMCHAK & C. A. ROSEN. 1989. Virology **173**: 335–337.
26. GENTZ, R., C.-H. CHEN & C. A. ROSEN. 1989. Proc. Natl. Acad. Sci. USA **86**: 821–824.
27. LEGRICE, S. F. J. & F. GRÜNINGER-LEITCH. 1990. Eur. J. Biochem. **187**: 307–314.
28. LJUNGQUIST, C., A. BREITHOLTZ, H. BRINK-NILSSON, T. MOKS, M. UHLÉN & B. NILSSON. 1989. Eur. J. Biochem. **186**: 563–569.

Integration of Cell Culture with Continuous, On-Line Sterile Downstream Processing

P. GRANDICS, S. SZATHMARY, Z. SZATHMARY, AND
T. O'NEILL

Sterogene Bioseparations, Inc.
140 E. Santa Clara Street
Arcadia, California 91006

Mammalian cell lines are becoming a method of choice for the production of complex proteins that require extensive posttranscriptional modifications for bioactivity. The commercial success of these protein products depends on the development of efficient and economical production methods. The product must be of the highest quality and obtained at a high yield. It should be produced in reproducible fashion over long periods of time, and the entire process should be automated. This, however, is a difficult task because of the complexity of current bioprocess technologies.

To meet these objectives, a technological synthesis (integration) of the fragmented upstream and downstream processing is necessary. The individual unit operations must be modified and organized so that they can be linked together into a continuous chain of operation.

The integration of upstream and downstream processes should also allow continuous, on-line product recovery. Because cell culture requires sterility, downstream processing should also be operated sterile in an integrated, automated bioprocess. The product is removed from the cell culture as it is being produced, which can eliminate degradation. This profoundly affects the quality of the product and simplifies downstream processing by minimizing the number of processing steps. With highly selective affinity chromatography the culture medium can be recirculated, improving media utilization. Also, it is important to use affinity separation which does not change the composition of the cell culture medium.

High protein (serum)-containing media could also be used in the integrated system to produce a "more" natural product without affecting the efficiency of operation and the final yield. The culture conditions profoundly affect the quality of the product.[1] Cell culture media affect the glycosylation of proteins which in turn causes pronounced changes in circulatory life/therapeutic index of the product.[2] Culture media containing serum preserve the natural glycosylation pattern of proteins,[3] therefore, it should be considered the medium of choice. Proteolytic or other degradation during cell culture can also be reduced using serum-containing media and more importantly by removing the product from the catabolic environment of cell culture as it is being produced.

If protein is purified as it is being produced, on-line, real-time process monitoring can be included in the system. This allows optimization of the process and provides a high degree of reproducibility of production that is difficult to achieve in biological

322

systems using current bioprocess technologies. On-line, real-time process monitoring will extend the longevity of continuous culture, thus greatly improving process economy.

The integration of cell culture with on-line protein purification requires the development of new affinity technologies. It requires stable, chemically inert, sterilizable (autoclavable) activated affinity chromatography resins. The resin must be compatible with mammalian cell culture. As mammalian cells are very sensitive to a wide range of chemicals even at ppm concentrations, it is important that the resin does not leach toxic chemicals or introduce bacterial/viral contamination into the cell culture.

The affinity/immunoaffinity chromatography resin must also withstand the conditions of prolonged cell culture. This includes elevated temperature, oxygenation, and the presence of proteolytic enzymes in the culture medium. Methods for elution of protein from immunoadsorbents need to be developed which do not cause denaturation of eluted proteins and at the same time preserve the binding capacity of the immunoadsorbent for long periods of time.

Current-activated resins are not suitable for integration because they are not autoclavable or chemically inert. Even if the problem of sterilization is overcome, toxic leaving groups of activated media poison cells in culture. An example of this is isocyanate groups leaching from cyanogen bromide activated matrices. Current immunoaffinity chromatography methods would not allow integration because of the rapid loss of immunosorbent capacity[4] on 10–30 cycles. Leached ligand also complicates downstream processing, requiring extra purification steps.

We have integrated the production and purification of complex proteins, such as monoclonal antibodies or recombinant proteins, into a single unit, automated process. The integration of cell culture and protein purification was made technically and economically feasible by the development of the stable, nontoxic, autoclavable, activated resin, Actigel-ALD, as well as the nondenaturing elution medium, Acti-Sep.[5,6]

MATERIALS AND METHODS

A mouse x mouse hybridoma (HB57 obtained from ATCC) producing an IgG monoclonal antibody was used. The cells are grown in RPMI 1640 medium supplemented with antibiotics and iron-supplemented bovine serum. Cells were stained with trypan blue dye and counted using a hemocytometer to determine cell density and viability. Glucose and lactate levels in the cell culture medium were measured using kits based on enzymatic assay from Sigma.

Monoclonal antibody (mAb) concentration in the cell culture medium is determined using ELISA. The purified mAb concentration is determined by OD_{280} absorbance measurement.

The hollow fiber bioreactor (13 ml extracapillary volume) and the hollow fiber oxygenator were from Microgon. Actigel-ALD Superflow activated resin, ActiSep Elution Medium, and CleanWash wash medium were from Sterogene Bioseparations, Inc. Cell culture medium RPMI 1640 was from Mediatech. Iron supplemented newborn calf serum was from Hyclone Laboratories. Anti-mouse IgG antiserum was obtained from Antibodies, Inc. The specific antibody to mouse IgG was purified by immunoaffinity chromatography as reported.[5]

RESULTS AND DISCUSSION

In the integrated cell culture/protein purification system, the desired protein product is produced using mammalian cell culture technology in the bioreactor loop. Highly specific affinity chromatography separation is used in the separation loop to purify the protein product from the cell culture medium on-line. The operation of these two loops is integrated and automated. A schematic diagram of the system is shown in FIGURE 1. The system comprises two subunits, the bioreactor and the product recovery modules.

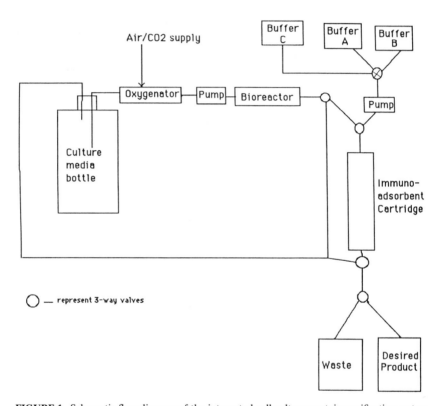

FIGURE 1. Schematic flow diagram of the integrated cell culture-protein purification system.

General Description

The cell culture unit employs a hollow fiber microfiltration membrane bioreactor to grow and maintain the mammalian cells in a perfusion mode. The bioreactor consists of closely spaced coaxial fibers in a cylindrical housing. The fibers have an inner diameter of 610 μm with 0.2 μm pore size. The space within the fiber is called intracapillary space. The cells are localized in the extracapillary space and the nutrient or culture medium flows through the intracapillary space. For oxygenation of the cell culture medium, a hollow fiber membrane oxygenator is used; 5% CO_2 in air mixture is passed through the oxygenator. The pore size of the fibers in the oxygenator is 0.02 μm. The bioreactor loop is kept in the incubator at 37°C. The cell culture medium is in a glass bottle equipped with sterile vent. A peristaltic pump is used for media circulation. The flow rate of culture medium through the bioreactor loop and the immunoaffinity column was 200 ml/min.

The bioreactor loop interfaces with the separation or product recovery loop through the immunoaffinity column (FIG. 1). The separation loop consists of a chromatography column, buffer and reagent bottles, a peristaltic pump, and an on-line UV monitor to detect absorbance in the column effluent. The direction of flow of the buffers, product, and waste fluids is regulated by appropriate solenoid valves linked to a controller and a PC. Separation of the desired product from the culture medium is achieved by using immunoaffinity chromatography. The culture medium is continuously passed through the affinity column specific to the product to be isolated. Periodically, the bioreactor is bypassed and the affinity column is eluted to collect the purified product. The product concentration in the eluate is determined by on-line UV spectrophotometry.

Cell Line Characterization

The hybridoma duplication time in stationary culture is 22 hours in RPMI 1640 supplemented with 10% bovine serum. The hybridoma is a low producer and requires a high serum concentration in the medium for optimal viability and production. We selected this low producer cell line to test the capabilities of the system. The growth, productivity, and cellular metabolism of HB57 hybridoma in stationary culture was first studied (FIGS. 2 and 3). As the cells reach confluence, the viability drops (FIG. 2). Monoclonal antibody production continues to increase even after the cells reach confluence (FIG. 3). The glucose and lactate levels are also monitored (FIG. 4). The changes in the glucose and lactate levels reflect a normal metabolic pattern.

The Immunoadsorbent

In this study, we cultured a hybridoma in the bioreactor, and the secreted IgG monoclonal antibody was purified on an anti-mouse IgG affinity column. The affinity-purified anti-mouse IgG, preadsorbed to eliminate cross-reactivity with bovine serum components, is coupled to Actigel-ALD Superflow at 4 mg/ml. The immunoadsorbent is packed into a 20 ml bed volume column.

It is preferred to use anti-antibody immunoaffinity resin over immobilized protein A or protein G for several reasons. These F_c receptor ligands bind bovine IgG along with the desired monoclonal antibody, necessitating the use of defined, low protein media or the development of additional purification methods to separate the product from the contaminating bovine antibody. However, serum-free media may cause aberrant glycosylation of the product[2] which needs to be examined in every case. F_c receptor ligands may bind essential growth factors from the cell culture medium that would hamper cell productivity. Another limitation is that only IgG antibody can be purified on F_c receptor ligands.

An anti-antibody immunoaffinity column overcomes all of these limitations and allows selective purification of low concentrations of the desired mAb product against high concentrations of irrelevant antibody. In the culture medium, the bovine IgG concentration is approximately 1 mg/ml (10% serum supplement), while the

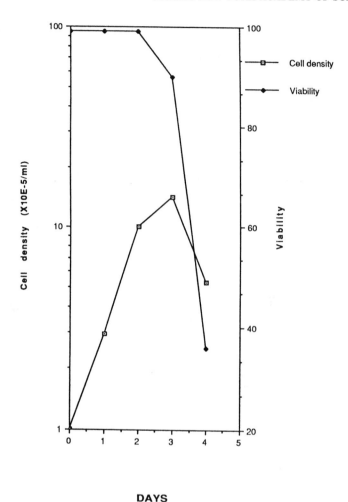

DAYS

FIGURE 2. Growth of mouse hybridoma HB57 in stationary culture.

monoclonal IgG concentration is approximately 10 μg/ml for this cell line. Also, antibody other than IgG can readily be purified on antibody-specific immunoaffinity columns.

SYSTEM OPERATION

The system is configured so that all the components coming into contact with the cell culture medium or buffers are autoclavable. Assembled components of the bioreactor and product recovery loops, including the chromatography column filled with the activated resin, Actigel-ALD Superflow, are sterilized with autoclaving.

The column bed is then conditioned and capture antibody immobilized by injecting a mixture of the antibody and the coupling reagent, $NaCNBH_3$, onto the column through a sterile port. The coupling is allowed to proceed for 6 hours at room temperature, and then the column is extensively washed with sterile saline solution.

To start up the system, the culture medium vessel as well as the buffer vessels are filled aseptically and the bioreactor loop is perfused with cell culture medium for several hours after which the medium flask is changed. The bioreactor is seeded with 50 million HB57 cells. The medium is replaced as soon as the nutrients are exhausted. After the first week, the culture medium bottle in the system is replaced on a daily basis with fresh medium. It took approximately 10 days for the cells to

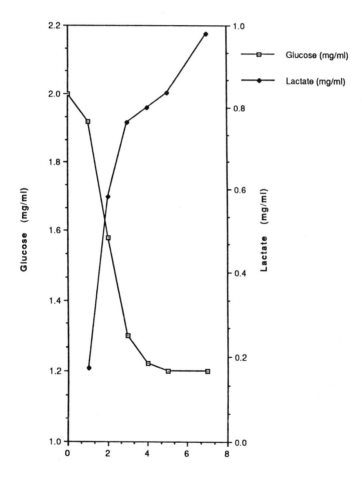

DAYS

FIGURE 3. Growth and mAb production of the mouse hybridoma HB57 in stationary culture.

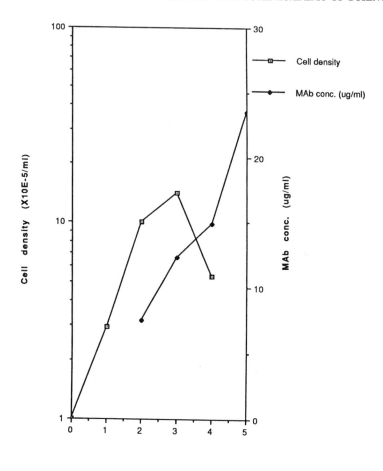

DAYS

FIGURE 4. Glucose consumption and lactate accumulation during the growth of mouse hybridoma HB57 in stationary culture when seeded at 1×10^5 cells/ml.

populate the bioreactor and secret 10 μg/ml mAb into the cell culture fluid. The mAb concentration in the spent medium is determined by using ELISA.

To monitor the metabolism of hybridoma cells in the bioreactor, the glucose consumption and lactate accumulation rate are measured (FIG. 5). The glucose consumption and lactate accumulation rate increase with time initially as the bioreactor becomes increasingly populated with hybridoma cells. After the 35th day of the run, we attempted to adapt the cells to a 5% serum-containing medium which slowed down the metabolism of the cells. As the hybridoma could not adapt to this reduced serum level, the initial 10% serum concentration in the culture medium was restored, leading to the restoration of normal metabolic functions of hybridoma cells. The high density cell culture still requires serum for normal metabolic functions.

After the fifth day, the product recovery loop was turned on and product purification commenced. When the immunoaffinity column nears saturation, the product recovery cycle is initiated. A typical product recovery cycle starts with bypassing the immunoaffinity column and directing the cell culture medium flow only through the bioreactor. From vessel A, a wash buffer, CleanWash, is applied to the column for 10 minutes. The wash fluid is directed into the waste container, while the OD_{280} of the effluent is monitored. When the absorbance falls to baseline, one bed volume of Actisep is applied and allowed to stay on the column for 15 minutes. Another wash buffer (0.5 M NaCl, 15 mM phosphate, pH 7.0) is then applied to displace the elution medium and the eluted product from the column. The absor-

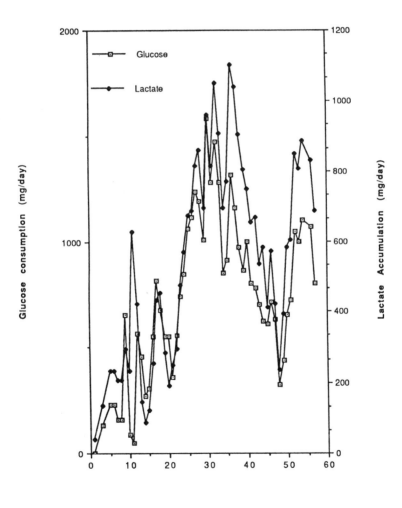

days

FIGURE 5. Glucose consumption and lactate accumulation in the integrated system.

bance of the column effluent is monitored, and the UV-absorbing material collected in the product vessel.

The column is then regenerated for the next product recovery cycle by removing traces of the elution medium by a short (5-minute) wash with 0.5 M NaCl, 15 mM phosphate, pH 7.0, followed by RPMI 1640 medium. The culture medium flow is again directed over the affinity column, and product adsorption from the culture medium continued. The productivity of the cell line required that the column be eluted twice a day. Practically, all the antibody ($>98\%$) is captured from the culture medium by the immunoadsorbent. The column binding capacity changed little over the 2 months of continuous operation (FIG. 6).

The purity of isolated antibody was tested by SDS-polyacrylamide gel electro-phoresis (FIG. 7). The purity of the antibody is approximately 98%. The production of mAb increases with time initially as the bioreactor gets populated with cells (FIG. 8). After the third week, the cells have populated the bioreactor and antibody production is stabilized. We have continuously harvested purified monoclonal anti-body from the system over a 2-month period of time after which the production run was terminated. Over 800 mg of pure antibody was collected from the system. This is in agreement with the productivity of the cell line. The productivity of the HB57 cell line in the bioreactor is found to be similar to the productivity observed in stationary culture. The purified monoclonal antibody reacts on ELISA with the pure antigen, human IgM. The activity of the purified antibody is $>95\%$.

CONCLUSION

We have achieved the integration of production and purification of secreted proteins from mammalian cell culture into a single unit operation. The integrated

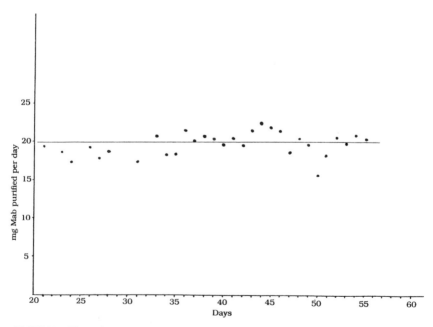

FIGURE 6. The stability of the operation of the immunoaffinity column; antibody recoveries.

heavy chain

light chain

FIGURE 7. Purity of isolated antibody on SDS-polyacrylamide gel electrophoresis.

system has been operating as a closed, sterile, automated production unit of pure monoclonal antibodies. This system is based on a novel activated resin technology that is nontoxic to mammalian cells and makes sterilization (autoclaving) of the resin possible, allowing sterile operation of immunoaffinity chromatography. This immunoaffinity column is stable under the conditions of mammalian cell culture for long periods of time. ActiSep elution medium allows long-term, stable operation of the immunoaffinity column (over 110 cycles) and preserves eluted antibody activity.

The high selectivity of immunoaffinity chromatography allows the replacement of fetal calf serum or defined media with inexpensive bovine serum, resulting in significant (approximately fivefold) cost reduction in the cell culture media cost.

The integrated system produces sterile, pyrogen-free purified antibody. The high specificity of immunoaffinity chromatography along with the optimization of washing steps allows the production of monoclonal antibodies at a purity of approximately 98% with activity > 95%. Proteolytic degradation of purified antibody is eliminated. The use of the integrated system can be extended to the production of other mammalian cell culture derived proteins.

The integration of upstream and downstream processing also has a significant impact on process development. More efficient (high yield) processes can be developed and optimized in less time. Because the protein product is purified as it is being produced, the pure product can be analyzed on-line, thus obtaining real-time data on how the cell culture conditions affect product characteristics.[7] This information could be used to optimize culture conditions at different stages in the cell culture. With this information, more efficient cell culture media can be developed, such as media optimized for the right glycosylation. An on-line, rapid, automated analytical instrumentation system developed around the integrated system would lead to a fully automated production system, resulting in a highly consistent production of cell culture derived proteins.

The capital expenses of a biopharmaceutical production plant could be reduced significantly if integrated production systems were used. Integration reduces the size of the production units because of the continuous operation of the system. The integrated system can use small bioreactors with high density, high productivity cell culture. Downstream, because of the continuous recycling of the columns, small chromatography columns suffice. Thus, besides reducing capital equipment costs, the disposable costs are also minimized. As the product is quarantined, in the event contamination occurs, the losses will be minimal as only the small columns, bioreactor, and the quarantined product will be discarded and the previously manufactured product is saved.

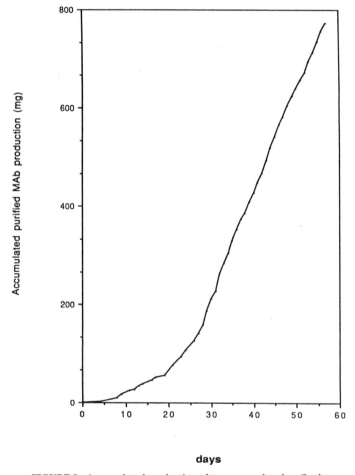

days

FIGURE 8. Accumulated production of pure monoclonal antibody.

The sizing down of the whole production system will greatly expand the application range for the integrated system. It is estimated that the current cost of a generic multipurpose biopharmaceutical facility (currently approximately 50 million dollars) may be reduced by an order of magnitude if the production was done in an integrated system. With this kind of integrated system, development and production of biopharmaceuticals can be achieved at a relatively low cost, approaching that of synthetic pharmaceuticals. Also the development of pharmaceuticals to treat diseases with a smaller patient population may become economically feasible with the integrated bioprocess system.

SUMMARY

The development of an integrated system for the continuous, automated production of pure cell culture derived proteins is discussed. The system comprises a cell culture subunit for the continuous culture of mammalian cells and a purification subunit linked on-line to the cell culture subunit. The cells are compartmentalized and continuously perfused with culture medium. The cell culture medium leaving the bioreactor is perfused through a sterile immunoaffinity column that instantaneously removes the product from the culture fluid. This results in improved product quality because the product is quickly removed from the cell culture, thus minimizing contact with degradative enzymes. The culture medium, stripped from the secreted product, is recirculated into the bioreactor. The system allows simple, automated, and economical production of purified proteins with higher quality than that possible with current production methods. The integration also allows on-line, real-time process monitoring, thus simplifying process development and allowing more consistent production of biologics.

REFERENCES

1. GOOCHEE, C. F. & T. MONICA. 1990. Biotechnology **8:** 421.
2. MAIORELLA, B. *et al.* 1990. Annual Meeting of the American Chemical Society. Washington DC. Abstr. 111.
3. DWEK, R. A. 1990. IBEX Scientific Conference, San Mateo, CA, p. 16.
4. EVELEIGH, J. W. & D. E. LEVY. 1977. J. Solid-Phase Biochem. **2:** 45.
5. GRANDICS, P., Z. SZATHMARY & S. SZATHMARY. 1990. Ann. N.Y. Acad. Sci. **589:** 148.
6. THALLEY, B. S. & S. B. CARROLL. 1990. Biotechnology **8:** 934.
7. SZATHMARY, S. & P. GRANDICS. 1990. Biotechnology **8:** 924.

Rational Design of Purification Processes for Recombinant Proteins

J. A. ASENJO, J. PARRADO,[a] AND B. A. ANDREWS

Biochemical Engineering Laboratory
University of Reading
Reading RG6 2AP, England

A critical element of modern process biotechnology is the separation and purification of a recombinant or mammalian protein from a fermentation or cell rupture supernatant. As it represents the major manufacturing cost, competitive advantage in production will depend not only on innovations in molecular biology and other areas of basic biological sciences but also on innovation and optimization of separation and downstream processes.[1]

The design of a process to economically purify a recombinant protein, maintaining a high yield, yet obtaining a virtually pure product while minimizing the cost, requires three main considerations: (1) clearly defining the final product requirements, (2) characterizing the starting material, and with these two pieces of information in hand (3) defining possible separation steps and constraints regarding operations and conditions to be used.[2]

It is necessary to define the final product and have information on its uses. How is the product going to be used (if therapeutic, what size doses, how many, and how often?)? Questions regarding the purity required (e.g., 99%, 99.9%, or 99.99%) as well as allowable ranges of impurity concentrations are vital. With therapeutic proteins all impurities have to be minimized, whereas for the production of bulk industrial enzymes this is not the case. For instance, in vaccine production it is necessary to remove all traces of unwanted immunogens to prevent potentially catastrophic immunological side reactions.

Large scale process design will be mainly dependent on the physical, chemical, and biochemical properties of the contaminating materials in the original broth and those of the protein that will constitute the final product. The properties of the starting material will be partially determined by its fermentation source, namely, bacterial, yeast, or mammalian cell, the type of cultivation medium used (e.g., presence of albumin, calf serum, proteases, and solid bodies such as whole cells or cell debris), whether the product is intracellular or extracellular, and if it is present in inclusion bodies or other types of protein particles (e.g., virus-like particles, VLPs). To these factors we must add the actual physicochemical properties of the product (surface charge/titration curve, surface hydrophobicity, molecular weight, biospecificity towards certain ligands [e.g., dyes], pI, and stability) as compared to those of the contaminant components in the crude broth.

This paper reviews and discusses recent developments in the design of purification processes for recombinant proteins and the use of modern design tools including databases for rational process selection, simulations, and advanced computer systems used for this purpose.

[a]PRESENT ADDRESS: Departamento de Bioquímica, Universidad de Sevilla, 41012-Sevilla, Spain.

MATERIALS AND METHODS

Adsorption experiments were carried out in batch in stirred flasks using 5 ml of swollen gel with 10 ml of protein sample. Concentration of the protein sample was about 5 mg/ml. The sample was incubated until equilibrium was reached and the gel was settled by centrifugation. Protein adsorbed in the gel was estimated by the difference between the initial and the final concentrations of protein in solution. Total protein was measured by ultraviolet absorbance (280 nm) using the respective extinction coefficient for each protein.

The buffers used for each pH were prepared according to the method of Scopes[3] (e.g., 0.01M Tris HCl for anion exchange and 0.01M MES-KOH for cation exchange). Therefore, for anion exchange with DEAE-Sepharose the buffers used were diethanolamine, triethanolamine, imidazole, histidine, and pyridine at pH 9, 8, 7, 6, and 5, respectively, and for cation exchange with CM-Sepharose they were lactic acid, acetic acid, MES, MOPS, and tricine at pH 4, 5, 6, 7, and 8, respectively.

Before the adsorption experiments were carried out at each pH, the gels were equilibrated by washing 10 times with 10 ml of buffer at a higher ionic strength (0.5 M). They were then equilibrated at the lower ionic strength (0.01 M) by washing 5 times with 10 ml of buffer at the appropriate pH but the lower ionic strength.

Titration curves for the proteins (BSA and myoglobin were obtained from Sigma and thaumatin from Tate and Lyle, Reading) were obtained by dissolving 20 mg of pure protein in 50 ml of deionized water and the protein was titrated with 0.01M HCl or 0.01M NaOH.

PURIFICATION PROCESS AND UNIT OPERATIONS

The first step is to define realistic separation steps on the basis of all the information provided. The following five main heuristics or rules of thumb provide a good basis for initial process selection:

Rule 1: Choose separation processes based on different physical, chemical, or biochemical properties.

Rule 2: Separate the most plentiful impurities first.

Rule 3: Choose those processes that will exploit the differences in the physicochemical properties of the product and impurities in the most efficient manner.

Rule 4: Use a high resolution step as soon as possible.

Rule 5: Do the most arduous step last.

An important point that needs consideration is that once the purification procedure is set and regulatory approval of the product is underway, the purification procedure cannot be changed. Only a particular product obtained by a specific procedure obtains regulatory approval; hence, once this is given, the purification procedure is fixed. Therefore, even in the very early stages of protein purification, only laboratory procedures that can realistically be used in large scale, should be employed, that is, procedures for which suitable large scale equipment either exists or might be developed in the foreseeable future. Otherwise more rationalized and efficient processes will have to await the second or third generation of process and plant design, which can be a very wasteful exercise.

The number of necessary steps in a large scale protein purification procedure is usually not more than four or five, and they can be divided into two main subprocesses of protein recovery/isolation and protein purification:

Protein Recovery/Isolation

1. Cell separation
2. Cell disruption and debris separation (for intracellular proteins only)
3. Concentration

Protein Purification

4. Pretreatment or primary isolation
5. High resolution purification
6. Polishing of final product

It is important to consider the process of fermentation and downstream processing as a single system so that, for example, the effect of decisions about the fermentation conditions on subsequent purification stages is made clear. Product concentration will partly depend on the reactor system used (stirred tank/air-lift or hollow fiber). The presence of nucleic acids and proteases as well as bacterial contamination has to be minimized, which creates a need for rapid processing. The presence of calf serum will usually increase the number of purification stages required. Recombinant proteins in many cases are present in particles that need to be solubilized and refolded. In conclusion, it is important not only to discuss upstream processing in the light of all the protein purification stages but also to make the necessary decisions that will improve the recovery of the protein product early in the process development stages.

Protein Recovery

Recovery comprises removing the broth from the biochemical reactor system (e.g., a fermentor or a hollow fiber reactor) and processing it until a cell-free solution is obtained and the total protein concentration including the product is 60–70 g/L.[4,5]

For cell separation or harvesting the variety of equipment found in industrial practice is not very large (centrifuge, rotary vacuum filter, and membrane filtration[6]) and the decision depends on the microbial source, equipment availability, equipment efficiency, and economics. If the product of interest is secreted (extracellular), then the liquid part is kept (FIG. 1); if the product is intracellular, the solid fraction of the harvesting is kept (FIG. 2). When a mammalian cell culture is used, the product is usually extracellular. Typical harvesting operations used are centrifugation (mainly for yeast but also for mammalian cells and bacteria), rotary vacuum filtration (mainly fungi), and microporous filtration (bacteria, yeast, mammalian cells, and also fungi).

Cell disruption is required when the product is intracellular. The equipment is selected mainly on the basis of the microbial source and product. The choice of disruption technique determines the size of the resulting debris which in turn has an influence on subsequent operations. Typical operations used are pressure homogenization (most bacteria including *Escherichia coli* and yeast) and bead milling (gram-positive bacteria and specific yeast applications).[6] Mechanical disruption releases nucleic acids that need to be precipitated. There is one standard method to achieve nucleic acid precipitation, the use of polyethyleneimine. Separation of cell debris from the proteins in solution has to be undertaken once the cells are disrupted. As a result of this step, the product will be in a solution with other proteins but without solids.

If the intracellular product is manufactured in *E. coli*, high expression of heterologous proteins will usually accumulate in the form of insoluble inclusion bodies (FIG. 3). This necessitates the processing of the inclusion bodies into the native protein by denaturing and refolding. If the intracellular product is manufactured in yeast, in many instances the protein is present in homogeneous particulate form, typically 30–60 nm particles such as virus-like particles (VLPs) (FIG. 4). Although the processing of intracellular particulate recombinant proteins is an important aspect of downstream processing, not many satisfactory methods exist for large scale separation, denaturation, and refolding of the particulate proteins.

Concentration is usually required when the protein concentration of the harvested, disrupted, and separated stream is below 60–70 g/L. With some proteins it is very difficult to obtain higher concentrations without a serious increase in viscosity, which would then impose very poor transport characteristics on the system. If a membrane (ultrafiltration) is used for concentration, the resulting flux characteristics determine the highest possible concentration that could be obtained from the

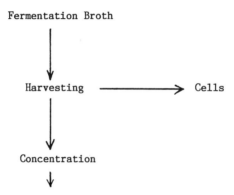

FIGURE 1. Recovery subprocess of extracellular product (yeast, mammalian, and bacterial). (From Asenjo.[18])

operation. If at the point where flux has dropped below an acceptable limit the concentration is below 60 g/L, then the proteins can be precipitated (e.g., with ammonium sulfate) to increase the final concentration.

Protein Purification

At this point the broth contains proteins and some other components such as lipids and/or wall or other polysaccharides, salts, and water in a concentrated solution. After the solution volume is considerably reduced in the previous concentration step, the total protein content suitable for chromatographic purification is 60–70 g/L.[4,5,9] Here, there will be a number of alternative combinations of purification processes (TABLE 1). For the recovery, resolution, and purification of a single protein, ideally one would like one step to extract almost 100% of the protein from this mixture with no contaminants. As this is virtually impossible, two or in some

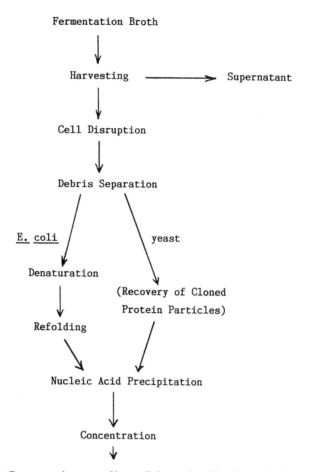

FIGURE 2. Recovery subprocess of intracellular product (*E. coli*, yeast). (From Asenjo.[18])

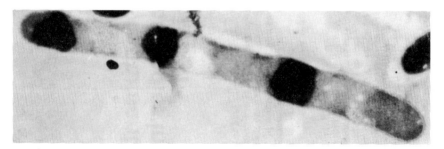

FIGURE 3. Recombinant *E. coli* cells showing cloned protein present in inclusion bodies (refractile bodies of protein). (From Kane *et al.*[7])

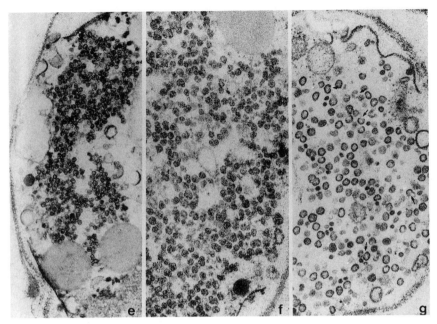

FIGURE 4. Electron micrographs of virus-like particles (VLPs) cloned in yeast cells *S. cerevisiae* (approximately 50,000 magnification). (From Muller *et al.*[8])

TABLE 1. Chromatographic Operations for Large Scale Purification of Proteins[a]

Physicochemical Property	Operation	Characteristic	Use
Van der Waals forces, H bonds, polarities, dipole moments	Adsorption	Good to high resolution, good capacity, good to high speed	Sorption from crude feedstocks, fractionation
Charge (titration curve)	Ion exchange	High resolution, high speed, high capacity	Initial sorption, fractionation
Surface hydrophobicity	Hydrophobic interaction	Good resolution, speed and capacity can be high	Partial fractionation (when sample at high ionic strength)
Biological affinity	Affinity chromatography	Excellent resolution, high speed and high capacity	Fractionation, adsorption from feedstocks
Hydrophilic and hydrophobic interactions	Reversed phase liquid chromatography	Excellent resolution, intermediate capacity, may denature proteins	Fractionation
Molecular size	Gel filtration	Moderate resolution, low capacity, excellent for desalting	Desalting, end polishing, solvent removal

[a]Adapted from Asenjo and Patrick.[9]

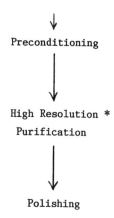

FIGURE 5. Purification subprocess. (From Asenjo.[18])

cases three or four stages will probably be needed to achieve the final purity required for the particular application (FIG. 5).

As most of the excess water has been extracted, a purification step of extremely high resolution should be used to minimize the number of stages used and hence maximize yield. However, in many cases this may not be possible at this stage as some of the contaminants still present may produce fouling of the affinity or high resolution ion exchange column and hence shorten its life. Therefore, a first step in protein purification from other contaminants will probably be necessary. This would constitute a clean-up step of pretreatment or primary isolation (FIG. 5). For this, a relatively inexpensive treatment to clarify the stream from suspended materials and nonprotein contaminants in addition to salts should be used. This step will not give a very high purity but must give a very high recovery yield in terms of target protein product recovered. Typical operations for this step would include inexpensive or disposable adsorption like a Whatman DE52 ion exchange cartridge, a hydrophobic interaction step, aqueous two-phase partitioning, or precipitation of the proteins using salt. After this step, a high resolution one will most probably be used, such as the high resolution protein purification in FIGURE 5. This stage should give a product of up to 99% (usually 95–98%) purity. Typical operations will include one or two high resolution ion-exchange chromatography steps or affinity chromatography. Although high resolution is the main concern in this stage, an adsorbent that will also give a high recovery yield should be chosen or designed.

After the high resolution step a polishing step is usually necessary to obtain ultra high purity. This will depend on the final use of the protein, and in some cases it is probably the most difficult task. If another physicochemical property cannot be exploited, gel filtration will be used which can separate dimers of the product (due to aggregation phenomena) or its hydrolysis products (due to action of proteases) solely on the basis of their different molecular weights. HPLC can also be used for polishing; however, this is an expensive technique for preparative purposes. It gives extremely high resolution but it may denature proteins.

Challenges with Recombinant Proteins

As shown in FIGURE 3, when the intracellular product is manufactured in *E. coli,* heterologous proteins will usually accumulate as large insoluble particles called inclusion bodies. If the intracellular product is manufactured in yeast, in a number of cases the protein is present in particulate form, typically 30–60 nm particles such as virus-like particles (FIG. 4).[10,11]

Regarding cell breakage, some of the advantages of chemical and enzymatic permeabilization and lysis methods recently discussed[6,10,12–14] should be investigated in greater detail, particularly because mechanical disruption techniques have several drawbacks related to obtaining high product yield (micronized wall materials, nucleic acids, high viscosity, complex mixture of contaminants, and partially damaged product) that are difficult to overcome. Release of recombinant intracellular proteins by chemical permeabilization and enzymatic lysis techniques as well as the release of recombinant particles from yeast[10] has been successfully achieved.[6] For this, however, greater availability of specialized reagents (e.g., wall lytic enzymes) will be necessary as currently these are almost only available as laboratory reagents.

Inclusion bodies have to be solubilized, in many cases chemically or enzymatically modified, and correctly refolded, otherwise the process will produce large quantities of inactive product. This is usually the case when bacteria are used for the manufacture of human proteins. A very recent study of a process with *E. coli*[15] showed that denaturation and solubilization of inclusion bodies with, for example, guanidine HCl are the steps that account for most of the cost of downstream raw materials. This is clearly shown in TABLE 2 (77% of the cost for guanidine HCl and carboxypeptidase only and 92% for the first four items). This clearly shows that currently there are no

TABLE 2. Downstream Processing (Raw) Materials[a]

Component	Annual Requirement (kg)	Price[b] ($/kg)	Annual Cost ($)	Percent
Guanidine HCL	1,007.00 tons	2.15	2,165,100	56.4
Carboxypeptidase B	0.8	1,023/g	818,400	21.3
Formate	262,280	1.25	327,850	8.5
Cyanogen bromide	22,848	11.00	251,330	6.5
Ammonium sulfate	484,448	0.14	67,823	1.8
Ethanol 95%	7,185 L	8.63/L (incl. 120% Fed. Tax)	62,000	1.6
Sodium tetrathionate	4,771	11.00	52,481	1.4
β-Mercaptoethanol	9,600 L	3.85/L	36,960	1.0
Tris-HCL	2,912	5.50	16,016	
Glycine	1,628	5.13	8,349	
Sodium sulfite	9,572	0.57	5,456	0.8
Zinc chloride	90	10.70	965	
Trypsin	1.5	361.00	542	
Ammonium acetate	50	1.10	55	
NaOH/NH₄OH/HCL/acetate	(Negligible) Amounts		27,673	
	Total		3,839,000	100.0

[a]From Datar and Rosen.[15]
[b]Prices were either obtained from the Chemical Marketing Reporter or estimated from existing retail prices.

satisfactory methods for large scale denaturation and refolding of particulate proteins. Recent developments in the use of reverse micelles for protein refolding[16] and of two-phase aqueous systems for separation of virus-like particles from yeast homogenates[10] appear particularly attractive.

Separation of inclusion bodies from debris can be achieved on a large scale by the use of centrifuges even if the material is small (about 1.0 μm) mainly because of the relatively large density of inclusion bodies (e.g., 1.3 g/ml).[15] However, flow rates have to be reduced severalfold compared to the separation of whole *E. coli* cells where flow rates are already low. This results in large capital requirements.

Aqueous two-phase systems are a very attractive alternative for the separation of cell debris from target product proteins.[9,10,17] The separation of recombinant particles from yeast has been demonstrated using this technique.[10,11] In the presence of debris and recombinant particles two stages were more appropriate, the first to separate the cell debris and the second to separate contaminant proteins.[10]

Selection of Operations

Selection of operations required for recovery/isolation is relatively straightforward if the product is extracellular. When a product is intracellular, however, no satisfactory procedures are available for large-scale processing of inclusion bodies into native proteins. Recent advances in cell permeabilization and differential product release as an alternative to disruption should show important developments in the next few years.[12,14]

Selection of purification operations, on the other hand, is more cumbersome, and choosing those operations that will give the best results is not an easy task, particularly with the high resolution purification operations that are carried out in one, two, or even three steps. To be able to separate one protein from another, a difference in physicochemical properties between them is exploited. To design an optimal separation process is to exploit these differences in the most efficient manner to accomplish the desired separation. Individual separations will generally depend on more than one property difference for their overall performance, but one property will usually form the primary basis for separation.

Physicochemical property information should be available for the target protein and also for the major contaminants. It is also useful to have some information on the fermentation supernatant from which the protein has to be separated and on some of the intermediate process streams such as those shown in FIGURES 1, 2, and 5 including thermodynamic equilibrium and transport properties such as density, total protein concentration, particle size distribution when these are present, and viscosity. Main sources of proteins in the modern biotechnology industry are few: *E. coli* (intracellular), mammalian cells (extracellular), and yeast (extra- and intracellular). The characterization of product protein and contaminants has to be carried out in terms of charge and titration curve of major proteins, molecular weight, hydrophobicity, pI, and available biospecific interactions. Determination of this information can be performed on a case-by-case basis for the individual product proteins. General distribution of physicochemical properties of the host cells just mentioned (*E. coli,* yeast, and mammalian cells) should be generated as it will allow selection of purification operations on a much more rational basis. This is shown in a succeeding section in this paper on implementation of protein properties in a prototype expert system.

High resolution purification is usually carried out by chromatography. Selection of these purification operations is based on the efficiency of different chromatographic techniques to separate the target protein from the contaminating ones.

Different techniques exploit different physicochemical properties, and some are much more efficient than others in exploiting these differences. Ion-exchange chromatography will separate the proteins based on their difference in charge. The charge of a protein changes with the pH following the titration curve.[18] Hence, if carried out at considerably different pHs at which the difference in charge of three or more proteins is very different, this technique can be used twice to purify a protein from different protein contaminants.[18] Ion exchange can use small differences in charge to give a very high resolution and hence is an extremely efficient operation to separate proteins.

Affinity chromatography can have a very high specificity for a particular protein or a small group of proteins; therefore, it can also have a very high resolution. The matrix can be expensive, but it can be reused for long periods. Ligand leakage into the product can be a problem. Regarding cost, affinity chromatography will usually be more expensive than ion exchange.[19,20] Hydrophobic interaction chromatography (HIC) has been proposed only as a pretreatment step or as a first high resolution purification (HRP) step. The resolution is good but not always particularly high as the distribution of surface hydrophobicity in a protein can be random, thus giving only adequate resolution. Gel filtration for protein fractionation is normally not used as a high resolution operation in the large scale because of the low efficiency in exploiting differences in molecular weight.

PROCESS DESIGN

Process design and selection of operations is a complex procedure in which the design evolves from an initial stage to the final stage in a trial-and-error fashion, repeatedly revising and refining the initial assumptions and restrictions: (1) flow-sheet generation (qualitative/semiquantitative); (2) quantitative design of units; (3) revise flowsheet 1, then 2, and so forth until some objective is reached). An important aspect of process design involves the selection of operations and design of a process sequence (process synthesis). In the initial stages this process is more or less done using heuristics, using rules of thumb to arrive at a rapid (and reliable) specification of equipment type, size, and maybe cost.

The problems that have to be solved in process synthesis and optimization of downstream protein separations are: (1) choosing between alternative operations (e.g., homogenizer *vs.* bead mill or centrifugation *vs.* cross-flow microfiltration), and (2) designing an optimal chromatographic sequence with maximum yield and minimum number of steps (1, 2, or 3), a problem that is combinatorial in nature. The first type of problem can be adequately solved if appropriate mathematical correlations and models that can be used as simulation tools are developed. The second type of problem has been partially tackled in classical chemical process engineering (e.g., distillation and extraction), that is, finding a rigorous solution using numerical methods like mathematical programming techniques (e.g., resolution of "tree structure"[21]) or, more recently, using an expert systems approach. For the design of an optimal sequence, purely mathematical techniques have limited use in biotechnology because of a lack of useful design equations and databases. The second approach is more attractive because it allows the use of empirical knowledge that is not rigorous in nature and is typical of that used by experts. Computer-based expert systems are an important tool in the field of artificial intelligence. Efforts have been made to develop expert systems for this purpose[22-24] or to adapt existing software systems (called shells)[4] for the manipulation of heuristics, databases, and simple algebraic design equations.

Use of Mathematical Models

Mathematical models and mathematical correlations of individual operations will allow simulation of performance and also may be used for scale-up. Computer simulations are useful to optimize individual separations.[25] Examples of useful downstream process simulations and investigation of process conditions are microbial cell breakage and selective product release using enzymes[26,27] and investigation of the affinity and ion-exchange chromatography of proteins.[28,29] The first constitutes an example of cell breakage and intracellular product release, and the second one of protein separation by chromatography.

One of these examples is the modeling of cell lysis and product release using enzymes. Recently, substantial attempts to model enzymatic cell lysis and product release were carried out. Three models of yeast lysis developed by Hunter and Asenjo serve different purposes.[26,30,31] The simple model is a lumped, two-step model that follows the major features of the data and may prove useful for design of simple lysis reactors for total protein release.[30,32] The structured model, which accounts for the source of protein within a cell, was developed to gain a mechanistic basis for predicting the effects of untested process conditions and to follow the release of protein from different cellular locations.[26] A schematic diagram of the reactions of this model is shown in FIGURE 6. The structured mechanistic model of the kinetics uses differential equations, one for each cellular component. This model was used for process simulations in which the release of proteins from different cellular locations can be predicted, analyzed, and then improved.[26] Process simulations were performed to predict the release of proteins from the wall, cytosol, and intracellular particles such as organelles or recombinant protein inclusions.

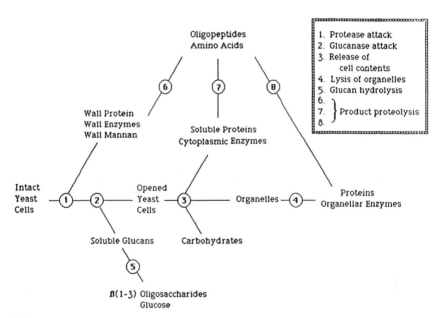

FIGURE 6. Mechanism of enzymatic cell lysis in structured model. (From Hunter and Asenjo.[30])

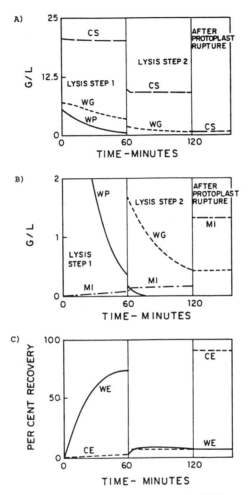

FIGURE 7. Enzyme recovery from subcellular structures. **(A)** Cell structure breakdown; — wall protein; ---- wall glucan; —— cytosol; —·— mitochondria. **(B)** Cell structure; closeup showing mitochondria. **(C)** Enzyme release, percent of original enzyme in cell; — wall enzyme; ---- cytoplasmic enzyme. (From Hunter and Asenjo.[30])

A simulation of site-linked product recovery is presented in FIGURE 7. In the first lysis step, using lytic enzyme and osmotic support, 93% of the wall protein was released from the cell wall. Some protein is hydrolyzed by the presence of some "destructive protease," but 74% of it survived to be recovered at the end of the first hour. As only 3% of the protoplasts burst during this step, little cytoplasm was released. The second digestion was included to decrease the amount of structural glucan from 50% to 13% of its original mass. This also made the cells more fragile for rupture in the third stage. Only a small amount of cytoplasmic protein is released from the protoplasts during the second stage, a desirable result since protein located inside the protoplasts is not attacked by "destructive protease." At the end of the second hour the protoplasts are easily broken by stirring or centrifugation. Almost all of the mitochondria (96%) are released during protoplast rupture.

This performance was analyzed, and the means of improving and maximizing the recovery of wall, cytosol, and mitochondrial proteins in three different stages were investigated. This resulted in the conditions shown recently[12] in which the results of computer simulations with the structured model (FIG. 7) were analyzed and improved and led to the concept of differential product release (DPR). This clearly represents one of the most important applications of mathematical modeling for process simulation, process investigation, and process development.

More recently a population balance model that takes into account the lack of homogeneity of microbial cell populations was described.[31]

A Prototype Expert System

Today there are well-developed expert software systems or "shells" that help develop an organized knowledge base from the domain knowledge and also provide the inference engine. Asenjo et al.[4] found that for prototype systems shells can be adequate, particularly if they are capable of evaluating uncertainties associated with the inference process. Expert knowledge was obtained mainly from industrial experts working on the large-scale separation and purification of therapeutic, diagnostic, and analytical proteins. It soon became apparent that the true bottleneck in the development of expert systems for protein purification was not in its implementation but in the acquisition, clarification, formalization, and structuring of the knowledge domain.

The knowledge was expressed in about 65 rules, some of which carry a degree of uncertainty.[4] The downstream process was divided into two distinct subprocesses, a first subprocess called recovery/isolation, after which the total protein concentration is 60–70 g/L, and a second subprocess called purification. Processing of recombinant proteins present in intracellular inclusion bodies or other particles was not considered in this prototype but has now been included in the expanded version (next sections).

The recovery subprocess comprises harvesting, cell disruption, separation of solid debris, and precipitation of nucleic acids (these last three steps are used only if disruption is required) and concentration. This subprocess is characterized by the objective of recovering the product from the production system. The purification subprocess takes the 60–70 g/L protein solution and purifies the individual protein product to a high purity with a high yield. It comprises preconditioning or cleaning (to obtain a "sparkling clear" solution), high resolution purification, which can be carried out in one or two steps, and polishing, when necessary (usually to remove traces of minor contaminants as for therapeutic applications).

The various parameters used to characterize the broth and the culturing system are:

Source (bacteria, yeast, fungus, and mammal)
Product of interest: Name
Cellular location (intracellular, extracellular, and unknown)
Titration curve (charge as a function of pH), isoelectric point
Surface hydrophobicity
Biospecificity database
Molecular weight
Two-phase aqueous systems database

The proposed process consists of a sequence of operations to obtain the stated design objective. Several different sequences of operations may accomplish the same objective. In those cases, a quantitative degree of performance (given by the "certainty factor") of each operation is assigned by the expert and carried by the system into the proposed design.

This prototype expert system (65 rules) did not have a database on physicochemical properties of main protein contaminants, so the selection of high resolution purification operations was rather empirical and was based on knowledge of the efficiency of the different techniques to separate a protein from its main contaminants. Efficiency was classified as high, medium, or low.[4]

In the development of the prototype expert system it was found that selection of operations in the recovery subprocess could be well structured. In the second subprocess (purification) the structuring of knowledge was more difficult. The main deficiency of available information was in the selection of high resolution purification operations (usually one or two chromatographic steps) that should be based on the physicochemical (molecular) properties of the proteins and those of the major contaminant proteins. This information is vital in selecting the right operations and in the best possible order according to their relative efficiencies (refs. 18 and 33 and next section). A considerable lack of information was also the case for the separation of minor contaminants present that are removed in the final polishing stage (also usually a chromatographic step).

For the selection of operations, information generated on a very small scale in

TABLE 3. Properties to be Exploited for the Separation and Purification of Different Proteins

1. Charge (Titration Curve)
2. Biospecificity
3. Surface Hydrophobicity
4. pI (Isoelectric Point)
5. MW (Molecular Weight)

terms of "efficiency" of separation or alternatively information on physicochemical properties (charge-titration curve, bioaffinity, surface hydrophobicity, pI, and molecular weight) is necessary. In this case the deviation of the value for the product protein from those of the main contaminants should be used. A factor for efficiency of the operation in exploiting this difference also has to be included in this evaluation.[18] This is discussed in detail in the next two sections (Protein Properties and Implementation).

Use of Protein Properties

Selection of actual operations is based on information generated on a small scale to determine performance and efficiency of particular separations. Alternatively, information on physical, chemical, and biochemical properties of product and contaminants can be used to predict such performance (TABLE 3). In this case the deviation (Dev) of the value of the protein product from those of the main contaminants should be found. A factor for efficiency (Eff) of the separation operation in exploiting this difference or deviation of physicochemical property has to be included in this evaluation. It is possible then to define a "separation coefficient" (SC) that can be used to characterize the ability of the separation operation to separate two or more proteins.[18]

$$SC = f(Dev, Eff)$$

In chromatographic separations such as ion exchange or affinity chromatography there are differences in the cost of matrices used (e.g., protein A affinity chromatography uses a much more expensive matrix than does CM-Sepharose ion exchange); however, most of the cost in such a process is associated with the hardware (columns, accessories, and control system) as most matrices can be reused for long periods of time resulting in lower matrix replacement costs. Differences in the cost of a purification operation can be taken into account by the use of a cost factor (CF), giving an expression for the economic separation coefficient (ESC):

$$ESC = f(SC, CF)$$

The values of the parameters in these two expressions should range between $0 \leq Eff \leq 1$ and $0 < Dev \leq 1$. As such values are relative, the maximum value for Dev for individual properties has to be defined within this range and the value for Eff given to a particular operation will also depend on the range (or maximum possible value for the deviation of a specific protein property), which has to be standardized for different operations. The value of the CF (cost factor) will be ≤ 1 or > 1 and a standard operation (such as ion exchange using CM-Sepharose) should be given a value of 1. We have made a first attempt to define a separation coefficient, SC as $SC = DF \cdot \eta$, where DF is the deviation factor and η the efficiency (Dev = DF = deviation factor; Eff = η = efficiency). This is shown in TABLE 4. We also recently suggested[33] the inclusion of a term for concentration, as this will affect the selection criteria because the contaminants in higher concentrations should be removed first (Rule 2). However, as concentration apparently does not intrinsically affect an actual separation coefficient, the suggestion of using the term "separation selection coefficient" (SSC) when including the concentration term θ has been preferred here ($SSC = DF \cdot \eta \cdot \theta$). The term that takes into account a cost element has been called the economic separation coefficient (ESC). Following the previous argument a more appropriate name could be the "economic separation selection coefficient."

The two parameters, η, and the cost factor, CF, are this far empirical and rather subjective. A more rigorous estimation is presently under study in our group. The cost factor is not based on a rigorous economic evaluation, as was recently carried out in our group,[20] but a very "approximate" evaluation of the cost involved in using such an operation. Many elements apart from direct variable or capital costs affect such a decision and hence this approximate evaluation of cost and therefore CF (e.g., availability of matrix in the pilot plant, reliability, robustness with variation in feedstock, speed of process implementation, or quality control). This is also partly related to the fact that the production cost of a therapeutic or diagnostic protein is still only a small fraction of the final price. Consequently, the cost differences found in a rigorous economic evaluation are much more marked than are those shown in the expression of TABLE 4. Values shown in TABLE 4 will be subjected to modifications as the rationale proposed is tested in real cases.

Implementation in Prototype ES; Hybrid and Evolutionary Systems

The rationale for selection of high resolution purification operations has been implemented into our prototype expert system. This was done by interfacing a program in PASCAL in which the main molecular properties of a target product protein were compared with those of the main protein contaminants and then used to select the most appropriate high resolution purification operations.[33] The rationale discussed in the previous section and in TABLE 4 was used.

The main sources used for the production of recombinant and mammalian proteins today are few. For the purpose of our prototype, only three main production systems were chosen for initial characterization of the main protein contaminants present in these sources. These are *E. coli* (intracellular proteins), yeast (intra- and extracellular), and mammalian cells (extracellular proteins). Initial results from our present work on characterization of main proteins in these sources are shown in

TABLE 4. Separation Coefficients

$$SC = DF \cdot \eta$$

DF = Deviation factor for hydrophobicity, molecular weight, and pI

$$DF = \frac{\text{Protein Value} - \text{Contaminant Value}}{\text{Max.[Protein Value, Contaminant Value]}}$$

DF = 1.0 for Affinity Chromatography

$$\eta = \text{Efficiency} = \begin{cases} 1.00 \text{ for Affinity Chromatography} \\ 0.70 \text{ for Ion Exchange} \\ 0.35 \text{ for Hydrophobic Interaction Chromatography} \\ 0.20 \text{ for Gel Filtration} \end{cases}$$

$$SSC = DF \cdot \eta \cdot \theta$$

θ = concentration factor

$$\theta = \frac{\text{Concentration of the Contaminant Protein}}{\text{Total Concentration of Contaminant Proteins}}$$

$$ESC = \frac{SSC}{CF}$$

$$CF = \text{Cost Factor} = \begin{cases} 1.0 \text{ for affinity chromatography} \\ 0.6 \text{ for gel filtration} \\ 0.3 \text{ for ion exchange} \\ 0.3 \text{ for hydrophobic interaction chromatography} \end{cases}$$

TABLES 5, 6, and 7.[34,35] This approach appears conceptually valid for molecular weight and for hydrophobic interaction chromatography, but care has to be taken in the selection of ion-exchange chromatography as a suitable method. Values of the isoelectric point, pI, of proteins are only useful for the selection of operation conditions when using an anion or a cation exchange matrix[3] but not for the selection of operations that will give better separation resolution between proteins. For this, data on the charge of the proteins as a function of pH[18]—the titration curves—or on its adsorption properties on the different matrices are necessary. Typical data on titration curves of proteins are shown in FIGURE 8. Data on adsorption properties of

TABLE 5. Characteristics of Main Protein Contaminants in *E. coli* Matrix (from Keeratipibul[34])[a]

Band No.	MW	φ	pI
8	120,200	0.02 M	5.0
5	145,000	Not seen	5.2
17	82,000	0.13 M	5.2
1	> 200,000	Not seen	4.6
33	13,804	0.64 M	5.2
34	27,200–22,900	0.26 M	4.6
28	39,500	0.13 M	5.7
25	39,500	0.64 M	4.6
27	39,500	0.13 M	4.6
24	120,200	0.02 M	5.5
14	145,000	Not seen	4.6

[a]Cell lysate was prepared by bead milling.
NOTE: φ = hydrophobicity; MW = molecular weight; M = molarity of $(NH_4)_2SO_4$; pI = isoelectric point. MW was measured by polyacrylamide gel electrophoresis (PAGE). pI was measured by isoelectric focusing. φ was measured by hydrophobic interaction chromatography (HIC) using a Phenyl-Superose gel in an FPLC and a gradient elution from 2.0 M to 0.0 M $(NH_4)_2SO_4$ in 0.1 M KH_2PO_4. Units used are the concentration of $(NH_4)_2SO_4$ at which the protein eluted.

the same three proteins on anion and cation ion-exchange matrices for the same three proteins are shown in FIGURE 9.

α (FIG. 9) is the partition coefficient defined as protein adsorbed on the matrix divided by total protein present (protein adsorbed plus protein in solution). The isoelectric point (pI) of thaumatin is about 10.5; hence this protein is positively charged at most pH values (below the pI) and thus adsorbs well on a cation exchange matrix (CM) and poorly on an anion exchange one (DEAE) at virtually all values of pH. BSA, on the other hand, has a pI around 4.8 and is thus negatively charged at pHs above this value (FIG. 8B). Its value of α is about 1 at pH values above 5 on DEAE-Sepharose and α decreases from about 0.9 at pH 4 to 0.2 at pH 8 on the

TABLE 6. Physicochemical Characteristics of the 10 Main Bands Present in the *S. cerevisiae* Lysate (from Noble[35])

Band Number	Molecular Weight	Isoelectric Point	Hydrophobicity φ [M][a]
25	38,020	6.1–6.4	0.42–0.27
35	60,260	5.7	0.34
41	154,880	7.0	0.34
47	14,790	4.9	0.04–0.0
31	8,510	5.1–5.7	0.42–0.27
30	12,300	6.7–7.0	ND[b]
33	41,690	6.4–6.7	ppt.[c]
22	41,690	5.4–7.0	0.64–0.57
43	54,950	7.6	EtOH[d]
46	3,390	8.1	ppt.[c]

[a]Units of hydrophobicity in Table 5.
[b]ND = not determined.
[c]ppt. = precipitated.
[d]EtOH = eluted only with ethanol.

cation exchange matrix, a behavior that follows the shape of the titration curve but not strictly the value of the isoelectric point. This will depend on the ionization of the matrix used (CM-Sepharose). For myoglobin which has a pI of about 7.0 the value of α increases on the positively charged matrix (DEAE) to a value of 0.7 at pH 9. This protein adsorbs well on the cation exchange matrix at pHs of 5.0–6.5. The low value of α (<0.2) at pH 8 which appears unexpected can be explained by the fact that CM-Sepharose loses its charge at low pH values (Levison[36]; information supplied by manufacturers). To summarize, the behavior shown in FIGURES 8 and 9 on behavior of three pure proteins indicates a good correlation in behavior between the titration curve of proteins and its adsorption properties on anion and cation exchange matrices provided the properties of the matrices are known. A more rigorous approach was recently carried out in which the titration curve of both the adsorbate (protein) and the adsorbent (gel matrix) was measured and a theoretical partition coefficient was calculated from Coulomb's law. This law states that the interaction force between charges e_1, e_2 is proportional to e_1e_2; therefore, the net interaction

TABLE 7. Physicochemical Characteristics of the 10 Main Bands Present in the Mammalian Cell Culture Supernatant (from Noble[35])

Band Number	Molecular Weight (Daltons)	Isoelectric Point	Hydrophobicity ϕ [M][a]
13	66,070	4.8	0.97–0
3	204,170–141,250	5.4–8.7	0.97–0.564
2	295,120	6.0	0.97–0.708
10	72,440	5.4	ppt.
11	52,950	5.2	ppt.
9	72,440	5.7	ppt.
12	169,820	5.4	ppt.
6	2,820	5.7	1.25, 0.564
15	6,460	5.2	ppt.
5	169,820	4.8	0.708 $[(NH_4)_2SO_4]$ (M)[a]

[a]Units of hydrophobicity in Table 5.
[b]ppt. = precipitated.

force between the protein and matrix should be proportional to the product of the degree of ionization of each component.[37] The theoretical partition coefficient obtained by this approach followed quite closely the behavior of the values obtained experimentally for the enzyme polygalacturonase and CM-Sephadex.[37]

The value of pI (TABLES 5, 6, and 7) is only directly relevant for the selection of chromatofocusing as a separation operation, but this technique is not a particularly feasible operation for large scale use. This is not only because of the high cost of a polybuffer but mainly because such buffers are unacceptable to be used for therapeutic proteins.

As more accurate and detailed information on the protein contaminants is made available, it will be appropriately implemented in the database. The "expanded" prototype with access to the databases and the more rational selection of high resolution separation operations discussed in this paper resulted in an expert system with approximately 130 rules in addition to the PASCAL interface. We are presently testing the use of this expert system for selection of purification operations.

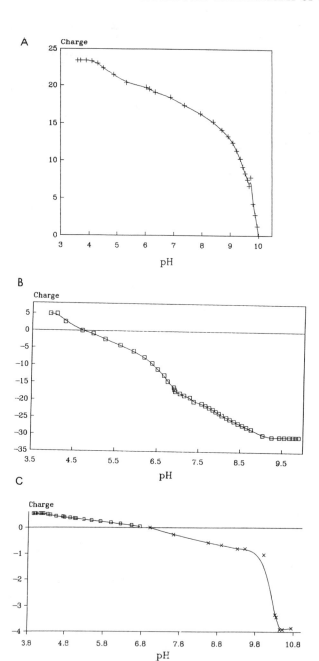

FIGURE 8. Titration curves of three proteins: **(A)** thaumatin; **(B)** bovine serum albumin (BSA); and **(C)** myoglobin.

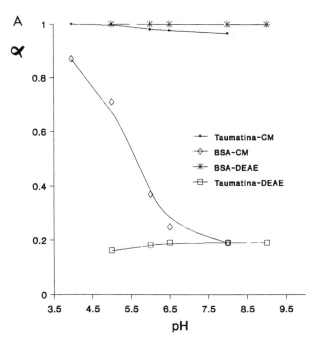

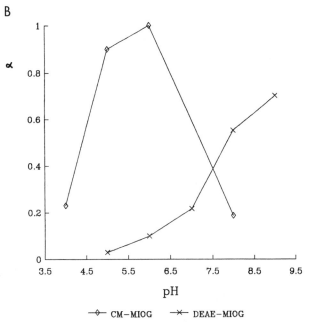

FIGURE 9. Partition (distribution) coefficient α (adsorbed protein/total protein) for three proteins: thaumatin, BSA, and myoglobin on an anion exchange (DEAE-Sepharose) and a cation exchange (CM-Sepharose) matrix as a function of pH. (**A**) Thaumatin and BSA; (**B**) myoglobin.

It is clear that heuristic methods (such as those that have been implemented in expert systems) are one end of the spectrum of available process synthesis techniques; the other end consists of rigorous methodologies such as mathematical programming techniques. A second stage in the development of expert systems should consider the introduction of quantitative models (mathematical correlations, design equations, and short-cut methods) for the design and evaluation of individual operations and their alternatives and for introducing basic cost calculations into the selection procedure of alternative processes. Such a hybrid system that should include heuristic rules, more rigorous information and design correlations in addition to databases is particularly attractive for process biotechnology.

CONCLUSIONS

The work described on rational design of protein purification processes clearly shows that properly developed expert systems can be a vital tool to assist with solving the knowledge-intensive and heuristic-based problem of process synthesis in biotechnology. Rigorous methods will not be appropriate to solve the overall synthesis problem, as rigorous information and mathematical correlations are not readily available as they are in chemical process engineering. The overall downstream process synthesis problem in biotechnology does not have a strict combinatorial nature, whereas the high resolution purification stages within the purification subprocess (1, 2, or even 3 purification stages in which several alternatives in different order combinations can be used) do.

Rigorous models, however, have a very important role in the simulation of individual operations (e.g., for process evaluation and comparison of performance and cost of individual operations). It is clear that the limiting factor in the development of expert systems for protein purification is not the implementation of new artificial intelligence programs but the acquisition, clarification, formalization, and structuring of the domain of expert knowledge.

With recombinant proteins a crucial issue is the processing of inclusion bodies or recombinant particles into the native protein as by denaturing and refolding. Clearly not many satisfactory methods exist for large-scale processing of these particulate proteins. The latest developments in the field are discussed.

The interfacing of the prototype expert system with a database of physicochemical and thermodynamic properties of main protein contaminants in typical production streams is an important improvement. This will allow the selection of high resolution purification operations on a much more rational basis, resulting in a very improved process selection and thus rational process design. To advance the further development of this field, there is an important need for generating more detailed databases for protein products, fermentation streams, and contaminants. These include databases on surface hydrophobicity, molecular weight, isoelectric point, but also, most importantly, titration curves for proteins.

SUMMARY

Recent developments in the rational design of purification processes for recombinant proteins are discussed. A review of the main issues involved in process design for protein separation and purification is presented with particular emphasis on the challenges posed by recombinant proteins. This includes physicochemical characterization of target protein and main contaminants, the use of rigorous mathematical modeling and process simulation as well as the development of an expert system and the application of this technology for optimization and design of large scale processes.

An expert system for selection of optimal protein separation sequences will give the user a number of alternatives chosen on the basis of extensive data back-up on proteins and unit operations.

REFERENCES

1. WHEELWRIGHT, S. M. 1987. Bio/technology **5:** 789.
2. ASENJO, J. A. 1989. The Rational Design of Large Scale Protein Separation Sequences. 32nd International IUPAC Congress, Stockholm, August 2–7.
3. SCOPES, R. K. 1987. Protein Purification, 2nd. ed. Springer-Verlag. NY.
4. ASENJO, J. A., L. HERRERA & B. BYRNE. 1989. J. Biotechnol. **11:** 275–298.
5. PHARMACIA. 1983. Scale Up to Process Chromatography. Guide to Design. Pharmacia. Uppsala, Sweden.
6. ASENJO, J. A. 1990. Cell disruption and removal of insolubles. *In* Separations for Biotechnology II. D. L. Pyle, ed.: 11–20. Elsevier.
7. KANE, J. F. & D. L. HARTLEY. 1988. Trends Biotechnol. **6:** 95.
8. MULLER, F., K.-H. BRUHL, K. FREIDEL, K. V. KOWALLIK & M. CIRIACY. 1987. Mol. Gen. Genet. **207:** 421–429.
9. ASENJO, J. A. & I. PATRICK. 1990. Large scale protein purification. *In* Protein Purification Applications: A Practical Approach. E. L. V. Harris and S. Angal, eds.: 1–28. IRL Press. UK.
10. HUANG, R.-B. 1990. Ph.D. Dissertation, University of Reading.
11. RIVEROS-MORENO, V. & J. E. BEESLEY. 1990. *In* Separations for Biotechnology II, D. L. Pyle, ed.: 227–237. Elsevier.
12. ANDREWS, B. A., R.-B. HUANG & J. A. ASENJO. 1991. Differential Product release from yeast cells by selective enzymatic lysis. *In* Biologicals from Recombinant Microorganisms and Animal Cells–Production and Recovery. M. White, S. Reuveny & A. Shaffermann, eds.:307–321. VCH Publishers.
13. NAGLAK, T. J. & H. Y. WANG. 1990. In Separations for Biotechnology II, D. L. Pyle, ed.: 55–64. Elsevier.
14. NAGLAK, T. J., D. J. HETTWER & H. Y. WANG. 1990. Chemical permeabilization of cells for intracellular product release. *In* Separation Processes in Biotechnology. J. A. Asenjo, ed.: 177–205. Marcel Dekker. NY.
15. DATAR, R. & C.-G. ROSEN. 1990. Downstream process economics. *In* Separation Processes in Biotechnology. J. A. Asenjo, ed.: 741–793. Marcel Dekker. NY.
16. HAGEN, A. J., T. A. HATTON & D. I. C. WANG. 1990. Biotechnol. Bioeng. **35:** 955–965.
17. ANDREWS, B. A. & J. A. ASENJO. 1989. Aqueous two-phase partitioning. *In* Protein Purification Methods: A Practical Approach. E. L. V. Harris & S. Angal, eds.: 161–174. IRL Press. Oxford.
18. ASENJO, J. A. 1990. Selection of operations in separation processes. *In* Separation Processes in Biotechnology. J. A. Asenjo, ed.: 3–16. Marcel Dekker. NY.
19. DUFFY, S. A., *et al.* 1988. Optimal Large Scale Purification Strategies for the Production of Highly Purified Monoclonal Antibodies for Clinical Application. 196th ACS National Meeting, MBTD division, Los Angeles, CA, Sept. 25–30, 1988.
20. KOSTI, R. 1989. Economic Evaluation of Large Scale Protein Purification Operations. M. S. Thesis, University of Reading.
21. PROKOPAKIS, G. J. & J. A. ASENJO. 1990. Synthesis of downstream processes. *In* Separation Processes in Biotechnology. J. A. Asenjo, ed.: 571–601. Marcel Dekker. NY.
22. SILETTI, C. A. & G. STEPHANOPOULOS. 1986. Computer Aided Design of Protein Recovery Processes. 192nd ACS National Meeting, Anaheim, CA, Sept. 1986.
23. WACKS, S. 1987. Design of Protein Separation Sequences and Downstream Processes in

Biotechnology; Use of Artificial Intelligence, M.Sc. Thesis, Columbia University, New York.

24. SILETTI, C. A. 1989. Computer Aided Design of Protein Recovery Processes, Ph.D. Thesis, MIT, Cambridge, MA.

25. HEDMAN, P., J. C. JANSON, B. ARVE & J. G. GUSTAFSSON. 1989. Large scale chromatography-optimization of preparative chromatographic separations. Proc. of 8th Int. Biotechnol. Symp. 1. G. Durand, L. Bobichon, & J. Florent, eds.: 612–622. Société Française de Microbiologie.

26. HUNTER, J. B. & J. A. ASENJO. 1988. Biotechnol. Bioeng. **31:** 929–943.

27. LIU, L. C., G. J. PROKOPAKIS & J. A. ASENJO. 1988. Biotechnol. Bioeng. **32:** 1113–1127.

28. CHASE, H. A. 1988. Affinity separations using immobilized antibodies. Symp. on Antibodies for Purification. SCI. London, March 1988.

29. ARVE, B. 1989. Simulation and Modelling of Chromatographic Processes. 32 International IUPAC Congress. Stockholm. August 2–7, 1989.

30. HUNTER, J. B. & J. A. ASENJO. 1986. *In* Separation, Recovery and Purification in Biotechnology: Recent Advances and Mathematical Modeling, J. A. Asenjo & J. Hong, eds.: 9–32. ACS Symposium Series 1986.

31. HUNTER, J. B. & J. A. ASENJO. 1990. Biotechnol. Bioeng. **35:** 31–42.

32. HUNTER, J. B. & J. A. ASENJO. 1987. Biotechnol. Bioeng. **30:** 481–490.

33. ASENJO, J. A. & F. MAUGERI. 1991. An expert system for selection and synthesis of protein purification processes. *In* Frontiers in Bioprocessing II, S. Sikdar, P. Todd & M. Bier, eds. ACS Books. Washington.

34. KEERATIPIBUL, S. 1989. Characterization of Proteins from *E. coli* and *S. cerevisiae* for the Design of Protein Separation Operations. M.S. Thesis. University of Reading.

35. NOBLE, I. 1990. Characterization of the Main Contaminant Proteins present in a Mammalian Cell Culture Supernatant and a *S. cerevisiae* lysate. B. S. Dissertation. University of Reading.

36. LEVISON, P. R. 1990. Whatman U. K. Ltd., Maidstone, Kent, personal communication at Seps. for Biotechnology II Conference, Reading, UK. September 1990.

37. HARSA, S., D. L. PYLE & C. A. ZAROR. 1990. *In* Separations for Biotechnology II, D. L. Pyle, ed.: 345–354. Elsevier.

Design of a Multipurpose Biotech Pilot and Production Facility

M. B. BURNETT, V. G. SANTAMARINA,
AND D. R. OMSTEAD

Ortho Pharmaceuticals
Raritan, New Jersey 08869

Increasing competitiveness within the biopharmaceutical industry demands that manufacturers take measures to minimize the time required to bring products to market. One such measure is to provide facilities that can support the processing of multiple products from early process development through initial manufacturing stages. Such multipurpose facilities have many advantages, including concentrating development efforts, easing the transfer of technology between development stages, and providing interim manufacturing capabilities.

BACKGROUND

The first biotechnology product was approved for market in 1980. Currently, more than 8 biopharmaceutical products are approved, and an additional 80 are estimated to be approved in the next 2–3 years.[1] The rate of emergence of new products creates a demand for facilities to produce product to support, first, clinical trials and, later, sales markets. Because biologically derived pharmaceuticals have different processing needs from those of conventional pharmaceuticals, the need for specially designed facilities is apparent.

Developing and manufacturing biologically derived pharmaceuticals require facilities designed to support both clinical development and manufacturing efforts. Often, the individual needs of the development and manufacturing groups are seen as distinct, thereby requiring facilities of different design. Development groups require the freedom to experiment with and analyze different processing techniques to optimize and scale-up processes; therefore, a development facility requires more processing flexibility than that normally found in a dedicated manufacturing facility. However, clinical and manufacturing processing protocols have many parallels; this suggests that a common facility design could be used to support both efforts.

The design, construction, and validation of a facility to produce biologically based pharmaceuticals can take in excess of 3 years (FIG. 1). Usually, construction of a dedicated manufacturing facility is not started until phase III clinical trials are nearly completed. This approach ensures that the product is efficacious, that commercial feasibility has been established, and that the manufacturing process is well defined. At this point in the product development cycle the risk associated with investing capital in a facility is reduced. However, by waiting until this point to construct a production facility, the time required to gain approval of a product after discovery can be in excess of 10 years (FIG. 2).

To shorten the time required to bring a product to market, several options exist: first, the construction of a dedicated manufacturing facility may be initiated in the early phases of clinical trials. This option, however, is economically risky because product efficacy, commercial feasibility, and process technology may not be fully

357

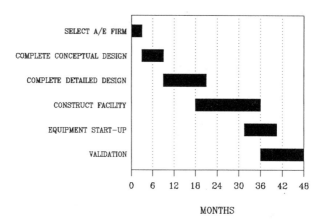

FIGURE 1. The time required to design, construct, and validate a biopharmaceutical facility can take up to 4 years (assumes a discrete design and discrete build project format).

established. As a second alternative, a contract manufacturing agreement at a contractor's existing facility could also be obtained. Contract manufacturing in this manner has risks because the transfer of technology from development to the contract manufacturer is often difficult, and the client loses a certain amount of direct control over the manufacturing process.

A third alternative, the implementation of a multipurpose, nondedicated bioprocessing facility, also exists. Such a facility can be designed to support simultaneous process scale-up and development, clinical production, and interim or small market manufacturing. In general, the use of multipurpose facilities allows for a reduction in the time required to bring a product to market, eases the transfer of technology between development and manufacturing areas, and offers an opportunity for greater processing control and good manufacturing compliance during the clinical phases. Additionally, such a facility could be used to supply interim manufacturing needs prior to constructing a dedicated facility or alternatively the entire manufac-

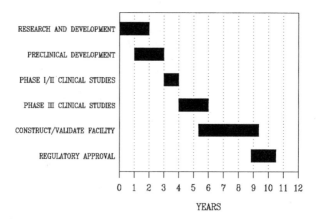

FIGURE 2. Developing and licensing a biopharmaceutical can take in excess of 10 years.

turing needs for small volume products. This paper discusses the key factors required for successful operation of such a facility.

DESIGN BASIS

The scope of a multipurpose development and bulk manufacturing facility should include: process development, preparation of material to meet research needs, preparation of material to support phase I, II, and III clinical trials, and the preparation of material used to launch or support a licensed product. To meet these needs the facility typically contains five general areas: (1) laboratories to support bench scale development and quality control efforts, (2) production suites to support scale-up, clinical, and manufacturing efforts, (3) utility areas for the production and distribution of purified waters and other utilities, (4) receiving and shipping areas to warehouse raw materials and the finished product, and (5) administrative offices.

To optimize the facility's productivity and flexibility, the ability to process multiple products simultaneously should be provided. The production suites, utility, and receiving/shipping areas should be designed to comply with current FDA guidelines. Flexibility for complying with future regulatory requirements should also be addressed. The incorporation of new or improved technologies should be straightforward, and the flexibility of the processing equipment and facilities should be maximized.

The aforementioned considerations directly impact the design of a facility and its processing equipment. In general, the greatest effects on design concern those portions of the building (and process) requiring regulatory approval: the production suites, utility areas, and shipping/receiving areas. The impact of regulatory design guidelines can be minimized in nonmanufacturing (i.e., laboratory and administrative) portions of the facility by providing a physical separation between the "controlled" and "noncontrolled" areas. The following sections address specific design aspects that concern the controlled (i.e., production, utility, and shipping/receiving) areas.

FACILITY DESIGN CONSIDERATIONS

The key aspects of physical design that need to be considered include the floor plan, heating and ventilation (HVAC) system, waste containment, and waste treatment.

The floor plan of the facility needs to address all of the design requirements. In our interpretation, FDA guidelines require the prevention of cross-contamination between different product lines and between different lots of a given product line. Prevention of cross-contamination is typically obtained by separating different processes either spatially or temporally. To operate multiple processes simultaneously, spatial separation is required. Spatial separation is achieved by physically providing unique areas for specific processes or process steps. For example, inoculum development, fermentation, and purification can be identified as unique bioprocess operations. Separating these operations into distinct areas using controlled access corridors and airlocks allows unique processes to be performed in each of the areas simultaneously. Furthermore, the access corridors and production areas must be configured to provide a logical flow of personnel, materials, and product through the facility. These flows should be designed to eliminate cross-contamination or adulteration of the product while maintaining accessibility and flexibility.

To further expand the productivity and utility of the facility, several similar production areas or modules can be included. The incorporation of multiple production modules allows for greater diversity in simultaneous processing. Additionally, one module can be dedicated to manufacturing, another to clinical production, and so on. Again, each of the modules would have to be separated and accessed via corridors and airlocks and the concepts of personnel and material flows incorporated into the design. The benefits of a modular design include maximizing design creativity (one good design can be used for many discrete areas), reducing the risk of cross-contamination, and ease of future expansion. Furthermore, by designing each production module as a self-contained unit, a logical boundary can be established between production areas that will aid in validating and/or licensing all or a portion of the building.

The HVAC system plays an important role in biopharmaceutical processing. The required environment can range from warm (37 ± 0.5°C for tissue culture work) to cold (4–8°C for downstream processing). The close temperature tolerances (and simultaneous need for maintaining humidity) used in biologics processing require the use of specialty environmental rooms. Additionally, airborn particles can add to the potential for cross-contamination (especially when working with microbial fermentations). Air entering a production area is normally filtered using a high efficiency particulate air (HEPA) filter. Maintenance of particulate-free air entering the room minimizes exposure of product and removes contaminating particles created within the room. The factors that determine the cleanliness of the air entering the room include the volume of air introduced into the room, the amount of air recycled back into the room, and the nature of the operations being conducted in the room.

Air normally enters the room through ducts or terminal filters mounted in or near the ceiling. In biopharmaceutical facilities the flow of air is downward and the air is exhausted from the room through ducts near the floor. This downward flow of air establishes a washing action, flushing particles away from the work areas and equipment. Consequently, the rate of air flow into the room (expressed in terms of room volume changes per hour) directly affects the cleanliness of the environment. Higher volume changes per hour result in a cleaner environment. However, high flow rates are not always warranted. The nature of the operations within an area also affects the required cleanliness of the environment. For example, initial inoculum development for a fermentation batch requires a very clean environment to minimize the chances of contamination by airborn microorganisms. These activities are generally undertaken on a small scale and can be conducted within a HEPA-filtered hood. In addition, the bioreactor-based fermentation activities are also subject to contamination via microorganisms, but the scale and nature of the reactor operation are such that a very clean environment is not possible. However, contamination can be avoided by properly designing the fermentors and associated aseptic transfer equipment.

As the product proceeds from fermentation through purification, the environment should be made progressively more clean to protect the integrity of the product. Consequently, the airflow rate to the inoculum and fermentation areas is generally moderate (i.e., 20–40 room volumes per hour), and in the final purification areas, the air flow rate can approach 100 changes per hour.[2]

For all areas, exhaust air can be either vented to the atmosphere or partially recirculated. The amount of recirculation used will depend on the processes and areas involved. To avoid cross-contamination, the air from different processing areas should not be mixed; recirculated air should only be returned to the room from which it was exhausted. Production areas with a high potential for creating airborn contamination should use little or no recirculated air. Areas that require very clean environments (and that do not create many particles) can benefit from recirculating the air; recirculating air helps maintain constant temperature and humidity and provides a cleaner feed to the filter. The amount of recirculation has a great impact on capital and operating costs, and the design decisions should be carefully made.

Finally, the static pressure of air within each room should be adjusted (by changing input and exhaust rates) to achieve a defined net flow of air "leakage" from one area to another. Areas that are "dirty" (e.g., exit corridors) or contain processes that create airborn particles (e.g., fermentation) are typically assigned the lowest pressure and act as a "sink" for air leakage, thereby minimizing the potential for cross-contamination. Areas that require the cleanest environment (e.g., final purification) have the highest pressure; an over pressure of "clean" air is used as a barrier to contamination with "dirty" air. The dynamic control of air pressure within a multi-room facility requires a sophisticated system of controls as well as a carefully designed floor plan.

Integrating the HVAC system into a modular design approach normally requires the use of multiple air handling units, one for each of the processing modules. Using multiple units aids in defining and licensing a specific production area. To prevent cross-contamination between modules or production areas, common access areas should have air pressures more positive (for access corridors) or more negative (for exit corridors) than the adjacent processing areas. The inoculum areas should have moderate pressure. The fermentation area should have low pressure to prevent airborn cross-contamination. The purification area should have the highest pressure to prevent entry of particles from other areas and thus preserve the purity of the product.

When working with recombinant or pathogenic organisms, National Institutes of Health guidelines[3] indicate that certain containment safeguards be incorporated into the process to prevent release of active organisms into the environment. The level of containment required depends on the specific nature of the organism being used. If recombinant/pathogenic organisms are to be used in the facility, then specific provisions outlined in the NIH guidelines should be made for the containment of these organisms.

The first step in developing a containment system is to define the areas where the active organisms will be used. Certainly areas of inoculum development and fermentation will require the use of active organisms. The next step in developing a containment system is to define the level of containment required, based on the intended microorganism, processing scale, and so forth. When trying to predict the needs of an unknown future project, as is the case in designing an R&D/pilot facility, this second step is not straightforward; a prediction will have to be made about the future containment requirements and guidelines. Usually, some degree of isolation is necessary, but the highest degree of containment may not be warranted.

The guidelines/requirements for process containment are similar but not identical to those for current good manufacturing practices (cGMPs). Process containment is intended to protect the environment and personnel from the "product," whereas cGMPs are intended to protect the product from the environment. The two design approaches overlap at the early stages of production but begin to diverge at later stages of purification. Fermentation areas are generally kept at "negative" air pressure levels to prevent cross-contamination; fermentation equipment is designed for aseptic containment. These concepts parallel the guidelines for handling biohazardous materials. The purification area, however, is generally maintained at "positive" pressures to provide as clean an environment as possible. Purification equipment can be, but is not always designed for, *in situ* sterilization or containment.

On the basis of these factors, a logical design approach would be to define the biohazardous containment area (i.e., the area held at low pressure) to include the inoculum and fermentation areas, but to exclude the purification areas. Active organisms would only be processed in the inoculum and fermentation areas, and all material leaving these areas could be chemically or thermally deactivated. In establishing a logical flow of process materials through the facility, the inoculum and fermentation areas would most likely be in close proximity to, if not adjacent to, one another, and the containment system for each module would therefore, be confined to these areas of the facility.

Again, the level of biohazard containment depends on the intended process and microorganisms. The containment system should prevent release of active product or aerosols into the atmosphere as well as provide a means for decontaminating both solid and liquid wastes. Primary containment of liquids and aerosols can be addressed within the basic design of the processing equipment and operating protocols. Decontaminating small volumes of solid and liquid waste can be handled "manually" in an autoclave. Decontaminating large quantities of waste often requires a separate decontamination (i.e., "kill" tank type) system. In a modular facility, one decontamination system can be dedicated to each production module and can be located adjacent to the inoculum and fermentation areas to ensure that the active organisms are confined to a specific location in the facility and to prevent any cross-contamination between modules. To prevent biohazard release in the event of an accidental spill, provisions should be made within the suspect areas to contain all biohazardous material from the largest process vessel. The biohazardous material can then be decontaminated before disposal.

The shipping, receiving, and warehouse areas of the facility need to be designed to comply with cGMP requirements. Staging areas for received raw materials and outgoing product need to be established. Areas for sampling and quarantine storage of raw materials are required. Storage equipment for perishable goods need to be provided with the required alarms, temperature monitoring devices, and the like. Although these requirements are the same as those for other pharmaceutical manufacturing facilities, they are more restrictive than those normally required for research facilities. It should also be noted that the high volume and diverse nature of the materials required for research will affect the size of the warehouse area.

PROCESS DESIGN CONSIDERATIONS

In designing a multipurpose production facility, considerable weight must be given to the range of intended processes, the utility needs, the design of the processing equipment, and the required process scale. These process needs will have an impact on the facility layout in terms of size and number of specific rooms, the initial capital expense for the facility, and the future utility of the facility. The process needs can be grouped into the following areas: utilities, material preparation, process, transfer, and cleaning equipment.

A significant portion of the expense in constructing and operating a GMP bioprocessing facility is in providing the necessary validated utilities: deionized water, water for injection, and clean steam. These types of systems are typically expensive to build (sanitary piping and valving are used) and require both validation and routine monitoring of the fluid quality. Economy can be realized by "sharing" these utilities between multiple production modules. The cost of multiple utility systems is prohibitive, and proper design of common systems eliminates the potential for cross-contamination. Of course the entire system must be validated and established operating protocols adhered to even in non-GMP service areas. Any attendant inconveniences in operation of shared utilities are compensated for by providing consistent production quality utilities to all stages of product development. However, risk is involved, for if the utility system becomes inoperable or the fluid is not within specification limits, processing within the whole facility is affected.

An area within the facility must be designated for preparing the raw materials for use in processing. For bioprocesses, these materials include growth media, nutrients, and buffer salts. This capability can be dedicated within each processing subarea, or a common preparation capability can be used to service all areas within a processing module. The preparation area will require equipment for weighing and dispensing solid and liquid raw materials, metering water and tanks for batching the combined raw materials. Depending on the scale of operation, the tanks can be fixed or portable. Utilizing a common preparation area has its benefits (i.e., economy, localizing activities, more straightforward floor plan), but the size of the equipment and the needs of each processing area will dictate whether dedicated preparation capabilities are required.

The design and type of processing equipment depend on the intended process. To meet the demands of future processes, equipment design needs to allow for future process changes or enhancements as well as for flexibility in physical arrangement. In a modular facility, the fixed equipment could be duplicated within each module to increase the interchangeability of processes and areas. This approach, however, will become expensive and somewhat limits overall flexibility. An alternative is to use, where possible, portable or skid-mounted equipment. Skid-mounted equipment is self-contained and can be moved to accommodate processing needs; consequently, less duplication and more diversity would be obtained. The disadvantage of skid-mounted equipment is that it is difficult to obtain fully integrated and automated processes; often, skids from different manufacturers will use different control components and philosophies. However, this is not as important in the development stages of process as it is in the final manufacturing process. Utility and diversity are more important than integrated automation during the early stages of process development. As the process approaches the manufacturing stage, equipment specific to the process needs can always be obtained.

The majority of processing equipment will be used in the fermentation and purification areas of the modules. A battery of fermentors or bioreactors, ranging in size, should be available. Smaller fermentors can be skid mounted, but the larger ones will most definitely be fixed. The fermentors should be designed to incorporate future needs, but initially they should not be so complicated as to make them economically or operationally unfeasible. The configuration of the bioreactors chosen (i.e., stirred tank, air lift, hollow fiber) should be considered carefully. In the absence of defined process needs, the equipment with the best overall suitability and diversity should be selected.

Other equipment used in the fermentation area includes homogenizers, micro- and ultrafiltration equipment, and centrifuges. Except for the largest processing scales, this equipment can be skid mounted. Utility "stations" to provide power and other services need to be installed in the production area to support the operation of the skid-mounted equipment. These utility stations can be designed to allow the use of any skid-mounted equipment at any station within the facility.

The purification areas can be designed so that all operations are undertaken with portable equipment. The facility can then provide necessary rooms and utility stations with required skid-mounted equipment (ultrafiltration, chromatography, etc.) for use as required by the process. However, when implementing larger process scales, tankage and other fixed equipment may be required.

To provide for fluid transfer from room to room and between fixed equipment within rooms, a transfer network needs to be established. Because this network will transport product and/or raw materials, it will have to be of sanitary design and consequently will be expensive. The transfer network should allow flexibility with the origin and destination of fluids (e.g., media, inoculum, and cleaning solutions) without being excessively complex. Transfer panels or nodes are well suited for providing the required flexibility; several transfer lines can meet at a node, and jumper connections are made from one line to another. With an appropriate configuration of a node and jumpers, specific transfer line connections can be "allowed" or "not allowed." Individual, valved manifolds are not often incorporated in bioprocessing areas, mainly for reasons of cost and validation complexity.

Transfer lines and operations must be designed for the special needs of the multipurpose facility: isolation of processes, containment, and validation. A system of transfer nodes should be established throughout the processing areas to allow for logical separation. In this way, one room need only be "connected" to another when a transfer operation is taking place; from a processing point of view, by blocking off the connecting transfer lines, the rooms can be isolated from one another. Similarly, lines that are used to transfer processing fluids within "containment" rooms or areas can be isolated from "noncontainment" areas. The physical layout of process piping used to connect different rooms and areas is not generally straightforward. This piping will be used for different products at different stages of production. Aseptic transfer will be required, especially in transferring inoculum between fermentors. The piping normally needs to be cleaned in place and sterilized in place between uses. Consequently, the transfer piping system should be designed to have a minimum of branching and no dead legs. Drains and steam traps should be provided at low points, and valves and gasketed connections should be kept to a minimum.

Cleaning multiproduct processing equipment is a very important aspect of operations. Because the same equipment will be used to process different products, repeatable and reliable cleaning is required. Automated clean-in-place (CIP) systems are well suited to this need. A cleaning procedure for a specific piece of equipment can be established and validated. Once programmed into the CIP system, the cleaning regimen will be repeated exactly from run to run, guaranteeing that cleaning will be consistent. The CIP system is connected to the processing equipment via supply and return lines. These lines can be routed directly to specific equipment or can be incorporated into the transfer piping and node design.

Several options exist for incorporating a CIP system into the multipurpose facility. A single, fixed system can be used to service the entire facility. This approach is beneficial in that the equipment is centrally located and capital cost is decreased. However, it will be difficult to certify process isolation. Another approach is to use several fixed CIP systems dedicated to specific areas of the facility. This allows the required isolation, simultaneous processing and containment, but the equipment costs would be elevated. A third approach is to use portable CIP equipment. This equipment provides the same automated cleaning procedures as the fixed equipment, but it can be shared between areas. With a portable CIP unit, suitable utility stations are required to provide the utilities for the CIP equipment as well as to allow the connection of CIP supply and return lines to the processing equipment. This method would provide the necessary isolation and containment at a reduced capital cost. However, operational complexity is substantially increased.

PROCESS MONITORING AND CONTROL

Monitoring and control of production processes as well as subsequent analysis of collected data are as important in bioprocess operation as in other areas of the pharmaceutical industry. These capabilities can be implemented using several design philosophies. The most common options include both centralized or distributed approaches.[4] In the 1980s the use of distributed control became common primarily due to the intrinsic reliability of commercially available systems. In recent years, however, the extreme cost of distributed systems, coupled with the improved reliability of less expensive, centralized systems, have led to a resurgence in the use of centralized process monitoring and control. For biologics processing this trend is attenuated by the perception that centralized computer systems, particularly those used to simultaneously support R&D and manufacturing operations, are difficult or impossible to validate.

To address the need for validation, a hybrid approach that maintains the required flexibility and reliability without undue expense has come into common use. In such a system, processing equipment (e.g., bioreactors) containing fully distributed local loop controllers is employed. In turn, these controllers can be linked to centralized control systems that can implement both batch and continuous control using a remote setpoint approach. For a multipurpose facility, individual centralized control systems can be implemented in each processing module. These systems can then be linked to each other and to an analogous laboratory research and development system using a conventional local area network. This design approach allows for free exchange of all collected information among all portions of the facility. It also allows for data isolation at either the control system or the equipment level, if that is required to meet computer validation requirements.

SUMMARY

With proper design, the integration of development and manufacturing efforts within one facility can be accomplished. A modular facility with isolated production areas will allow for the simultaneous production of development, clinical, and/or manufacturing material. Such a modular facility will help shorten the time required to bring a product to market, concentrate many related process activities, and help reduce facility, equipment, and overhead redundancy. This type of design, however, will impose some restrictions on development activities and personnel: rigorous manufacturing protocols will have to be adopted in certain areas to preserve manufacturing integrity.

Facility design should be developed based on the expected needs of the process. Separate processing areas should be provided to allow for isolation and/or simultaneous processing. How each room relates to another, in terms of operation requirements, potential for cross-contamination, and accessibility, will affect the design basis for the floor plan, access hallways, HVAC system, and containment system.

Services provided to all production areas (i.e., deionized water, water for injection, and clean steam) can be distributed from centrally located generating equipment. The risk of cross-contamination occurring via service utilities is minimal, and economics favors using central equipment.

Fixed-in-place processing equipment can be duplicated among modules to ease the transfer of a process from development to manufacturing groups and areas.

Fixed equipment should be based on a design that will offer the most processing flexibility and allow for future upgrade or incorporation of new technology.

Where possible, equipment should be designed around a skid-mounted or portable configuration. This type of configuration allows the equipment to be shared between modules and results in lower initial costs without restricting productivity. Initial portable equipment should be designed for general purpose use. Upgrade of existing equipment or purchase of new equipment specific to the manufacturing needs can be accomplished when the processing needs have been defined. The factor that has the greatest impact on whether equipment is portable or fixed is the scale of operation. An appropriate scale should be chosen that will support the majority of research, clinical, and interim manufacturing needs. Prescribing too large a scale will require more and larger fixed equipment, which will have a great impact on the size and cost of the facility.

REFERENCES

1. Genetic Engineering News. Vol. 10, March 1990, p. 4.
2. DEL VALLE, M. A. 1989. HVAC systems for biopharmaceutical manufacturing plants. BioPharm 2: 26–42.
3. NIH. 1986. Guidelines for research involving recombinant DNA molecules. Fed. Register 51: 16957–16985.
4. OMSTEAD, D. R. 1990. Computer Control of Fermentation Processes. CRC Press. Boca Raton, FL.

Plant Design and Process Development for Contract Biopharmaceutical Manufacture

S. P. VRANCH

Celltech Limited
Slough, Berkshire SL1 4EN, UK

Companies that employ a contract manufacturer have a variety of motives and requirements. Usually, they need the product quickly but do not have spare capacity in their own development or manufacturing groups, or they do not have the technology. A contractor can increase the chance of success by developing and manufacturing the product, using a systematic approach that is based on experience.

The customer expects from the contractor a confidential service, with high priority and rapid completion according to an agreed schedule and product specification. The contractor should be financially secure for continued business, have a proven scientific track record, have premises and equipment in compliance with CGMP, and produce material of the required quality. Quality includes the need for documentation in compliance and validated methods. In this paper I describe in general terms how Celltech achieves these needs, concentrating on process development and plant design, so that useful discussion points may emerge. Although we have facilities for the exploitation of microbial and yeast expression systems, I will concentrate on the manufacture of recombinant proteins and monoclonal antibodies derived from animal cells.

PROCESS DEVELOPMENT

Cell Line Evaluation and Development

Reception of Cell Line and Initial Testing

A hybridoma or recombinant cell line is received from the customer, taken to a dedicated suite outside the production building, and there placed in quarantine. The cells may already be well characterized by the customer, but the first objective is to test the line to show the absence of microbial contamination, mycoplasma, and sometimes viruses. For example, if a murine cell line is shown to contain a murine retrovirus, which is likely to affect the ability of the customer to market a product, then he is informed. Additional viral testing is also done to ensure that any contaminating virus is not a potential pathogen. The presence of mycoplasma precludes the further use of a cell line, and fresh cells are requested.

Cell Transfer and Cloning

The cells are transferred to the cell evaluation laboratory suite where a small "back-up" cell bank is prepared with all manipulations done under class 100 conditions. This is the beginning of a process to evaluate the potential of the cells for adaptation to optimal growth conditions and to assess their productivity, using a program developed with hundreds of lines. Re-cloning and screening can also be used, if required, to increase productivity. In addition, Celltech operates a service for raising and selecting hybridoma cell lines for producing homogenous clonal and stable antibodies.

Media Optimization

Normally, the cells are adapted to grow in a proprietary serum-free medium, a process that typically takes 8 weeks and involves monitoring of productivity and stability.

Since 1985, we have been progressively introducing synthetic media into production. The elimination of fetal bovine serum reduces costs, minimizes the risk of contamination, and assists in optimizing the process through to purification. We have succeeded in adapting all rodent hybridoma lines so far tested to grow in a serum-free medium, containing 1 g protein/liter, without a significant loss in yield compared to that with serum-supplemented media.[1] Media development for the manufacture of recombinant proteins is similarly addressed. All the media development work at this stage is carried out in shake flasks or roller bottles and spinner vessels. Once a stock of cells has been adapted, it is frozen in liquid nitrogen. An ampule is taken and cells are revived for a 15-generation cell stability study.

Fermentation Development

The adapted cells and optimized media are used for fermentation development, typically in a 5-L airlift fermentor. In the production plant there are 100, 200, 1,000, and 2,000 L airlift fermentors (FIG. 1). Conditions in the 5-L airlift fermentors have been scaled down to simulate cell growth in the production fermentors. For example,

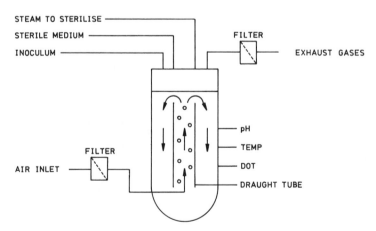

FIGURE 1. Principle of an airlift fermentor.

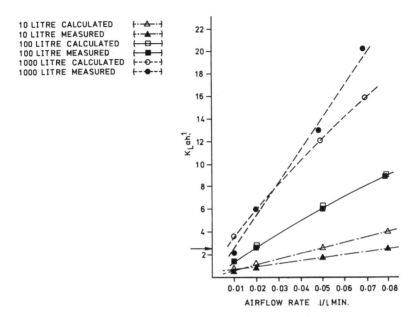

FIGURE 2. Effect of airflow rate on oxygen transfer coefficient. *Arrow* indicates oxygen transfer coefficient required to meet the oxygen consumption rate of 2×10^6 cells/ml.

the effect of airflow rate on the oxygen transfer coefficient for a range of airlift fermentors has previously been reported (FIG. 2). Specific air flowmeters are used to achieve the required oxygen transfer rate.[1]

Yields in roller bottle culture can be used to predict final yields. Suspension cultures were shown to be about 4.6 times more productive than were roller cultures in trials with eight murine hybridoma lines.[2] The range of antibody concentrations in airlift fermentors is 40–500 mg/L with an average of over 100 mg/L. This compares with 10–100 mg/L for static flasks or roller bottles. These increased yields have resulted from process optimization studies, especially media development and adjustment of fermentor parameters.[1]

We have developed processes for the manufacture of recombinant antibodies using suspended CHO cell lines. A novel gene amplification system[3] has enabled cell lines to be established rapidly.

In some instances we have scaled-up alternatives to the airlift fermentor, such as a microcarrier process that has been developed for attached cells. The system comprises a 40-L perfusion vessel with spin filter and has been used to manufacture t-PA using a murine cell line that was initially grown in low serum medium and then perfused continuously in protein-free medium. For example, a t-PA concentration of 55 mg/L was obtained over 800 hours, making more than 90% single chain t-PA (FIG. 3).[4]

Chemostat systems have been used to study cell growth and antibody synthesis by hybridoma cells. Antibody synthesis is stable over a variety of nutrient limitations. The chemostat has the same advantages for study of cell physiology with animal cells as it does for microbes.

Downstream Processing Development

A typical train of purification involves the use of unit operations including cell separation, concentration, filtration, immunoaffinity purification, ion exchange, preparative HPLC, size exclusion chromatography, and formulation. The principal operations have been scaled down from our full-scale plant, so that we can quickly simulate the conditions for processing and then estimate productivity. Research and Development staff carry out this work and transfer the technology into Production. The adequacy of in-house assay systems can be determined during this period. A process can usually be developed in a period of 10 weeks that will provide material for assay development and indicate conditions for process parameters. Depending on the yield obtained, full-scale manufacturing runs can be scheduled and the preliminary validation batches processed if required. Appropriate process documentation will be assembled and issued by Quality Assurance for Production use (FIG. 4).

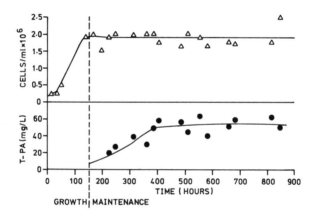

FIGURE 3. Perfused microcarrier culture of C127 cells expressing tissue plasminogen activator (t-PA). The cell line was grown in a serum-containing medium and maintained by perfusing with a proprietary protein-free t-PA production medium. Perfusion rate was 0.5 fermentor volumes/day.

In devising the purification process, the removal of traces of contaminating proteins, pyrogens, and DNA has to be addressed. Virus clearance studies have been performed for all significant unit operations. Celltech often uses immobilized protein A sepharose for protein purification, and although expensive, it gives a high yield and quality and can be used for successive batches of material. Reliability and consistency of column operations are improved with automatic control.[2]

MANUFACTURE

Master Cell Bank and Working Cell Bank

The intended use of the biopharmaceutical product at the current stage of its development dictates the standard of supporting documentation and process validation required. For example, the material may be used for toxicology or research or for clinical trial. Urgent supplies of material for research are sometimes needed, with

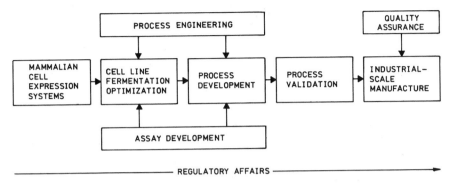

FIGURE 4. Process development scheme.

only minimal supporting documentation, namely, that which is sufficient to ensure good science and the ability to reproduce the same quality of material.

The establishment of a validated master cell bank before the start of manufacture, however, ensures that a reliable source of cells can be used, and development work does not have to be repeated if the protein becomes successful and a production license is required. Thus, a working cell bank is typically created, and cell stability studies are routinely performed (FIG. 5). Good manufacturing practice is therefore established even though it may not be mandatory for the product at this stage of its life cycle.

Inoculum Grow-Up

Inoculum is prepared in roller bottles (to about 2×10^5 cells/ml) and transferred to spinners and to a small scale fermentor at a concentration of 1×10^6 cells/ml. A

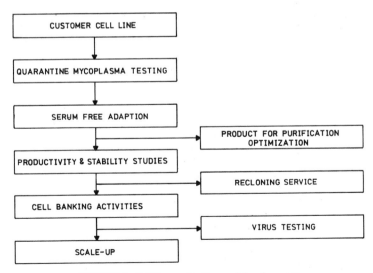

FIGURE 5. Cell line evaluation and development.

volumetric inoculum transfer ratio of 10:1 is most commonly used, but may vary depending on the cell line employed. Cell lines are handled in completely separate inoculum preparation laboratories, each equipped with class 100 laminar airflow units. Incubators are used for flats and roller bottles and spinner flasks.

Fermentation

Fermentations are carried out in airlift fermentors in which the mixing required for heat transfer is provided by the flow of gas. We have not experienced cell damage caused by sparging, and cell viability is the same as that seen in roller and spinner cultures. Experiments have shown that the change in hydrostatic head equivalent to 11 meters has no deleterious effect on the cells.[1]

The course of fermentation is controlled with measurement of pH, dissolved oxygen concentration, cell density, and so forth, and criteria for harvesting are

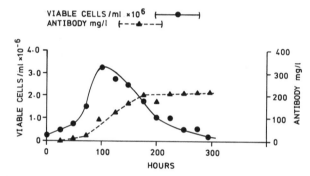

FIGURE 6. Batch culture of a mouse hybridoma cell line in serum-free medium. An IgG secreting mouse hybridoma cell line was inoculated into a 1,000-L airlift fermentor at a population density of 1.5×10^5 cells/ml. Viable cells were estimated by trypan blue dye exclusion on a hemocytometer and antibody concentration was measured by an HPLC technique.

chosen for each individual product and fermentation. A time course of some products is predictable. An example of the relation between growth and antibody production by a murine hybridoma cell line growing in serum-free medium in a 1,000 L airlift fermentor is shown (FIG. 6). Doubling time was 21 hours and cell growth reached a maximum concentration of 3.2×10^6 viable cells/ml. With this and many other cell lines, a large proportion of the antibody is synthesized during the stationary and decline phases of growth. In this example, 200 g of antibody was produced in 300 hours.

Results of process development studies are reviewed by production and development staff, and the size of the production fermentor is selected, between 100 and 2,000 L scale. Each fermentor train is equipped with seed fermentors, a media solution vessel and media filters, and a harvest vessel (FIG. 7). Each train is supplied with distilled water (WFI distribution system). Each piece of equipment is separately monitored and controlled by a computer control system that is dedicated to the train. The control system drives the CIP and SIP of the fermentor train, including the cell separation equipment, and this means that cleaning can be achieved consistently according to a validated protocol. The control system also provides adjustment of key parameters and produces in-process data.

Cell Separation

Although filtration can be used for cell separation, centrifugation is the preferred method of removing cells and cell debris. The Westfalia centrifuge is an automatic discharge disk stack machine that is automatically sterilized. We have chosen to strip and clean the centrifuge after every batch, although traces of cells are rarely found. Ultrafiltration is used to filter the centrifuge and concentrate the product, and filtration to 0.2 μm is carried out before further processing.

Protein Purification

The purification scheme is designed at the development stage and in compliance with the specification of the product so that the scheme will achieve the requirements of the certificate of analysis for the product. A pilot fermentation is advisable with some products to yield sufficient material for a tenth-scale purification train to be evaluated.

Precipitation, followed by ion exchange, has been widely used for monoclonal antibody purification, but ion exchange may not achieve the required purity as a single step, although it is satisfactory for some diagnostic and immunopurification reagents. Protein A Sepharose purification has proved more successful.[5]

In general terms a train of purification equipment is set up in the purification facility. The process is documented and transferred to production staff by those responsible for its development. The rooms are operated under class 10,000 conditions and are provided with separate change rooms as operators wear clean room clothing. When possible, the operations are contained. Rooms within the purification suite are dedicated to one product, and the room is cleaned between products. Columns and associated pipework and valves are also dedicated to particular products. Scheduling of purification is eased if products can be stored cool or frozen until an appropriate room becomes available. Any freeze-thaw process is validated at the development stage.

PLANT DESIGN FEATURES

In designing a multiproduct plant the importance of operating systems that avoid cross-contamination cannot be overemphasized. Appropriate CGMP procedures include features of containment and the use of operating systems by trained staff. Critical services to vessels, such as fermentor air supplies, media transfer lines, and vessel vents are provided with duplicate filters. Pipework links between process vessels are achieved via flow plates to ensure that segregation is achieved. The correct routing through the flow plates is checked by the control system.

Automatic control systems for the operation of SIP and CIP as well as for the production fermentor and associated vessels are programmed to ensure a correct sequence of operations. The control system generates prompts for the operator when intervention is needed, and the correct response has to be entered before the control sequence will proceed. Any intervention is automatically logged.

The process plant is designed to be contained to minimize the release of aerosols containing cells. The safe use of animal cells for large scale production of pharmaceuticals has been established for many years.[6] All of the containment features for the

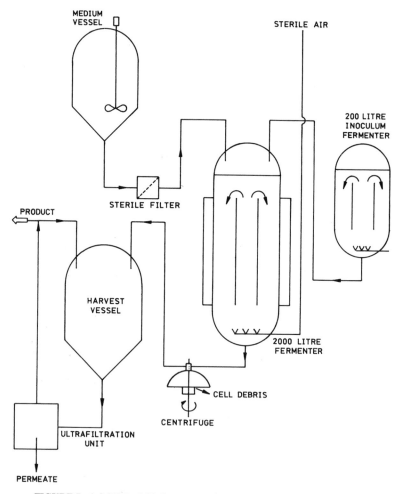

FIGURE 7. A 2,000-L airlift fermentor with associated process equipment.

protection of the product also contribute to the protection of the environment. We do not handle human pathogens in the production facilities, but containment is needed to minimize the release of viruses that may infect the cells. All recombinant cells are inactivated before they leave the production facility and the plant is in compliance with GILSP (Good Industrial Large Scale Practice).[7]

The process engineering of the plant is based on all welded, stainless steel pipework using sterile couplings where welds cannot be used. The plant is sterilized with clean steam and pipework slopes to ensure condensate removal. Vessels and purification laboratories are supplied with distilled water of WFI quality from a recirculating loop at high pressure and 80°C. The cooling water in the heat exchangers is automatically maintained at a pressure lower than that in the loop. All take-off points are regularly sampled and sanitized.

Routine environmental monitoring in the manned and unmanned state ensures that when "action limits" are attained, the cause can be investigated promptly.

CONCLUSION

The examples given herein demonstrate how process developments and plant design are implemented in a validated 40,000 ft^2 production facility used for the contract manufacture of biopharmaceuticals. For each product, process development of the cell line and of the manufacturing process is carried out to take advantage of the expertise built up in processing hundreds of cell lines. The airlift fermenter has proved a reliable and effective bioreactor that facilitates rapid scale-up. The evaluation of unit operations for protein purification is backed up with fundamental work, such as virus clearance studies, to ensure that an appropriate train of equipment can be installed for rapid manufacture. Documentation, quality control, and quality assurance are provided in keeping with commercial and regulatory requirements.

The production facility is designed into segregated fermentation systems and separate purification rooms. Provision for the validation of cleaning and the avoidance of cross-contamination are crucial elements in the engineering design.

SUMMARY

The development of processes for the contract manufacture of biopharmaceuticals is described, starting with the receipt of the cell line, through process development, and into manufacture in dedicated facilities. The plant is designed for rapid adaptation to each new product, and model systems used in the laboratory enable scale-up to be reliably predicted.

REFERENCES

1. BIRCH, J. R., K. LAMBERT, P. W. THOMPSON, A. C. KENNEY & L. A. WOOD. 1987. Large Scale Cell Culture Technology. Hanser Publishers.
2. RANSOHOFF, T. C., M. K. MURPHY & H. L. LEVINE. 1990. Biopharm, March: 20–25.
3. Published International Patent Application No. WO 87/04462.
4. RHODES, P. M. & J. R. BIRCH. 1988. Bio/Technology 6: 520–523.
5. KENNEY, A. C. & H. A. CHACE. 1987. J. Chem. Tech. Biotechnol. 1987. 39: 173–182.
6. PETRICIANNI, J. C. 1987. Swiss Biotech 5: 32–37.
7. VRANCH, S. P. 1990. ASTM STP 1051. W. C. Hyer, Jr., ed.: 39–57. American Society for Testing Materials. Philadelphia.

Design and Construction of a Bulk Biopharmaceutical Pilot Plant

DON BERGMANN

SmithKline Beecham Pharmaceuticals
King of Prussia, Pennsylvania 19406-0939

SmithKline Beecham recently opened their new Biopharmaceutical Development Facility in suburban Philadelphia. The facility is intended for the development and scale-up of manufacturing processes for the bulk manufacture of therapeutic proteins. The facility is also designed to accommodate the preparation of materials to support clinical trials for these agents. As such, the facility is designed to conform to the current Good Manufacturing Practices and National Institutes of Health/ Recombinant DNA Advisory Committee guidelines for work involving recombinant microorganisms.

The facility was designed to provide for the development and scale-up of processes for the production of recombinant proteins from bacterial, fungal, and animal cell sources and at culture volumes ranging from 5–3,000 L. The facility also provides for the recovery and purification of proteins from these sources and scale of operation. The areas have been divided into research and development and analytical laboratories, process development scale-up areas, and clinical production areas.

The facility consists of five floors that include three floors of development laboratories and pilot plant areas, a ground level mechanical area, and a rooftop mechanical penthouse. Because of the desire to separate microbial process areas from animal cell culture process areas and the protein purification areas from the up-stream processes, each function was provided a dedicated floor. The areas consumed by each type of function, in square feet, are as follows: process areas, 28,698; laboratories, 7,963; offices, 6,443; mechanical areas, 15,040; personnel services, 16,871; a total of 75,015.

Because of the changing technologies for biopharmaceutical processes, the facility was designed with the utmost of flexibility including the provision of a variety of services to the pilot plant areas and room for the addition of future equipment. The important services provided to the facility are as follows:

1. Deionized water
2. Water for injection
3. Clean steam/plant steam
4. House gasses
    Carbon dioxide
    Nitrogen (Liquid and Gas)
    Oxygen
    Ammonia
    Compressed air (instrument and process)
    Natural gas
5. House vacuum
6. Process-chilled water system
7. Controlled environment rooms ($-20°C$ through $37°C$)
8. Biological waste treatment system
9. Distributed control system (DCS)

Two areas of particular interest are the biowaste system and the distributed control system. The biowaste system consists of two two-tank systems connected to the biological process areas. All floor drains and process drains in these areas are connected to the system. Each tank system is independently connected to the process drains to provide the capability of independently handling specialty waste while still processing routine process wastes. Each tank system operates in an alternating batch mode, so that as one tank is collecting waste the other is either in stand-by or sterilizing waste. Sterilized wastes are normally discharged to the aqueous waste system that ultimately drains to the publicly owned treatment works (POTW). In cases in which waste cannot be discharged to the POTW, inactivated wastes can be transferred to a fiberglass holding tank for later drumming and off-site disposal.

A distributed control system is employed throughout the process areas to operate various pieces of process equipment including fermentors and chromatography columns. The system is capable of providing process control, real time monitoring and alarming, data logging, trending of process variables, and report generation.

The project was conceptualized in January 1986 and actual engineering started in November 1986. Construction began in September 1987 and mechanical completion was achieved in August 1989. Validation planning started in January 1989, and validation field work began in August 1989. By August 1990, 22 major building systems had been validated. The facility is now fully occupied and operational.

An Independent Pilot-Scale Fermentation Facility for Recombinant Microorganisms

LAWRENCE D. KENNEDY, DENISE E. JANSSEN,
MICHAEL J. FRUDE, AND MICHAEL J. BOLAND

Biochemical Processing Centre
Department of Scientific and Industrial Research
Palmerston North, New Zealand

The Biochemical Processing Centre is a research group within the Department of Scientific and Industrial Research (the major publicly funded research and development organization in New Zealand) that is operating a contained facility for use with recombinant microorganisms. Equipment available includes stirred-tank reactors of 40 and 1,000 L working volumes, a Westfalia CSA 8 steam-sterilizable contained centrifuge, biohazard hood, double-door autoclave, ultra-high temperature (UHT) flow sterilizer, and ancillary support equipment. This is housed in a purpose-built fermentation room, with entry through an airlock/control room. A floor plan of the room and equipment normally used is shown in FIGURE 1.

The facility has been approved by the New Zealand Advisory Committee on Novel Genetic Techniques (ACNGT) and by the Interim Assessment Group on Genetically Modified Organisms for large-scale work to Category 1 containment requirements, and is the first publicly owned approved facility in Australasia.

REGULATORY CONSIDERATIONS

Design and construction of the facility began before New Zealand guidelines for large-scale work with recombinant organisms were established by the ACNGT. However, because the ACNGT's small-scale guidelines had been based closely on those of Australia and the United Kingdom, adherence to the Australian large-scale guidelines was deemed appropriate. These define two categories, C1-LS and C3-LS, with C1-LS expected to be appropriate for most projects and C3-LS likely to be specified when "the donor DNA contains genes for potent toxins or pharmacologically active substances." Our facility complies with Australian C1-LS in all respects and also meets C3-LS in all except the provision of a shower near the exit and the airtightness of doors.

PRIMARY CONTAINMENT

The requirement for both levels 1 and 3 is that liquids be handled in closed piping systems and process equipment be designed to minimize the potential for rupture and to allow easy decontamination and maintenance. Steam sterilization is used for process equipment and transfer lines.

Cell harvesting is done with a Westfalia CSA 8 centrifuge fed directly from the fermentor using 50 kPa air pressure. Transfer lines used are pressure-rated braided stainless steel with Teflon lining, with connections via sanitary dairy unions and with diaphragm valves on product lines. The centrifuge is fitted to allow steam sterilization followed by cool-down under positive pressure from a sterile-filtered air supply for aseptic harvesting. Supernatant from the centrifuge is piped to a 1,000-L collection tank with a sterile vent filter and from there to the liquid waste treatment system.

Other primary containment systems include those for sampling and inoculation, which are designed to prevent release of viable organisms; details of the system have previously been described by Janssen *et al.*[1] All potentially contaminated drain lines, including those from steam traps, are piped to the liquid waste treatment system.

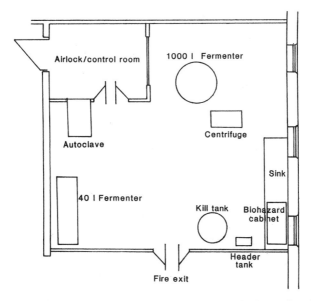

FIGURE 1. Floor plan of the fermentation area, showing major items of equipment.

SECONDARY CONTAINMENT

The fermentation room is designed to contain any release if primary containment is broken. Construction is inside an existing concrete building, with washdown paneling for walls and ceiling. An impervious epoxy-based floor is continued up the walls sufficiently to contain the contents of the 1,000-L fermentor in a catastrophic spillage. The floor slopes away from the exits, with the normal entrance being at the highest point. Both the normal and emergency exits are above the level of a 1,500-L spill. The room is entered through an airlock/control room divided by a "boot bar" (a physical barrier on which one sits to don protective footwear and too wide to step over readily) to prevent walk-through access. Distinctive footwear and protective clothing are required on the inner side of the boot bar.

The fermentor room is maintained at about −25 Pa relative to ambient pressure with the control room at about −10 Pa. The room is ventilated by three HEPA filters, with two for recirculation and one for exhaust. Ambient air enters the control room via a dust filter, passes to the main room via another dust filter, and finally exists via the exhaust HEPA filter.

All liquid waste leaves the room by way of a treatment system based on a 500-L pressure vessel with a schedule of 20 minutes at 121°C. Liquid from the primary containment system piping enters via a diaphragm pump which runs continuously. The floor slopes to a sump, and from there washdown water and any spillages enter the system via a Mono pump controlled by a level switch in the sump.

Sterilization systems in use, including the operation of the waste treatment system, have been validated with *Bacillus stearothermophilus* spore suspensions and are monitored with several commercially available indicator systems.

SAFETY CONSIDERATIONS AND SYSTEMS

As part of the 1,000-L fermentor systems design, a HAZOP study by Gurnsey *et al.*[2] was made to ensure that containment would not be compromised by any subsystems' failure. The computer-based control functions are fed from an uninterruptible power supply, with other loads such as heating, stirring, and solenoid valve operation designed for fail-safe operation.

Microbiological monitoring within the room shows no loss of primary containment to date. The only bacteria regularly recovered have been *Bacillus* spp. from the floor; these bacteria are known contaminants of the water supply used for cleaning purposes in the building.

REFERENCES

1. JANSSEN, D. E., P. J. LOVEJOY, M. T. SIMPSON & L. D. KENNEDY. 1990. Technical problems in large-scale containment of rDNA organisms. *In* Fermentation Technologies: Industrial Applications. P. L. Yu, ed.: 388–393. Elsevier. London.
2. GURNSEY, J. C., M. P. GURNSEY & I. S. LAIRD. 1990. Application of a hazard and operability (HAZOP) study to a biological containment facility. *In* Fermentation Technologies: Industrial Applications. P. L. Yu, ed.: 394–399. Elsevier. London.

Environmental Impacts upon Biotechnology Facility Design

A Review of Chiron's Recent Environmental Impact Report for a Biotechnology Facility

ROBERT F. KONOPACZ

Chiron Corporation
Emeryville, California 94608

In the 1990s the biotech facility designer will be required to become proficient in legislation dealing with environmental considerations of the location intended for development. Although the Food and Drug Administration, Washington, DC, currently regulates environmental issues for the industry through Establishment License Applications and New Drug Applications, the Environmental Protection Agency (EPA), Washington, DC, is becoming sensitive to increasing its regulatory administration on the biotechnology industry. Current biotech designs have dealt primarily with revisions of the guidelines from the National Institutes of Health, Bethesda, Maryland.[1] These guidelines advise the designer on containment of recombinant organisms for large and small scale processes. The facility designer needs to address increasing regulatory issues from agencies, such as the EPA, that may prevail during initial designs.[2] This may result in substantial cost savings by avoiding facility modifications during or after construction. These cost savings relate to process effluent and air emissions that may have not appeared significant in the prior decade because of the relatively small scale involved. As biotech processes mature, stricter regulation of design may evolve when communities become aware of increasing their role in safeguarding their controllable environments. The designer is faced with educating the general population near whom the facility may be located.

Chiron Corporation in 1987 elected to design a multipurpose biotechnology facility at its Shellmound site in Emeryville, California. This plant was projected to fulfill requirements for diagnostic, therapeutics, and vaccine development. It would be one of three buildings that would complete a biotech campus. The facility would have approximately 60,000 square feet devoted to several different scales of fermentation, recovery, and final purification operations.[3] The original schedule was designed around a modified fast-track premise. The design would move quickly to the field through coordinated design packages upon in-house approval. Working with a local developer, a preliminary design package was submitted to the Emeryville Planning Department for building permits. This required several public and closed city council hearings. The result was an agreement with the city that Chiron Corporation would abide by environmental conditions without the need for an environmental impact report (EIR). These conditions were described in a document referred to as negative declaration. It describes by item mitigation that are required by the company and the developer to abide to for construction and operation of the intended facilities. During the 30-day final open period after the negative declaration was filed, Chiron Corp., the local developer, and the city of Emeryville were sued by a resident of the city of Emeryville.[4] The suit alleged that proper procedures, as set up by California's Environmental Quality Act, had not been followed. The lawsuit was settled out of court with the requirement that the company and developer prepare an EIR.

The Chiron EIR was prepared by an independent agency that specialized in the preparation of this document. This document addressed the cumulative impact that the facilities would have on air, water, and traffic. Biological, radiological, seismic, hazardous wastes, and archeological specialists were hired to assist in reviewing the design and to prepare impact statements. A draft EIR was then provided to the public and to government agencies such as The U.S. Army Corp. of Engineers, San Francisco, California.[5] After a 30-day review period, formal open hearings were held to include public comment. These comments were combined with formal written replies from citizens and public and private agencies. Each comment was researched and given a formal written reply which was combined with a final analysis into the final EIR. This final EIR was then submitted to the Emeryville Planning Department for approval.[6] The EIR was approved with 74 separate mitigation items. A key component of the EIR was that the mitigation set forth in its review must be formally monitored. Failure to comply with the measures set forth can result in fines and can halt the construction or eventual operation of the facility. The requirement for an EIR will result in an overall schedule delay and additional costs to the facility in question. In the Chiron Corporation design this delayed the project by a year, and cost in excess of $200,000. For the biotech designer in politically sensitive areas, the effect of a required EIR should be calculated at the initial conceptual design. With regard to the rapid need for proper facilities the design engineer should prepare two estimates, one of which includes the EIR and its effects on costs and schedule.

The EIR affected the design of the plant by requiring that certain conditions be followed. Of key note was that only certain strains and species of organisms would be allowed to be used in the plant. The relaxation of certain strains of organisms to containment issues by the NIH did not matter. Heating, ventilation and air conditioning (HVAC) design and layouts for the fermentation areas were required to be designed according to Biological Safety II (Large Scale). The HVAC system was designed to allow for future installation of odor-adsorbent systems in case the fermentations created odor problems in the adjacent community. Personnel operating the plant were not to exceed a certain number, to decrease the traffic impact in the area. In this EIR, traffic impacts were more sensitive than biotech ones because of the dense urban traffic area in which the plant was to be located. From seismic studies, tanks used in the facility were designed as pressure vessels to decrease the possibility of collapse. The capacities and locations of storage tanks for acids and bases were determined by the city's fire department.

Inasmuch as each facility and location will be unique, it is imperative to the designer to address the political and educational characteristics of the community in determining site location. In the initial design, careful consideration to these outside influences will save the project from undesired delays in schedules and unforeseen overruns in costs.

ACKNOWLEDGMENTS

Grateful acknowledgments are extended to Tom Sanders, Chiron Corporation, Emeryville, California, and Susan Diamond, Brobeck, Phleger & Harrison, San Francisco, California, for their time and assistance in the preparation of the EIR.

REFERENCES

1. Federal Register. Vol. 53, No. 207, 1988. National Institutes of Health. Bethesda, MD.
2. GIBBS, J. N., I. P. COOPER & B. F. MACKLER. 1987. Biotechnology and the Environment: International Regulation. Stockton Press. New York, NY.
3. Chiron Corporation Biocenter 3 Emeryville, California Phase II-Preliminary Design. Vols. I & II. 1988. Jacobs Engineering Group. Pasadena, CA.
4. Flashman vs. City of Emeryville *et al.* 1988. Alameda County Municipal Court. Oakland, CA.
5. Draft Environmental Impact Report Chiron Pilot Manufacturing Plant. 1989. Vols. I & 2. EIP Associates. San Francisco, CA.
6. Final Environmental Impact Report Chiron Manufacturing Plant Response Document. 1989. EIP Associates. San Francisco, CA.

Index of Contributors